Nuclear Science: Engineering and Technology

Volume I

Nuclear Science: Engineering and Technology

Volume I

Edited by **Paul Patterson**

CL **LANRYE**
INTERNATIONAL

New Jersey

Published by Clanrye International,
55 Van Reypen Street,
Jersey City, NJ 07306, USA
www.clanryeinternational.com

Nuclear Science: Engineering and Technology
Volume I
Edited by Paul Patterson

© 2015 Clanrye International

International Standard Book Number: 978-1-63240-395-7 (Hardback)

Contents

Preface

The discipline of nuclear science deals with the study of the constituents and interactions of atomic nuclei. The origins of this discipline can be traced to the discoveries of atomic constituents such as the electron, neutron and proton. The electron model of JJ Thomson, experiments by Henri Becquerel and James Chadwick, Einstein's formulation of mass–energy equivalence have all played a crucial role in developing Nuclear Science and Technology.

The application of Nuclear Science can be found in fission and fusion of atomic nuclei, which form the fundamental premise of most nuclear experiments of contemporary times. Interaction and maintenance of systems and components like nuclear reactors, nuclear power plants, all involve nuclear science and technology.

There are numerous fields in which Nuclear Science is being utilized. Radiotherapy, which uses radiation to weaken or destroy particular targeted cells, uses a form of nuclear energy. In fact, millions of patients are treated with nuclear medicine each year. In industries, sealed radioactive sources are used for gauging applications and mineral analysis. Heat from nuclear reactors is used to generate electricity and also for desalination of plants, in order to extract drinkable water from seawater. Radioisotope thermal generators, such as Plutonium 238 are also used in space missions. Nuclear well logging is used for oil and gas exploration.

I'd like to thank all the researchers who've shared their studies with us.

Editor

Localization of Plastic Deformation in Copper Canisters for Spent Nuclear Fuel

Kati Savolainen[*], **Tapio Saukkonen, Hannu Hänninen**

Laboratory of Engineering Materials, School of Engineering, Aalto University, Espoo, Finland

ABSTRACT

Localization of plastic deformation in different parts (extruded and forged base materials as well as EB and FSW welds) of the corrosion barrier copper canister for final disposal of spent nuclear fuel was studied using tensile testing, optical strain measurement, scanning electron microscopy (SEM), and electron back-scatter diffraction (EBSD). Results show that in the base materials plastic deformation occurs very uniformly. In FSW welds the deformation localizes in the weld either at the processing line or at a line of entrapped oxide particles. In EB welds the deformation localizes to the equally oriented large grains at the weld centreline or at the steep grain size gradient in the fusion line.

Keywords: Copper; Deformation; Localization; Friction Stir Welding; Electron Beam Welding

1. Introduction

In Finland and Sweden the spent nuclear fuel will be encapsulated and deposited into long-term repositories deep in the bedrock [1,2]. The spent nuclear fuel will be placed in cast iron inserts surrounded by corrosion barrier copper canisters. The 50 mm thick copper canisters consist of an extruded tubular body and forged lid and bottom parts. The tubes and the lids/bottoms will be joined using either electron beam (EB) or friction stir welding (FSW). Different parts of the copper canister have inhomogeneous grain structures and the canister itself contains discontinuities and welds as well as possible minor defects, which cause stress concentrations. These regions may have a significant influence on the deformation of the canister during its lifetime, and their impact on the damage tolerance of the copper canister must be known. Plastic deformation of the canister caused by the repository conditions, such as hydrostatic pressure and bentonite swelling, will not be distributed evenly around the canister, but instead it will concentrate to certain locations. Microstructural defects, geometric discontinuities, as well as microstructural heterogeneity and residual stresses and strains of the welds, in particular, localize plastic deformation.

Residual stress and strain profiles in constrained thick-section copper welds on a local scale and their relaxation under creep and irradiation has not been assessed in detail. Also the relationship of strain localization and strain history to cold work distribution in copper canister has not been explored. The amount of residual plastic strain is known to have an effect on different long-term materials properties. For instance, effect of prestrain (dislocation density) on creep of copper has been shown [3]. In austenitic AISI 304 stainless steel the amount of cold work exceeding 20% is known to result in initiation of the stress corrosion cracking [4]. Thus, when wanting to ascertain and evaluate the integrity of the copper canister it is very important to know the material properties even in very small localized areas. Electron backscatter diffraction (EBSD) and electron microscopy have in recent years proved to be a promising method to study the plastic strain and other material properties in localized areas. However, due to time constraint when wanting to map large areas, it is beneficial to be able to concentrate only on the critical areas. The different regions of a weld are typically exhibiting large variations of residual plastic strain and various discontinuities in the material properties. The optical strain measurement using large test samples containing the whole weld area is an ideal method to detect the critical areas for strain localization, which can later be studied with electron microscopy in detail.

Welding of 50 mm thick copper is difficult and the two currently available methods are EBW and FSW. In EBW, a beam of high-energy electrons locally melts the material to be welded and as the beam passes, the material solidifies producing a very distinctive grain structure. Solidification initiates epitaxially on the base material on both sides of the weld [5]. The solidification forms long columnar grains, which turn to the welding direction (**Figure 1**). At the centreline of the weld the long columnar grains are almost parallel to the welding direction. The

[*]Corresponding author.

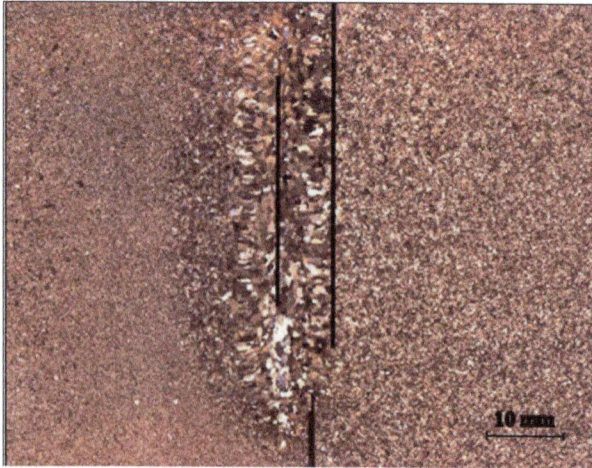

Figure 1. Macrograph of the transverse cross-section of an EB weld. Tube material is on the left side and lid material on the right side. The large grain size of the weld can be clearly seen. Fracture locations are marked with black lines. Thickness of the joint is 50 mm.

morphology of the grain structure can be controlled by several EB welding parameters, e.g. the acceleration voltage and the current of the electron beam and the welding speed. FSW is a solid state welding method, where a rotating tool with complicated features is plunged between the work pieces. The frictional heat between the tool and the work pieces plasticises the base material without melting it. The resulting microstructure has an equiaxed, small grain size and nearly random texture, similar to that of the base materials. Properties of the FSW welds have been reported to be very close to those of the base materials [6].

The aim of this study was to determine where the plastic deformation localizes in different parts of the copper canister. Different manufacturing and welding methods result in different microstructures. Samples of the extruded and forged base materials as well as EB and FSW welds transverse to the welding direction were tensile tested. An optical strain measurement system comprising of a CCD camera and a PC with an imaging software was used to measure and analyze the plastic deformation of the samples during tensile testing. The system records successive images of the sample surface and constructs the deformation field using advanced cross-correlation algorithms. Many studies have shown the usefulness of the OIC/DIC method in determining the mechanical properties of the welds [7-9].

2. Material and Methods

The base material was hot-worked phosphorus-doped (40 ppm) oxygen-free copper (Cu-OFP). The base materials of the forged lids and extruded tubes are slightly different depending on the manufacturing method. Therefore, the base materials of the studied welds were different. Due to

their similarity and restrictions in the amount of available test material, optical strain measurements were made only with tube material of the EB weld and lid material of the FSW weld. Forged and extruded base materials have a small grain size (appr. 75 µm).

Tensile testing was performed by a MTS 810 material test system at three strain rates: 10^{-4}, 10^{-5}, and 10^{-6} 1/s using flat dog-bone shaped tensile test samples. The tensile test samples were electro-discharge machined from full-size welding trial canisters with specific canister geometries for both welding methods. Gauge length of the samples was 40 mm, width 35 mm, and thickness 5 mm. Welds were located at the centre of the gauge length of the cross-weld specimens. Images for optical strain measurement were obtained at a rate of 0.3 Hz during tensile testing. Deformation fields were determined using Strain-Master optical strain measurement system by LaVision. Sample preparation for FEG-SEM/EBSD was made using SiC grinding papers, diamond polishing up to 1 µm, and electrolytic polishing. Zeiss Ultra 55 FEG-SEM and Channel 5 acquisition and analysis software by HKL Technologies were used in SEM and EBSD studies. For macroscopic studies the samples were etched using 50%/50% solution of distilled water and nitric acid for 60 s. Experimental methods to reveal the processing lines of FSW welds were developed using optimized values of electrolytic polishing.

3. Results

Figure 2 shows stress-strain curves for the test samples tested at strain rate 10^{-4} 1/s. It can be seen that the tensile strength of the EB weld is lower than that of the other samples. Tensile strength of the EB weld is 175 MPa while the other samples have a tensile strength of 200 MPa or higher. The extruded tube material has the highest tensile strength of 210 MPa. It can also be seen that the elongation to fracture of the EB weld is significantly lower compared to the other materials. Yield strength was 75 MPa for the tube material and EB weld, 95 MPa for the FSW weld, and 105 MPa for the lid material.

The effect of the strain rate in the studied range was negligible, as the tensile properties and fracture locations were the same for all corresponding test samples regardless of the strain rate. EB welds exhibited lowest tensile properties, while FSW welds showed tensile properties comparable to those of the base materials. **Figure 3** shows the final fracture of the cross-weld specimens of EB and FSW welds exhibiting the location of fracture in EB weld in the middle of the weld and partially along the fusion line between the weld metal and the lid material and in FSW weld fracture occurs along the line of entrapped oxide particles and the nearby processing line. In the base material specimens plastic deformation occurred very uniformly and fracture took place at the centre of the gauge

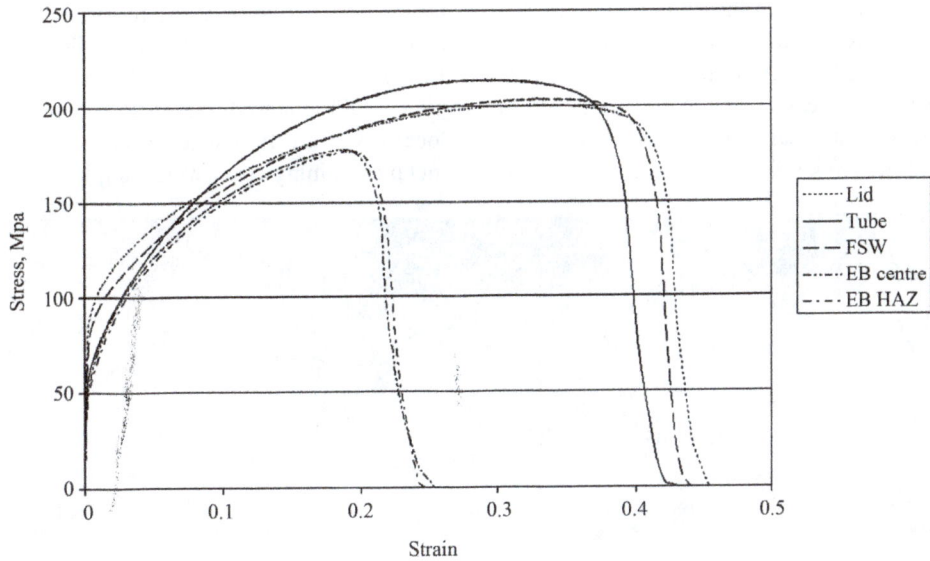

Figure 2. Stress-strain curves of test samples tested at strain rate 10^{-4} 1/s. Tensile strength of the EB weld is low compared to that of the other samples, 175 MPa as compared to 200 MPa or higher and the uniform elongation is about half of that of the FSW weld and the lid and tube base materials.

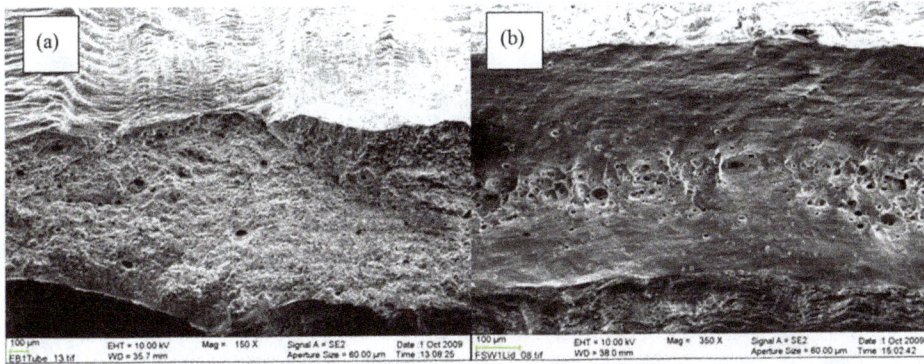

Figure 3. Examples of the fracture surfaces of cross-weld samples of (a) EB weld and (b) FSW weld showing the final ductile fracture mode and location of fracture in the weld metal.

length. Dimples and shear lips indicate a ductile fracture process. EB weld shows ductile shear fracture through the whole specimen. The appearance differs significantly from those of the FSW and the base materials where the shear lips formed on both side surfaces of the specimens.

Macrographs of the transverse sections of the studied FSW and EB welds are shown in **Figures 4** and **1**, respectively. Although both FSW and EB welds are heterogeneous, it can be seen that the grain size variation in the EB weld is significantly higher than that in the FSW weld. In FSW weld the grain size is slightly smaller than in the base materials (60 μm and 75 μm, respectively). The grain size in the EB weld can reach up to 12 mm. Fracture localizes in specific locations in the studied cross-weld specimens similarly at all strain rates used in the mechanical testing. The fracture locations are marked by black lines in the figures. FSW welds fracture either at a processing line or a line of the entrapped oxide particles (see details

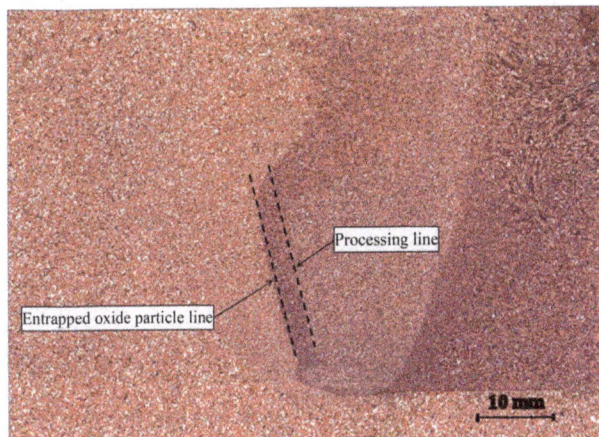

Figure 4. Macrograph of the transverse cross-section of an FSW weld. Tube material is on the right side and lid material on the left side. Different zones of the weld can be clearly seen. Fracture locations are marked with a black line. Thickness of the joint is 50 mm.

in **Figure 5**). Processing lines in the FSW welds are due to the etching effects revealed by the experimental electrolytic polishing method used. As can be seen from **Figures 5(a)** and **(b)** the processing line is not connected to any texture change or with local variation of residual strain in the weld metal. EB welds fractured either at the cen-

treline of the weld or at the fusion line between the weld metal and the lid material (see details in **Figures 6** and **7**). Based on the finite element (FE) modelling of an isotropic Cu-OFP material, the fracture occurs and deformation localizes at a different location than it does in the weld metal containing cross-weld samples.

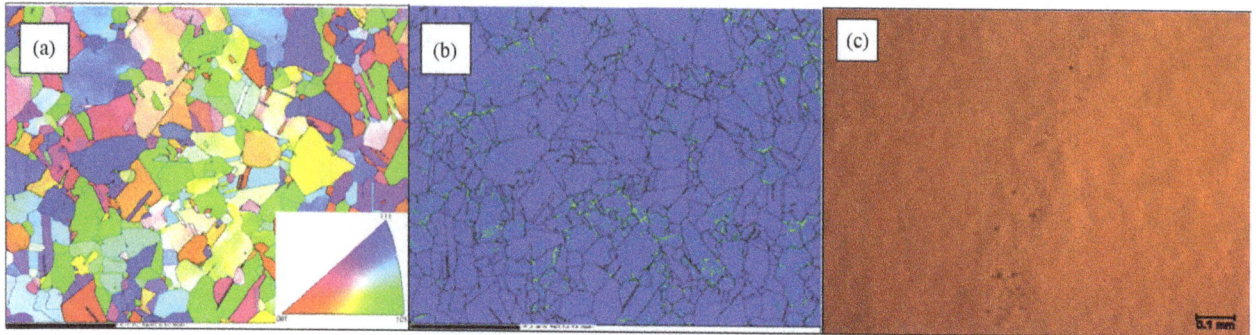

Figure 5. Details of the fracture locations in FSW welds. (a) Inverse pole figure EBSD map and (b) local misorientation map of the processing line (scale bar is 100 μ m) in the middle of the pictures. Colour key for the IPF map is shown in the insert and for the local misorientation map the colour key is the same as in Figure 6. There is no change in texture or distribution of residual strain (green colour) in the area of the processing line. Only a line of slightly smaller grain size can be vaguely discerned. (c) Optical microscopy image of the entrapped oxide particle line.

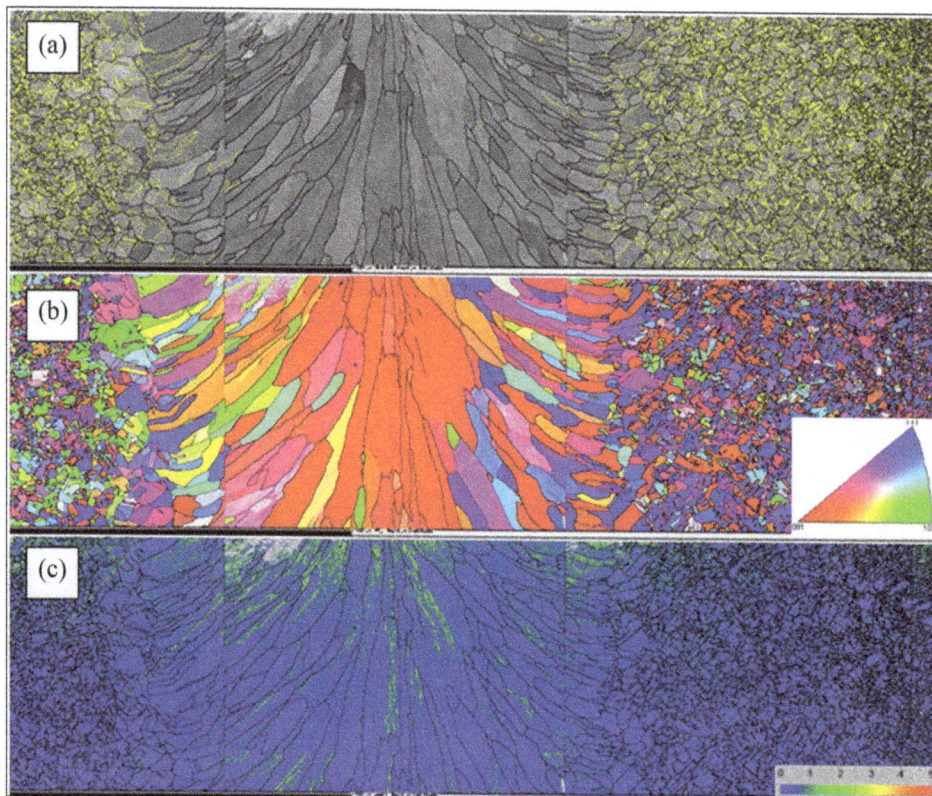

Figure 6. EBSD maps of a horizontal cross-section (top view) of the EB weld; lid base material is on the left side. (a) Pattern quality map shows the distribution of the grain size in the weld metal and the base materials. The yellow lines show twin boundaries inside the grains of the base materials. In the solidified weld metal twin boundaries are not present except in some grains close to the fusion line. (b) Inverse pole figure map (IPF) in the Z-direction shows that in the middle of the EB weld a highly oriented microstructure with <100> direction is dominating. (c) Local misorientation map shows the local concentration of higher strain (green colour) at the centreline of the weld and close to the fusion line between the weld metal and the base materials. Colour keys for the IPF and local misorientation maps are shown in the inserts.

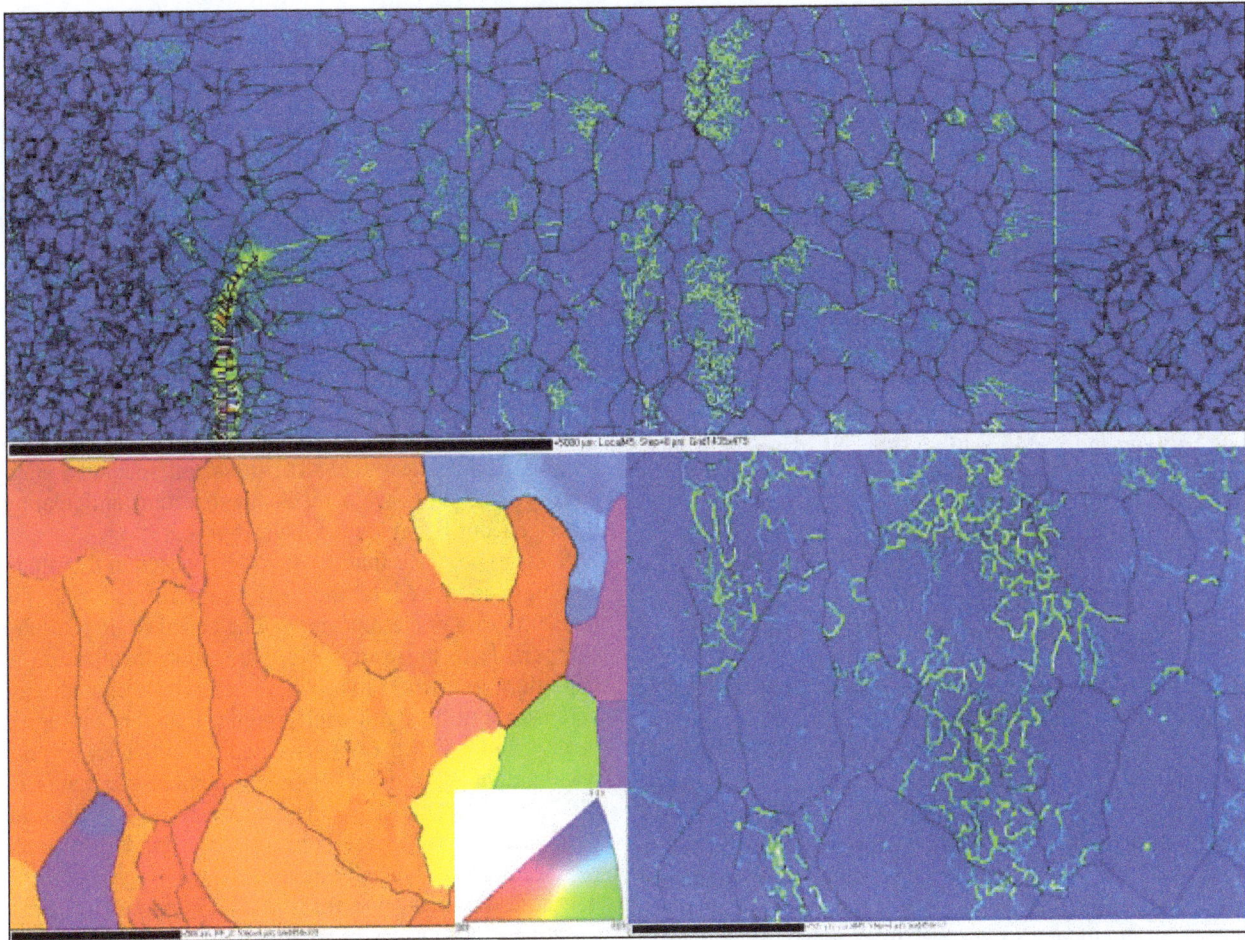

Figure 7. EBSD maps of the transverse cross-section of the EB weld. (a) Local misorientation map of the entire width of the weld; (b) High magnification IFP map of the centre line of the weld (colour key is shown in the insert); (c) Local misorientation map of the same area as in (b). Weld defect close to the fusion line in (a) does not show any marked residual strain accumulation. The colour key for the local misorientation maps is the same as in Figure 6.

Using the specially developed macro-etching technique and electrolytic polishing techniques, different areas and processing lines can be discerned by naked eye from the macro cross-section of the FSW weld. Especially the electrolytic polishing method reveals specific processing lines, which reflect light differently from the surrounding material. Even though these features and processing lines can be detected by naked eye or with small magnification stereo microscopy, they are very difficult to find in a SEM either in normal mode or using the EBSD mapping. This reveals that there are no easily discernible differences in grain size or orientation across these lines, as seen in **Figure 5(a)**. **Figure 5(b)** shows a typical local misorientation map of the dynamically recrystallized structure, in which some areas of the grains are just recrystallized (blue areas) and the continued deformation induces new strain (green areas) in the newly recrystallized grains. Also, in micro-hardness testing there are no differences in hardness across the processing lines. there are no differences in hardness across the processing lines. When using the intentionally too cold

processing parameters some localization of strain and increase in the nanohardness could be detected in connection with the processing lines in our earlier studies reported elsewhere [10]. Only when the processing lines are exactly marked on the FSW weld cross-section it can be surely stated, that the EBSD maps are run across the processing lines. Some of these processing lines are clearly associated with a concentration of oxide particles, but on other processing lines no features explaining their origin can be detected by FEG-SEM. In the lower part of the FSW weld, a change in the morphology of small Cu_2S precipitates has been observed across the processing line [11]. This can be a partial explanation to the optical contrast difference at the boundaries in the macro-section of the FSW weld. **Figure 5(c)** shows a detail of the entrapped oxide particle line.

Figures 6 and **7** show details of the regions of fracture localization in EB welds. In EB welds plastic deformation localizes mainly to the large grains in the middle of the weld or at the fusion line with a large grain size gradient between the weld metal and the forged lid material.

In the middle of the weld metal a highly oriented micro-structure is formed consisting of columnar grains oriented in <100> direction. Local misorientation map shows also that the residual plastic strain is highest in the middle of the weld and that the residual plastic strain distribution follows the elongated columnar grain structure of the weld metal showing highest strain close to the long grain boundaries of the elongated weld metal microstructure. The EB welds contained some small 0.5 mm welding defects (pores), but they had no effect on the localization of plastic deformation. The large grain size gradient between the base metals and EB weld metal localizes plastic deformation and causes sometimes fracture close to the fusion line between the weld metal and the lid material. Additional reason for cracking at the fusion line between the lid material and the weld metal is the high yield stress of the lid material causing a mechanical property gradient at this location.

4. Discussion

The effect of the strain rate in the studied range was negligible, as the tensile properties and fracture locations were the same for all corresponding test samples regardless of the strain rate. EB welds exhibited lowest tensile properties, while FSW welds showed tensile properties comparable to those of the base materials. The solid state nature of FSW is most likely the reason for the excellent mechanical properties of the FSW welds. Contrary to EB, there is no melting, no large solidified columnar grains or grain size gradients, and no local anisotropy or residual plastic strain gradients. Microstructure of the FSW welds is similar to that of the hot-worked base materials and there is also very little difference in the mechanical properties despite the processing lines and entrapped oxide particle lines of the FSW welds [12]. Also the creep testing results of Andersson *et al.* [13,14] at temperatures ranging from 20°C to 175°C at various testing configurations show that FSW welds have similar creep life and ductility as those of the base materials.

Laboratory mechanical testing of cross-weld specimens using small standard test samples across narrow, a few millimeter wide welds may give a wrong view of the localization of plastic deformation and ductility of the welds. It is important to use full-thickness samples covering the whole weld to have a realistic constraint of the weld structure with different microstructure and mechanical property gradients. For example, Lee and Jung [15] studied the transverse tensile strength of 4 mm thick FSW welds of commercially pure copper with small dog-bone cross-weld specimens. They noticed that fracture occurred outside the weld nugget in the HAZ in the area of lowest hardness. Ollonqvist [16] measured the tensile properties of the prototypical EB welds using the proportional cross-weld tensile specimens with rectangular cross-section (10 mm

× 8 mm). In the cross-weld testing failure occurred always in the centreline of the EB weld, but the measured elongations to fracture were high and comparable to the base material testing results. Specimens comprising the whole weld area used in this study give a different view of the ductility of the EB welds. The uniform elongation and the elongation to fracture are much smaller than the corresponding values of the base materials and the FSW weld. This is due to constraints caused by the large gradients in the grain size and orientation not covered by small standard specimens. Therefore, the present kind of mechanical testing of heterogeneous weld microstructures gives more realistic view of the strain localization and fracture in the welds of the thick-wall copper canisters of spent nuclear fuel than the small specimen standard type of cross-weld mechanical testing anticipating homogeneous behaviour of the samples.

It has to be noted that the strain rates used in this study are high compared to the real strain rates during the 100,000 years in the repository. The actual strain rates are comparable to those of creep. The strain localization at extremely low strain rates combined with elevated temperatures will be significantly higher than that presented in this study. Creep of EB welded 50 mm thick copper was studied by Aalto [17] in the temperature range of 20°C - 150°C. The EB welds are clearly weaker than the base material. The creep failure occurred close to the weld centreline especially in the low stress region of creep. It was also noticed that fracture was intergranular in the weld metal. The cause of the creep strength reduction was thought to be due to the cast microstructure and the strong anisotropy of the EB weld. Andersson *et al.* [13] studied creep of copper EB welds at temperatures ranging from 75°C to 175°C. Also their results show that the main creep strain accumulation and fracture occurred in the weld metal. Thus, the present mechanical tests give also indications of the strain localization tendency in various heterogeneous microstructures of thick-wall copper welds in creep.

Welding of thick-wall copper canister results in the development of heterogeneous microstructures and, since microstructure and properties are related, welds will exhibit variations in strength and ductility. Laboratory testing using standard cross-weld small specimen geometries anticipating homogeneous plastic deformation cannot give critical information on strain localization and damage development in the variable heterogeneous microstructure of the welds where deformation is heterogeneous. Strain localization and local damage development are significant in assessing the in-service mechanical failure processes, failure criteria and behaviour of the copper canister.

5. Conclusions

Based on the results obtained in this study, the following

can be concluded about localization of plastic deformation in different parts of the copper canister for spent nuclear fuel:

1) In FSW welds fracture occurs mainly at a processing line and occasionally on the line of entrapped oxide particles.

2) In EB welds fracture occurs in the middle of the weld or at the fusion line between the weld metal and the lid material.

3) In base materials deformation occurs uniformly.

4) Tensile strength and elongation to fracture of the EB welds were considerably lower compared to the base materials and FSW welds, which exhibited similar mechanical properties.

5) The microstructure of the EB weld shows large variable grain size with high anisotropy of the large columnar grains as well as residual strain accumulation in the middle of the weld and high grain size gradient at the fusion line between the weld metal and the base materials. The FSW weld shows similar hot worked microstructure with low localized residual strain accumulation as the base materials.

6. Acknowledgements

The authors wish to thank KUUMA, KYT2010, and KYT 2014 projects for funding, the Swedish Nuclear Fuel and Waste Management Co. (SKB) for FSW welds, Posiva Oy for EB welds, and Kim Widell for help in tensile testing and optical strain measurement.

REFERENCES

[1] B. Rosborg and L. Werme, "The Swedish Nuclear Waste Program and the Long-Term Corrosion Behaviour of Copper," *Journal of Nuclear Materials*, Vol. 379, No. 1-3, 2008, pp. 142-153.

[2] D. W. Shoesmith, "Assessing the Corrosion Performance of High-Level Nuclear Waste Containers," *Corrosion*, Vol. 62, No. 8, 2006, pp. 703-722.

[3] B. Wilshire and C. J. Palmer, "Strain Accumulation during Dislocation Creep of Prestrained Copper," *Materials Science and Engineering: A*, Vol. 387-389, 2004, pp. 716-718.

[4] U. Ehrnstèn, P. Aaltonen, P. Nenonen, H. Hänninen, C. Jansson and T. Angeliu, "Intergranular Cracking of AISI 316 NG Stainless Steel is BWR Environment," *Proceedings of the Tenth Symposium on Environmental Degradation of Materials in Nuclear Power Systems—Water Reactors*, Lake Tahoe, Nevada, 5-9 August 2001.

[5] S. Kou, "Welding Metallurgy," 2nd Edition, John Wiley and Sons, New York, 2003.

[6] L. Cederqvist, "FSW to Seal 50 mm Thick Copper Canisters—A Weld that Lasts for 100,000 Years," *Proceedings of the Fifth International Friction Stir Welding Conference*, Metz, France, 14-16 September 2004.

[7] W. D. Lockwood and A. P. Reynolds, "Simulation of the Global Response of a Friction Stir Weld Using Local Constitutive Behaviour," *Materials Science and Engineering A*, Vol. 339, No. 1-2, 2003, pp. 35-42.

[8] C. Genevois, A. Deschamps and P. Vacher, "Comparative Study on Local and Global Mechanical Properties of 2024 T351, 2024 T6 and 5251 O Friction Stir Welds," *Materials Science and Engineering A*, Vol. 415, No. 1-2, 2006, pp. 162-170.

[9] W. D. Lockwood, B. Tomaz and A. P. Reynolds, "Mechanical Response of Friction Stir Welded AA2024: Experiment and Modelling," *Materials Science and Engineering A*, Vol. 323, No. 1-2, 2002, pp. 348-353.

[10] K. Savolainen, T. Saukkonen and H. Hänninen, "Banding in Copper Friction Stir Weld," *Science and Technology of Welding & Joining*.

[11] T. Saukkonen, K. Savolainen, J. Mononen and H. Hänninen, "Microstructure and Texture Analysis of Friction Stir Welds of Copper," In: A. D. Rollet, Ed., *Materials Processing and Texture*: *A Collection of Papers Presented at the 15th International Conference on Textures of Materials*, John Wiley & Sons, Hoboken, 2008, pp. 53-60.

[12] K. Savolainen, T. Saukkonen and H. Hänninen, "Optical Strain Measurement of Plastic Strain Localization in Nuclear Waste Copper Canisters," *Proceedings of Baltica VIII—International Conference on Life Management and Maintenance for Nuclear Power Plants*, VTT, Espoo, Finland, 9-14 August 2009, pp. 163-177.

[13] H. C. M. Andersson, F. Seitisleam and R. Sandström, "Creep Testing of Thick-Wall Copper Electron Beam and Friction Stir Welds at 75, 125 and 175°C," SKB Report TR-05-08, SKB, Sweden, 2005.

[14] H. C. M. Andersson, F. Seitisleam and R. Sandström, "Creep Testing and Creep Loading Experiments on Friction Stir Welds in Copper at 75°C," SKB Report TR-07-08, SKB, Sweden, 2007.

[15] W.-B. Lee and S.-B. Jung, "The Joint Properties of Copper by Friction Stir Welding," *Materials Letters*, Vol. 58, No. 6, 2004, pp. 1041-1046.

[16] P. Ollonqvist, "Microstructural Characterization and Mechanical Properties of Electron Beam Welded Thick Phosphorous Microalloyed Oxygen Free Copper (Cu-OFP)," M.Sc. Thesis, Helsinki University of Technology, Helsinki, Finland, 2007.

[17] H. Aalto, "EB-Welding of the Copper Canister for the Nuclear Waste Disposal," Posiva Report 98-03, Posiva, Finland, 1998.

2

External and Environmental Radiation Dosimetry with Optically Stimulated Luminescent Detection Device Developed at the SCK·CEN

Reinhard Boons, Mark Van Iersel, Jean Louis Genicot*
Radiation Dosimetry and Calibration-Environmental Health and Safety
SCK-CEN, Belgian Nuclear Research Center, Mol, Belgium

ABSTRACT

The laboratory of Radiation Dosimetry and Calibration of the Belgian Nuclear Research Centre (SCK·CEN) is using thermoluminescence dosimetry for more than thirty years for routine measurements and for R & D investigations. In 2002, it has developed an experimental device based on the optically stimulated luminescence (OSL) technique. This device is working with AL_2O_3:C crystals stimulated by the green line (488 nm) emitted by a 150 mW argon laser. This paper describes this device, its characteristics, some applications in space dosimetry and the R & D works initiated in this field during the next few years.

Keywords: External Dosimetry; OSL; Optically Stimulated Dosimetry

1. Introduction

Luminescence applied to radiation dosimetry is used in research in different application fields (geology, dating, radiotherapy, retrospective dosimetry, space dosimetry ···). SCK·CEN is using thermoluminescence detectors (TLD) for more than 30 years for routine control of the personal professionally exposed to ionizing radiations. The optically stimulated luminescence (OSL) technique also has been initiated in 2002 at the SCK·CEN for the same applications. A facility, using an argon laser has been developed to ensure the most versatile capabilities in the measurement of doses in different area with the best detection limits. This prototype device has been conceived in its first design by Dr. O. Goossens.

The main purpose of this R & D study was to extend the capabilities of the laboratory by taking profit of the different characteristics of the OSL detectors: their large dynamic range, their better sensitivity and reusability, their low fading and particularly the easy way they are used in comparison with the TLD techniques. The Al_2O_3:C crystal was chosen as first material to study the OSL techniques.

Taking part in different space research programs, the SCK·CEN enforced the laboratory to enlarge the number of detector types to optimize the measurement of absorbed and equivalent doses in the radiation environment found in the International Space Station [1].

2. Description of the Laboratory for External Dosimetry

When the first Belgian Reactor BR1 became operational in 1956, the film dosemeter was used to measure the effective doses to the body, the skin and the fingers in case of irradiation with γ-, X-, β-rays and neutrons and also in case of criticality accident. After modification of the film badge design, this technique has been used routinely until 1986 when it was replaced by the thermoluminescence detector (TLD) using three LiF crystals in a one-inch diameter plastic box. The main advantages of the TLD was a larger dynamic range, extending the irradiation period to one month instead of one week and an easier technique to read the detectors. The TLD technique is used today for routine measurements and accepted as legal mean for external dose assessments.

Each of the three LiF detectors (model TLD-100 also called MTS type) is measured by counting the light emitted when the crystal is heated to a constant temperature of 190°C during 13 seconds. The measurements are performed with an "Automatic TLD reader 4000" developed by NRG-Re in Arnhem, Netherlands. More than 6600 detectors batches are measured every month.

The TLD techniques have some limitations: the information is completely lost after the reading and the temperature required to read a detector depends on the detector type and can be high (>400°C). The measurement must be done with a photomultiplier in a special place

---

*Corresponding author.

External and Environmental Radiation Dosimetry with Optically Stimulated Luminescent Detection Device Developed at the SCK·CEN

9

protected from ambient light. The Teflon™ TLD and the MOSFET active detector are also used for special applications.

3. Requirements for OSL Dosimetry

The development of optically stimulated luminescence (OSL) techniques has been initiated at the SCK·CEN in 2002. The main purpose of this R & D work was to extend the capabilities of the laboratory by taking profit of the different advantages of this technique: larger dynamic range, better sensitivity and reusability (it was observed that TLD's can lose their sensitivity after repeated use, a default attributable to a redistribution of the impurities in the lattice after several thermal treatments of the crystal). The OSL technique is also easy to use.

The first crystal for OSL proposed by A. Romanovski in 1955 [2] was SrS:Eu,Sm. The Al_2O_3:C crystal has been chosen as a study material because of different advantages. It can be used in TL mode and is 30 to 60 times more sensitive than TLD-100 [3]. Al_2O_3:C is several times more sensitive in OSL than in TL [4].

Other parameters required for OSL dosimetry are the reproducibility, the stability, the linearity and the dose-rate independence [5]. In OSL materials, the dose dependence is first linear, followed by an approach to saturation. Assuming no dose-rate effects, one can calibrate the sample at high dose-rates and deduce the dose collected at a much lower rate. The stability of the stimulation light source and of the light sensitive device (e.g. the photomultiplier tube) is also important.

OSL presents other advantages over TL: the heating of the sample is not necessary, avoiding the blackbody radiation; they do not exhibit thermal quenching of luminescence and their sensitization effect is significantly less in OSL. A problem is present with Al_2O_3:C: each reading remove 0.2% of signal sensitivity so that the detectors are always recalibrated after each measurement.

4. Descriptions of the Experimental Set-Up

4.1. General Description

An Ar^+ laser (model Melles Griot 35 LAP 431) with a light output power of 150 mW has been chosen as stimulating light source (**Figure 1**). The optimum stimulation wavelength for α-Al_2O_3:C is 480 nm and the Ar^+ laser is able to generate an output at 488.8 nm and another one ay 515.70 nm. The second (green) wavelength has been selected because special effects may appear when blue light is used in OSL dosimetry [6]. The light output power of the laser is constant (150 mW) but the light stimulating the detector can be reduced by placing a neutral filter just before the crystal to minimize the scattering.

An electrical shutter is placed just after the laser module (part A) followed by a long pass filter. The laser beam

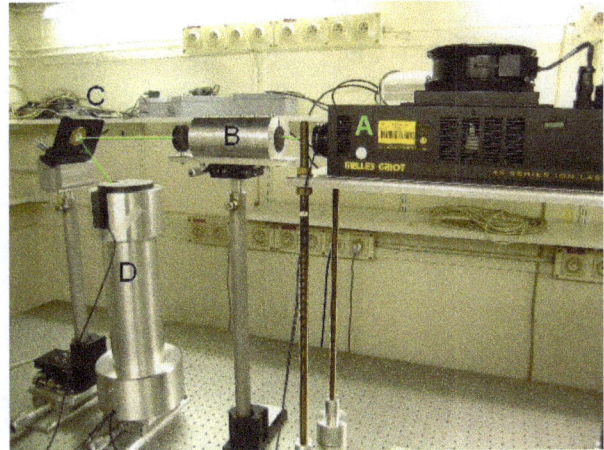

Figure 1. Layout of the OSL facility (A: Ar laser—B: Beam expender C: Mirror—D: tube with detector support, optics and PMT).

diameter, too small for the purpose (0.72 mm), is increased to 3.05 mm with a plano-concave lens followed by a confocal plano-convex lens.

The laser beam is reflected by a surface mirror to the sample placed on a quartz window above the detection module: the detector crystal emits a quite narrow light spectrum centered at 420 nm which is measured by a photomultiplier tube (PMT) with appropriate filters.

The photomultiplier tube is placed in a vertical aluminum tube to be protected from the external light except from the quartz window on top of the aluminum tube. Between the quartz window and the PMT, plano-concave and plano-convex lenses modify the shape and the size of the stimulated light emitted by the detector sothat the photocathode of the photomultiplier is completely illuminated. A filter pack eliminates the parasitic light and an electric shutter, placed before the PMT, protects it between measurements (**Figure 2**). With the aid of a microcontroller (ST62E20), this shutter is first opened and 0.3 s later the laser's shutter is opened to stimulate the detector. When the preset counting time is reached, both shutters are closed together.

A parabolic mirror, above the detector crystal reflects the useful light emitted upwards by the detector back to the PMT. This mirror is slightly ex-centered to avoid redirecting the light to the detector. The typical result of an OSL measurement is presented in **Figure 3**.

4.2. Detection Limits

The detection limit (a posteriori) is defined as:

$$LLD = \frac{3 \cdot \sigma}{C} \cdot D$$

where σ = standard deviation on the background count; C = maximum counts of the measurement during the calibration; D = Dose used for the calibration (Gy).

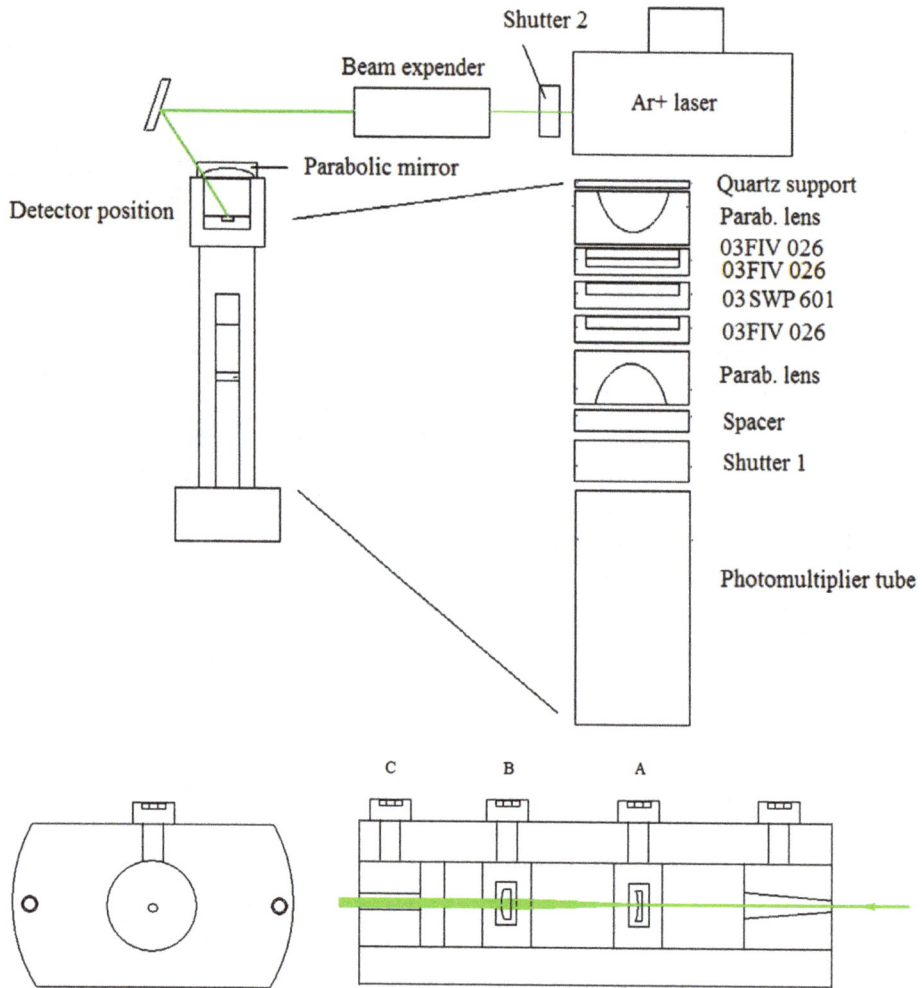

Figure 2. Layout of the OSL device using an Ar⁺ Laser and details of the beam expender.

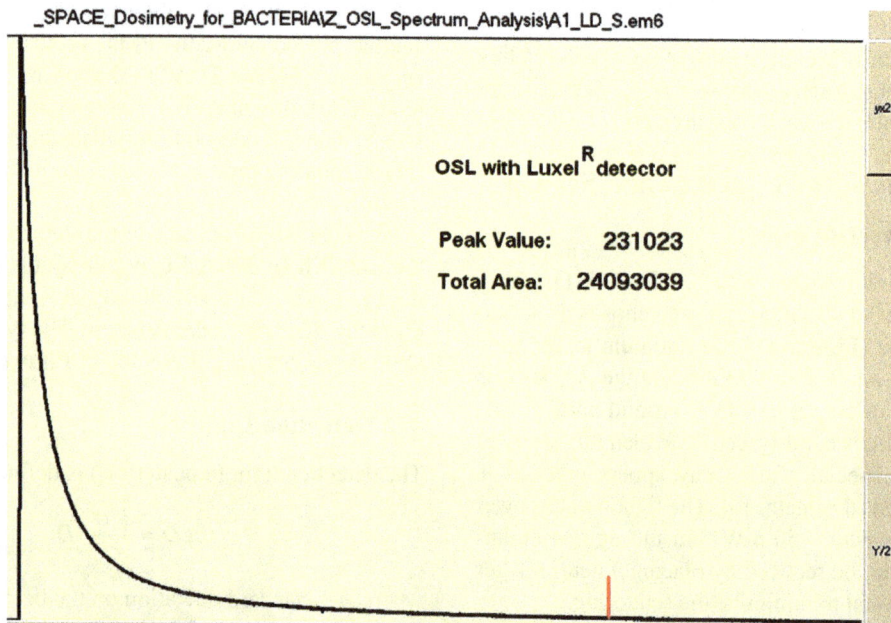

Figure 3. Result of OSL measurement (Luxel™ detector, Gamma irradiation from ¹³⁷Cs source).

This *LLD* is determined with the method developed by McKeever *et al.* [7].

4.3. Detectors Used in OSL

Two types of detectors are presently used in the laboratory: TLD-500 and Luxel™. The TLD-500 is a α-Al$_2$O$_3$:C mono-crystal. Before irradiation, this detector is annealed at 400°C during 60 minutes. After growth, the ingot of α-Al$_2$O$_3$:C is cut in crystal about 1 mm thick. This leads to discrepancies in the thickness, weight and transparency of the different detectors. The individual mass and transparency of 78 different crystals have been measured (**Figures 4** and **5**). These discrepancies lead to the necessity to use an "individual factor" defined as:

$$IF_i = \frac{S_i}{\frac{1}{n}\sum_{i=1}^{n} S_i}$$

S_i is the count measured during the illumination of detector *i* and n is the number of detectors in the lot. Measurements show that the IF can vary from 0.55 to 2.04 for TLD-500.

The detector, with the trademark Luxel™, is a polycrystalline form of the same material glued between two sheets of plastic (Ertalon™). This Luxel™ is easily bleached with exposition to daylight: the residual dose information is reduced to 1.2% after 12 hours and to 0.21% after 24 hours. The light transmittance of the Luxel™ is shown in **Figure 6**.

The transmittance of the detector is defined as the ratio between the measurement of a monochromatic light beam (from an L.E.D.) in a defined geometry and the same measurement when the detector is placed between the light source and the detector. The device to measure the transmittance is shown in **Figure 7**. An L.E.D. selected for its wavelength is connected to a constant current generator controlled by the computer via a USB connection card (K8055 from Velleman®). The transmitted light is measured with a TSL230 (fromTAOS) through a Propeller microcontroller array from Parallax®. The value for Al$_2$O$_3$:C is calculated from 50 measurements for each wavelength.

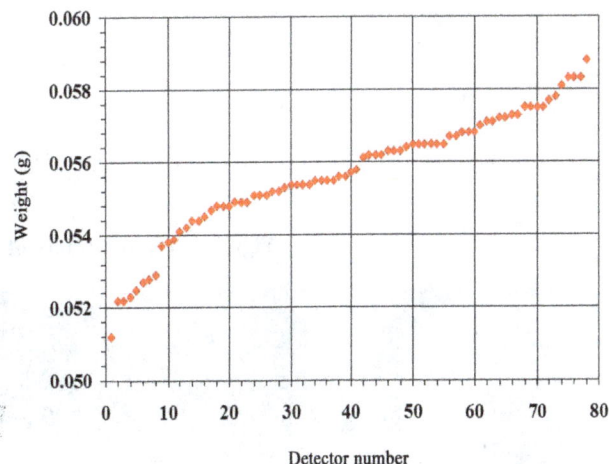

Figure 4. Masses (in g) of 78 different TLD-500 detectors.

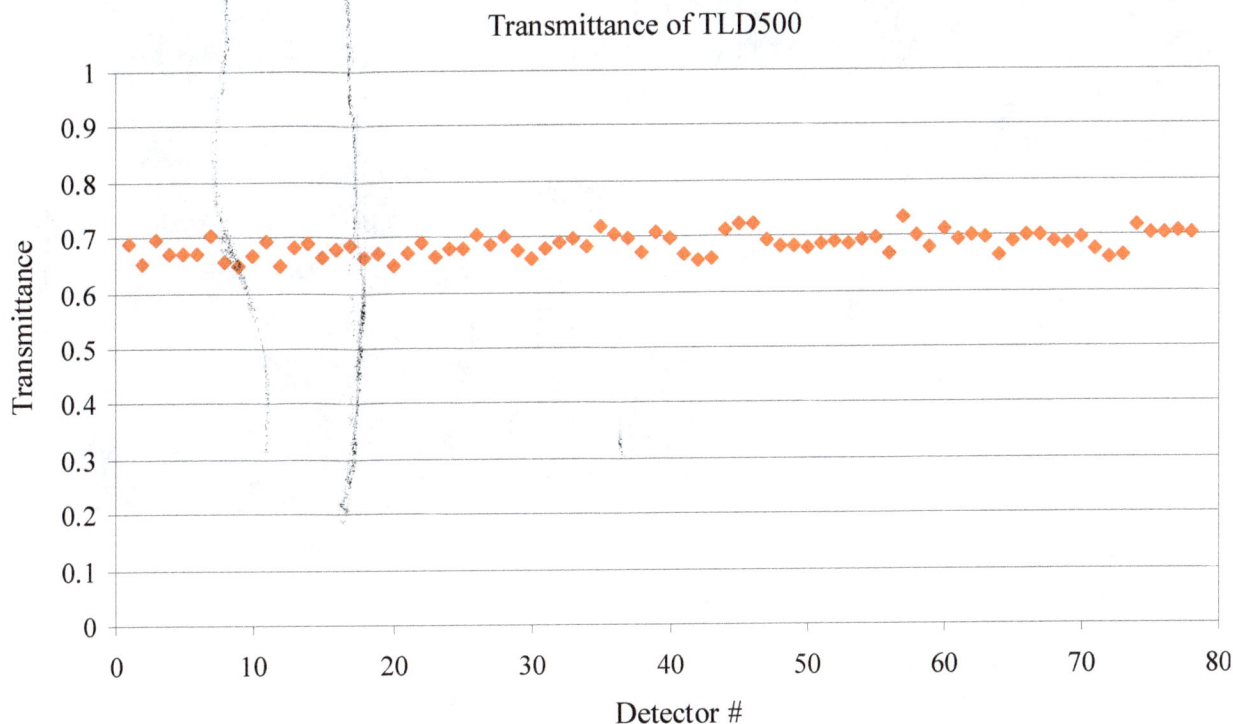

Figure 5. Transmittance of 78 different TLD-500 detectors.

Transmittance of Luxel$^{(TM)}$

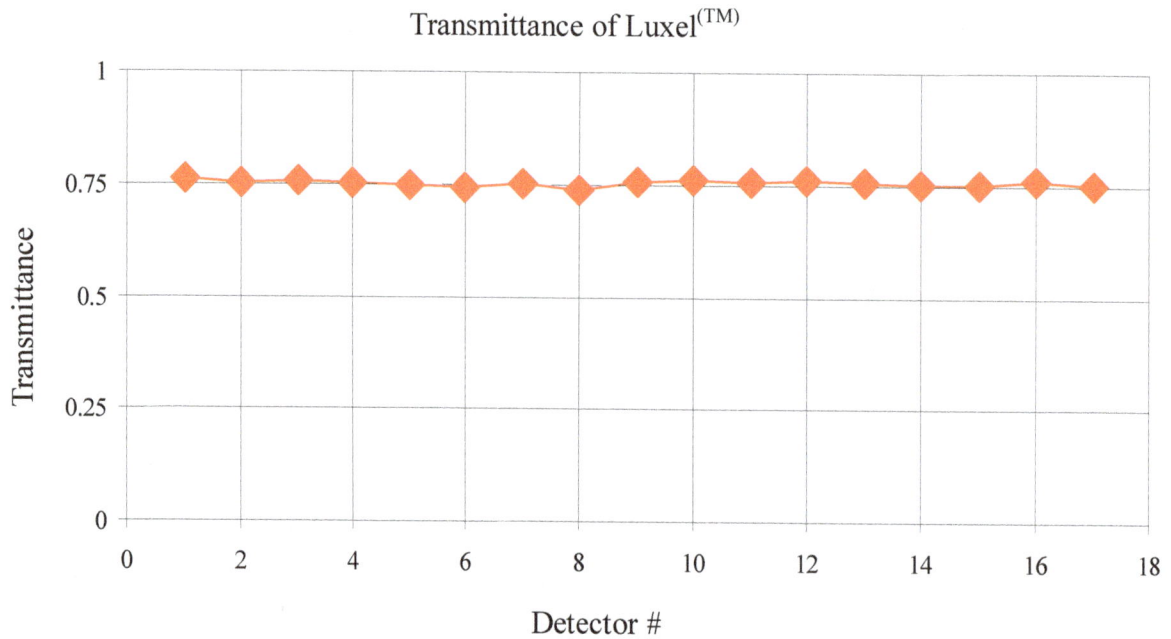

Figure 6. Transmittance of 17 different Luxel™ detectors.

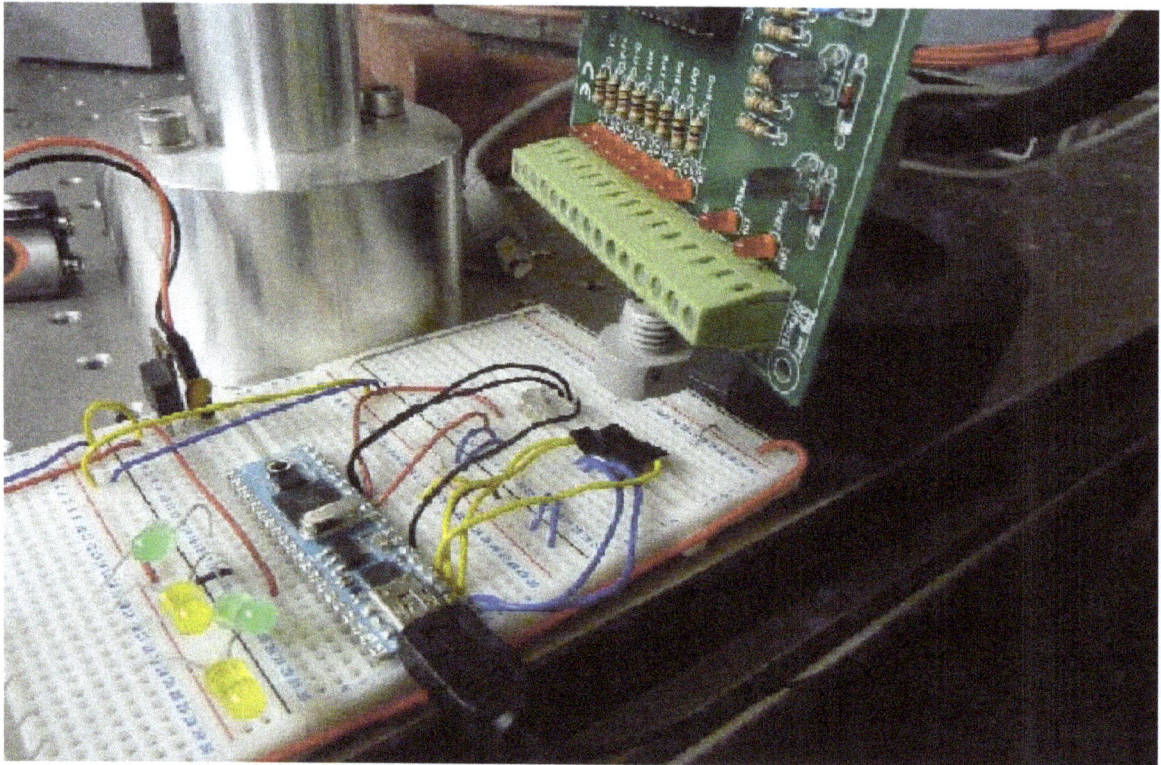

Figure 7. Mockup for the measurement of the attenuation coefficient of detectors used in OSL dosimetry.

The higher thickness of the TLD-500 with respect to the Luxel™ leads to another difference between the two detectors when they are read in transparency mode. **Figure 8** shows that when a TLD-500 is reversed during the measurement, more signal can be collected from the other side of the crystal. This effect is not visible with the Luxel™

detector (upper curve). These results show that there is an optimal thickness of the detector when it is read by transparency. This thickness is calculated from the attenuation coefficients of the material, for the two wavelengths implicated in the OSL process, with the following relation:

External and Environmental Radiation Dosimetry with Optically Stimulated Luminescent Detection
Device Developed at the SCK·CEN

13

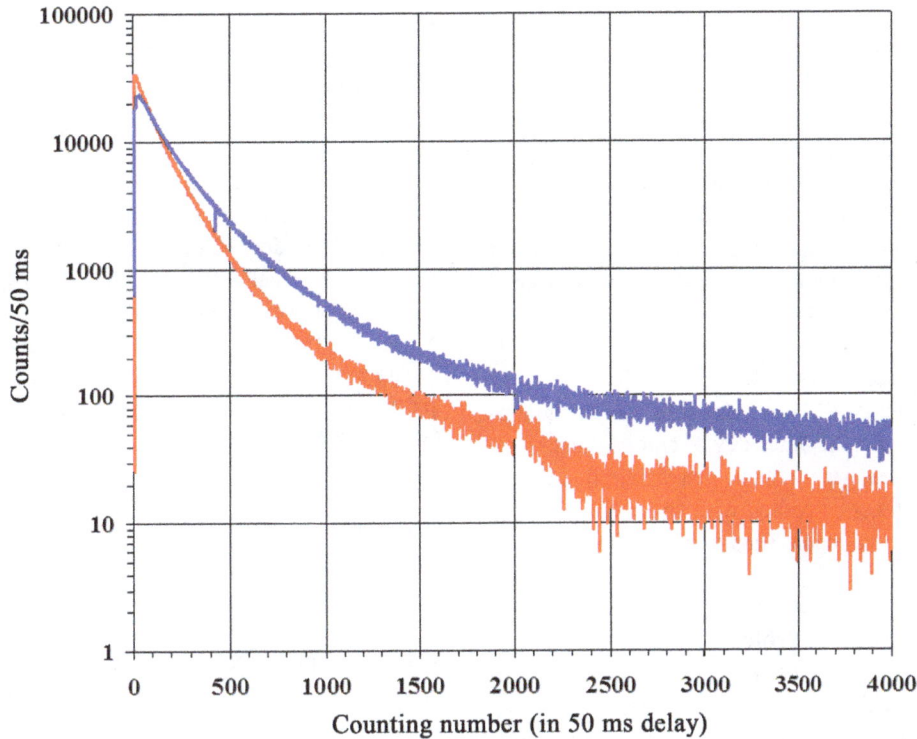

Figure 8. Effect of Recto/Verso measurement on TLD-500 detector (lower curve) and on Luxel™ detector (upper curve).

$$I = \frac{I_0 \cdot k}{\mu - \mu'}\left(e^{-\mu' \cdot t} - e^{-\mu \cdot t}\right)$$

where: I is the measured intensity of light emitted by the detector; I_0 is the stimulation light intensity; k a factor considering all other losses; μ, μ' the extinction coefficients for stimulation light and for stimulated light respectively; t the thickness of the detector.

Fifty detectors have been measured with the device described in **Figure 7** using two different wavelengths. The measured extinction coefficients are 2780 m^{-1} and 3070 m^{-1} for blue (465 nm) and UV (395 nm) respectively. These values introduced in the previous relation give the curve I/I_0 for different thicknesses presented in **Figure 9**. The optimal thickness for α-Al$_2$O$_3$:C is 3.4 μm. This value explains the effects seen in **Figure 8** and the advantage of the Luxel™ type.

4.4. Characteristics of the OSL Device

LiF:Mg,Ti (TLD-100) detectors used at the SCK·CEN has a detection limit L$_D$ of 50 μGy and a 7 decades dynamic range. Al$_2$O$_3$:C is able to detect doses as small as 20 μGy and presents an 8 decades dynamic.

Experiments showed that with this device, the detection limit is 7.2 μGy for the Luxel™ and 6.5 μGy for the TLD-500. The Luxel™ is linear between 0.1 and 100 mGy, becomes sublinear between 100 mGy and 1 Gy afterwards it saturates [6]. The reproducibility is 3.1% for Luxel™ and 8.4% for TLD-500 in OSL measurements. Each read-

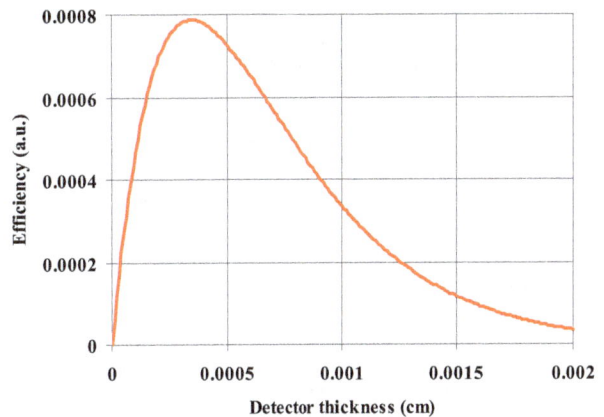

Figure 9. Evolution of the measured stimulated light from an Al$_2$O$_3$:C detector versus its thickness for a constant stimulation light (Optimum thicknesses: 3.4 μm).

ing of an Al$_2$O$_3$:C detector removes 0.2% of its sensitivity. For this reason and to avoid systematic errors due to loss of powder, the Luxel™ detector is preferably used only one time. This is possible because of its low price. The detector is first annealed with daylight through a window for about one day then used for the dose measurement and calibrated afterwards with a known dose. The detector is then discarded.

4.4.1. Stability of the Measurement Device

Different points are to be checked concerning the stability in an OSL measurement device: the stimulation light source, the stimulated light measurement device and the ambient light protection.

4.4.2. The Stimulation Light Source

The Ar^+ laser used in the device is characterized by an excellent stability of the output light intensity (better than 3%). This stability has been controlled frequently and presented a problem only one time due to a defect in the power supply leading to oscillations in the light output power as shown in **Figure 10**.

4.4.3. The Photomultiplier Tube

It is well known that a PMT is quite sensitive to the ambient temperature. The Ar^+ laser device and its power supply are important heat generators so that the confined room of the laboratory must be cooled with an air conditioning system to guarantee a stable sensitivity of the PMT and accurate measurements.

4.4.4. The Ambient Light Protection

An OSL measurement requires a complete protection from the ambient light which is not easy to guaranty when the measurement of light is involved with a big laser which must be kept outside the confined volumes of the detector and the PMT for cooling.

5. Applications of OSL Dosimetry

The laboratory of Microbiology at SCK-CEN, in collaboration with different universities, participates in several

ESA programs related to bacterial experiments in space: research program MESSAGE (Microbial Experiments in the Space Station About Gene Expression) studied the effect of space conditions on micro-organisms in general using some well-known bacteria. During another experiment, MESSAGE 2, the samples and detectors stayed in space for ten days, of which eight were in the service module of the ISS. This dosimetry experiment was a collaboration between different institutes (School of Cosmic Physics, Institute for Advanced Studies, Dublin, Ireland, Johnson Space Centre Houston, USA, Department of Radiation Dosimetry, National Physics Institute, Czech Republic Department of Physics, Oklahoma State University, Stillwater, USA and SCK-CEN, Mol, Belgium), so that the doses could be estimated by different techniques. For the high LET doses (>10 keV/μm), two types of track etch detectors were flown. The low LET part of the spectrum was measured by three types of thermoluminescent detectors (^{7}LiF:Mg,Ti; ^{7}LiF:Mg,Cu,P; Al_2O_3:C), and by the optically stimulated luminescence technique with Al_2O_3:C detectors, both in continuous and pulsed mode. The high LET results were of the order of 0.13 mGy or 2.4 mSv and LET spectra were obtained. For low LET radiations, small differences between different techniques and detectors were observed, ranging between 1.5 and 1.9 mGy, but a general agreement was observed. The differences may be generally understood from the different efficiencies of the different methods to heavy charged particles. OSL can find useful applications in retrospective dosimetry by measuring the tooth enamel for instance [8,9].

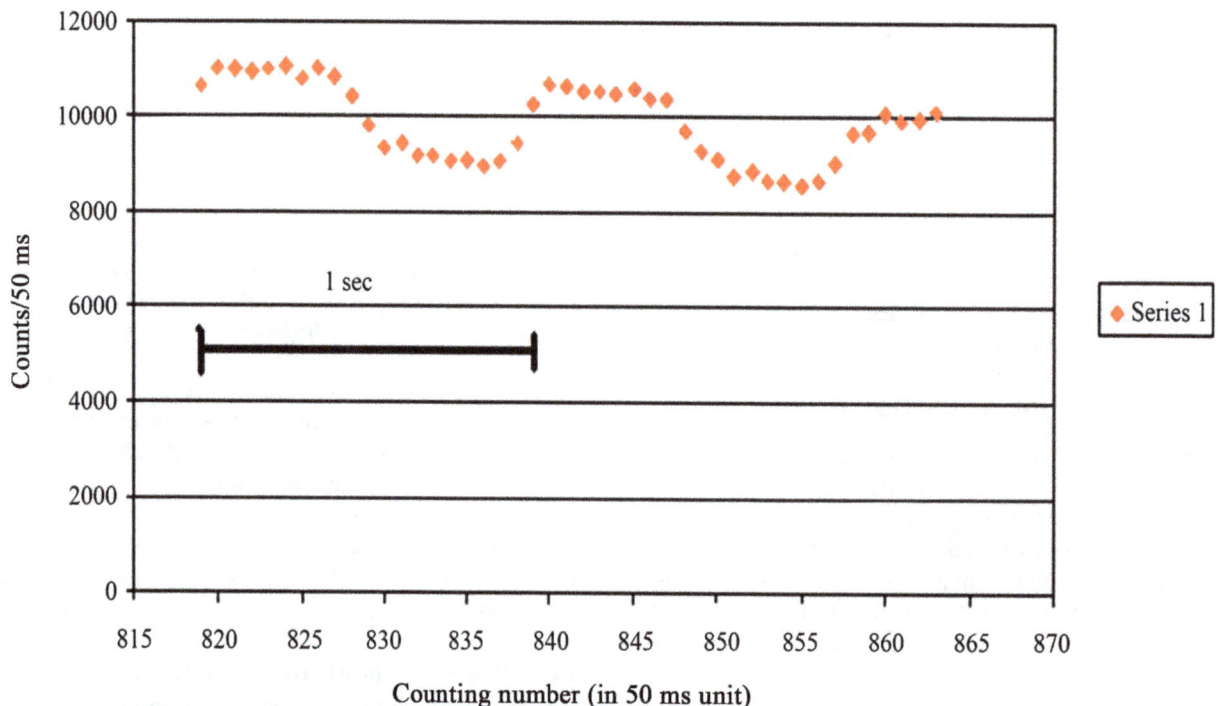

Figure 10. Instability of the Ar^+ laser power output.

External and Environmental Radiation Dosimetry with Optically Stimulated Luminescent Detection
Device Developed at the SCK·CEN

15

6. Future Developments

The device using the Ar^+ laser has shown real interest as an OSL dosimetric tool but requires small improvements. Protection to ambient light and protection to mechanical vibrations has been the first points to improve. A future step will implement a simplification of the operations to change the detectors between measurements. The Ar^+ laser will also be replaced by a cheaper YAG laser with a frequency doubler or by a diode pumped solid state (DPSS) laser. This DPSS laser comprises a laser diode emitting in the infra-red domain followed by a frequency doubler to generate the green color best suited for OSL applications. A prototype of the modified device is shown in **Figure 11**. This laser is characterized by a short term instability of the beam intensity (**Figure 12**) followed by a very stable emission after about 10 minute (**Figure 13**). Peltier cells will be added to the reading device around the PMT holder to reduce the leakage current and keep the efficiency constant. The microcontroller driving the shutters will also be replaced by another one with other duties. The mechanical shutters have shown their utility also in pulsed OSL techniques [10] but they should be advantageously replaced by a Pockel cell or by a pulsed L.E.D. to improve the speed and allow different dynamical studies

7. Conclusions

The innovation brought by OSL techniques in the external dosimetry has extended the capabilities of the SCK·CEN in term of personal dosimetry, environmental surveys, space and retrospective dosimetry. The OSL dosimetry has been initiated with an argon laser with excellent detection limits and a very good reproducibility of measurements. It could be improved and easily extended to other excitation light sources (green laser diodes or high power LED's). The use of the AL_2O_3:C in the form of powder between two sheets of polyester (Luxel™ detectors) will improve the precision of the measurements. The cheap price of this type of detectors will allow the single use technique with appreciable reduction of interpretation errors.

In spite of the quality of the present device, it will be necessary to develop a device, more easy to use with selected green or blue L.E.D.s.

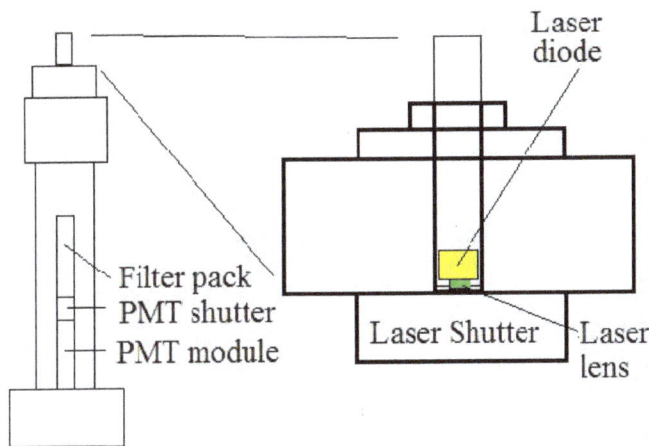

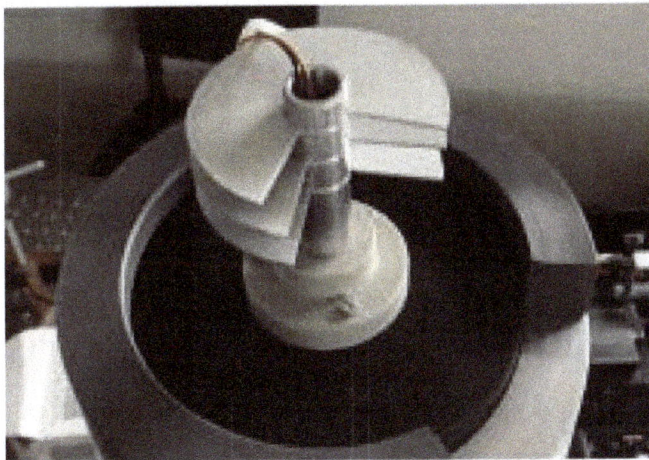

Figure 11. Description of the modified OSL device for stimulation with a DPSS laser.

DPPSS Laser Stability

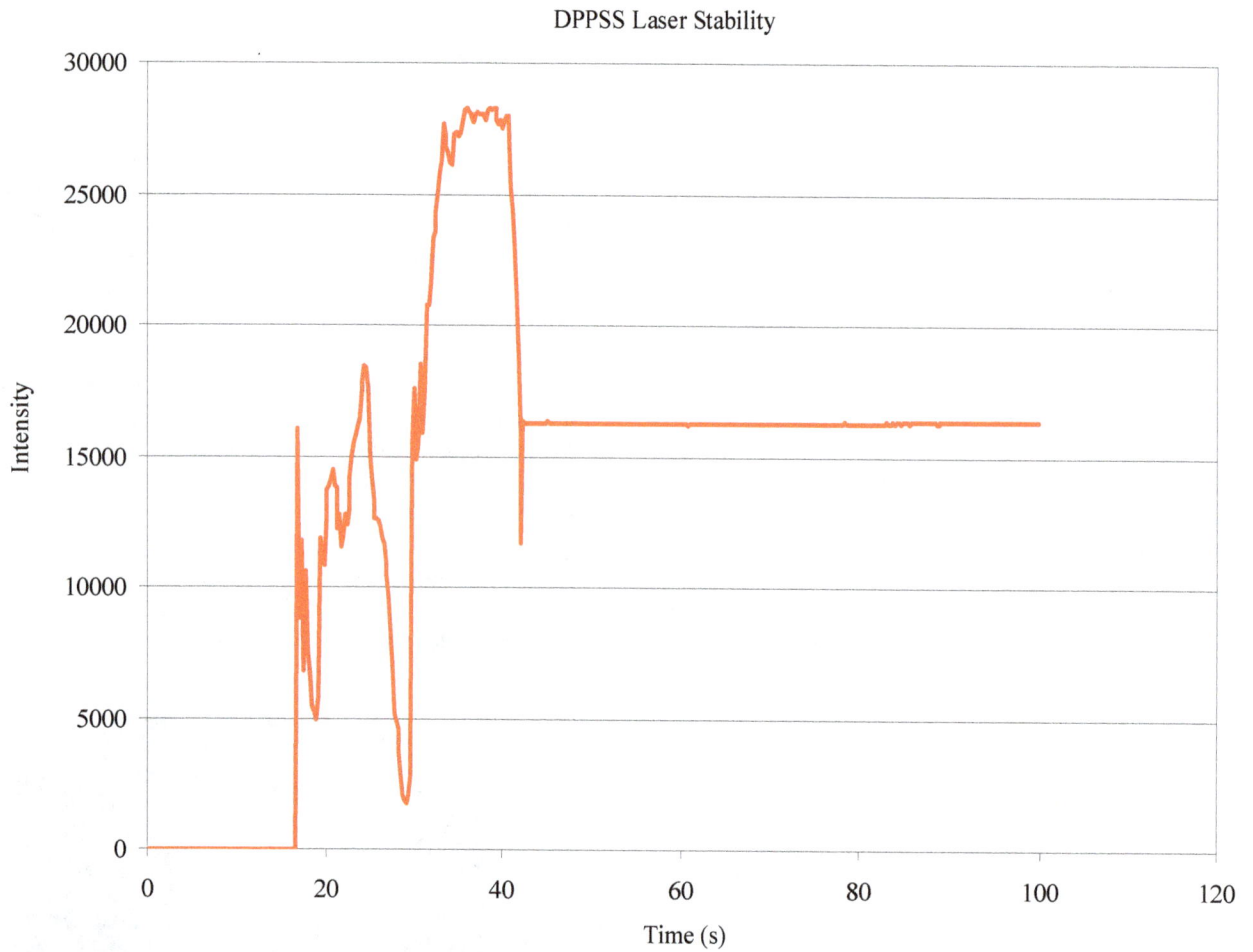

Figure 12. Laser output power after start up. The beam presents different multi-mode transverse before reaching a stable transverse mono-mode.

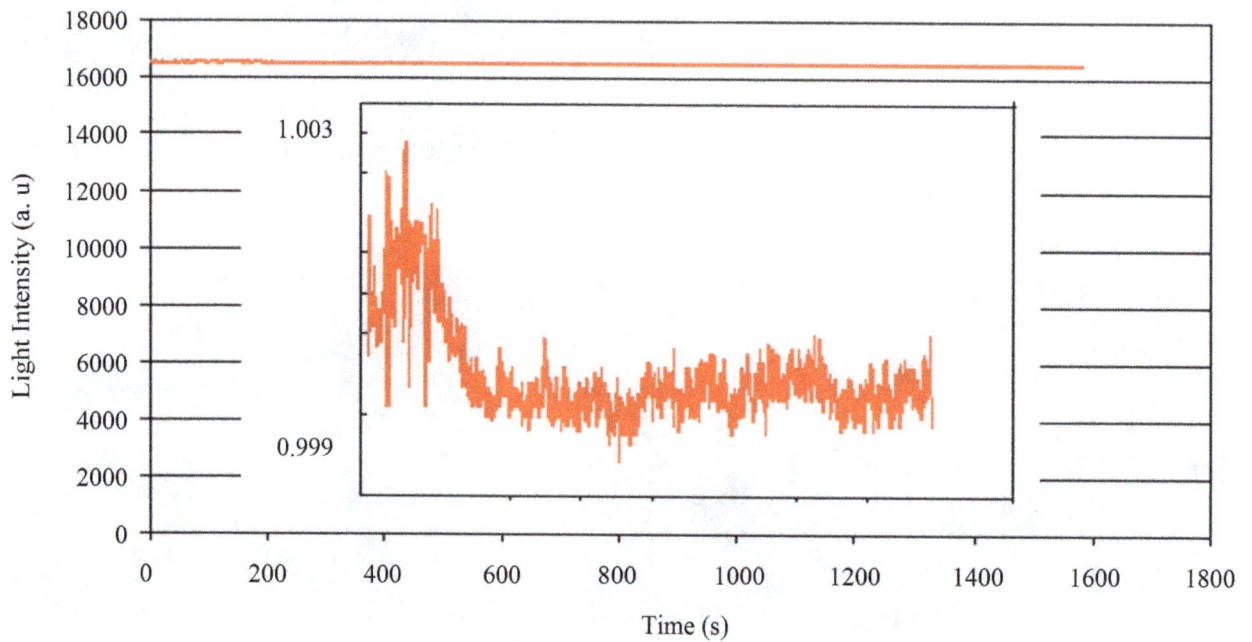

Figure 13. Stability of the DPSS laser 15 min after start-up and with cooling.

External and Environmental Radiation Dosimetry with Optically Stimulated Luminescent Detection
Device Developed at the SCK·CEN

17

8. Acknowledgements

For this work, we want to thank Dr. O. Goossens for his contribution in the first design of the OSL facility, B. S. T. Roggen for the different measurements in the characterization of the two types of sapphire detectors mentioned in this paper and measured with the Ar facility. We want also to thank M. J. Van Doninck for his continuous effort in different aspects in the development of the devices.

REFERENCES

[1]   S. W. S. McKeever, M. S. Akselrod and B. G. Markey, "Pulsed Optically Stimulated Luminescence Dosimetry Using α-Al$_2$O$_3$:C," *Radiation Proection Dosimetry*, Vol. 65, No. 1-4, 1996, pp. 267-272.

[2]   V. V. Antonov-Romanovskii, I. F. Keirum-Marcus, M. S. Poroshina and Z. A. Trapeznikova, "Conference of the Academy of Sciences of the USSR on the Peaceful Uses of Atomic Energy," USAEC Report AEC-tr-2435, Part 1, Moscow, 1956, pp. 239-250.

[3]   H. Y. Goksu, E. Bulur and W. Wahl, "Beta Dosimetry Using Thin-Layer α-Al$_2$O$_3$:C TL Detectors," *Radiation Protection Dosimetry*, Vol. 84, No. 1-4, 1999, pp. 451-455.

[4]   O. Roy, "Etude de la Luminescence Stimulée Optiquement (OSL) pour la Détection des Rayonnements: Application à un Capteur à Fibre Optique de Rayonnement γ," CEA Direction de l'Information Scientifique et Technique, 1998.

[5]   R. Chen, "Advantages and Disadvantages in the Utilisation of Thermoluminescence (Tl) and Optically Stimulated Luminescence (OSL) for Radiation Dosimetry," *IRPA Regional Congress on Radiation Protection in Central Europe*, Dubrovnik, Croatia, 20-25 May 2001, pp. 1-8.

[6]   S. W. McKeever and M. Moscovitch, "Topics under Debate—On the Advantages and Disadvantages of Optically Stimulated Luminescence Dosimetry and Thermoluminescence Dosimetry," *Radiation Protection Dosimetry*, Vol. 104, No. 3, 2003, pp. 263-270.

[7]   N. K. Umisedo, E. M. Yoshimura, P. B. R. Gasparian, E. G. Yukihara, "Comparison between Blue and Green Stimulated Luminescence of Al$_2$O$_3$:C," *Radiation Measurements*, Vol. 45, No. 2, 2010, pp. 151-156.

[8]   S. W. S. McKeever, M. W. Blair, E. Bulur, R. Gaza, R. Gaza, R. Kalchgruber, D. M. Klein and E. G. Yukihara, "Recent Advances in Dosimetry Using the Optically Stimulated Luminescence of Al$_2$O$_3$:C," *Radiation Protection Dosimetry*, Vol. 109, No. 4, 2004, pp. 269-276.

[9]   D. J. Huntley, D. I. Godfrey-Smith and M. L. W. Thewalt, "Optical Dating of Sediments," *Nature*, Vol. 313, 1985, pp. 105-107.

[10] D. I. Godfrey-Smith and B. Pass, "A New Method of Retrospective Radiation Dosimetry: Optically Stimulated Luminescence in Dental Enamel," *Health Physics*, Vol. 72, No. 5, 1997, pp. 744-749.

[11] J. L. Genicot, M. Moyaerts and W. Schroeyers, "Description of a Portable Devices Developed at SCK·CEN for OSL and TL Dosimetry," *Radiation Measurements*, Vol. 46, No. 12, 2011, pp. 1578-1581.

3

Reduction of Systematic Error in Radiopharmaceutical Activity by Entropy Based Mutual Information

Palliyakarany Thirumani Krishna Kumar, Toshikazu Takeda
Research Institute of Nuclear Engineering, University of Fukui, Fukui-shi, Japan

ABSTRACT

The quality of the radiation dose depends upon the gamma count rate of the radionuclide used. Any reduction in error in the count rate is reflected in the reduction in error in the activity and consequently on the quality of dose. All the efforts so far have been directed only to minimize the random errors in count rate by repetition. In the absence of probability distribution for the systematic errors, we propose to minimize these errors by estimating the upper and lower limits by the technique of determinant in equalities developed by us. Using the algorithm we have developed based on the technique of determinant inequalities and the concept of maximization of mutual information (MI); we show how to process element by element of the covariance matrix to minimize the correlated systematic errors in the count rate of ^{113m}In. The element wise processing of covariance matrix is so unique by our technique that it gives experimentalists enough maneuverability to mitigate different factors causing systematic errors in the count rate and consequently the activity of ^{113m}In.

Keywords: Random and Systematic Errors; Covariance Matrix; Limits for Correlated Elements by Determinant Inequalities; Mutual Information; Reduction of Systematic Errors by Maximizing Mutual Information

1. Introduction

Radiopharmaceuticals are used in nuclear medicine to study the functioning of organs and tissues. One of the major objectives of these radiopharmaceuticals in radio therapy is to ensure precise delivery of dose to a tumour. As radiation dose is proportional to the radioactivity, any error in measured activity will affect the dose deposited to the organ. Hence, it is important to estimate the amount of radiation delivered by these radiopharmaceuticals both for optimization of image quality and for radiation protection purposes.

The quality of the dose depends upon the gamma count rate of the radionuclide used [1]. The error in the number of counts at the gamma camera is a measure of the error in dose deposited at the human body tissues, which is a cumulative effect of the errors due to instrumentation and nuclear characteristics of the radiopharmaceuticals. Thus, any reduction in error in the count rate is reflected in the reduction in error in the activity and consequently on the quality of dose. The total error in the count rate estimate should include both the random and the systematic errors. To quantify the errors, covariance matrix has been identified as the error matrix as per international recommendation [2]. The diagonal and off diagonal elements of the covariance matrix represent uncorrelated random and the correlated systematic errors respectively. To our knowledge, so far many attempts have been made only to minimize the random error by repetition as it decreases by $\sqrt{N}$ if a measurement is repeated N times. On the contrary, the systematic error can never be reduced by repetition and is the main cause of correlation [3].

All the attempts made so far on systematic errors of SPECT imaging, either neglected them [4] or randomized them [5]. The only approach suggested to mitigate the systematic errors is to assume a statistical distribution for them like in Linearized Bayesian Update Procedure (LBUP) [6]. Assuming a statistical distribution for the systematic error requires knowledge of the second central moment or the variance. In the case of the systematic errors, the initial estimate is taken as the mean or the first moment of the probability density function. The estimate of the corresponding variance or the second central moment depends strongly on the availability of supplementary knowledge. As variance is seldom available for the systematic errors, their probability density function cannot be assumed and hence, LBUP cannot be applied in principle.

To circumvent the lack of knowledge of variance, we propose in this paper, estimation of limits for the systematic errors. Accordingly, if the upper and lower limit of the systematic error can be estimated and no additional information exists, then one can assume a constant probability density within these limits [2]. We have developed an algorithm based on the Technique of Determinant Inequalities (TDI) to estimate these limits. These limits are used to maximize the information theoretic concept of Mutual information (MI) to reduce the systematic errors. We demonstrate the utilization of our approach in reducing the systematic errors and consequently the total error in the gamma count rate of ^{113m}In.

2. Generation of Covariance Matrices from the Neutron Activation Analysis of ^{113m}In

^{113m}In is a diagnostic nuclear medicine agent for internal radiotherapy and is also used as a tracer in experimental studies [7]. In our laboratory, ^{113m}In is produced by neutron activation analysis in a standard neutron field of ^{252}Cf, by the reaction ^{113}In (n, n') ^{113m}In. ^{113}In foils with thickness 2.0 cm were irradiated for about 15 hours at a distance of 6 cm from the ^{252}Cf source. The californium ^{252}Cf is a needle type source, containing about 500 μg of ^{252}Cf encapsulated in a stainless steel cylinder having 5 mm diameter and 17 mm height. Induced gamma activities due to ^{113m}In were measured by a Ge (Li) detector having intrinsic efficiency ε and let the gamma count rate be C. The induced activity A of ^{113m}In depends upon its neutron absorption cross-section Σ, its atom density N and the induced neutron flux density φ of the ^{252}Cf. If m is the tissue mass, so that (A/m) is the activeity per unit mass, the absorbed dose rate D = k (A/m) E, where k is a constant whose value depends on the units used for other factors in the equation and E is the average energy released per transformation. Since, E is a constant for each radionuclide, k and E can be combined into a single constant Ω, hence, D = Ω (A/m). For a particular tissue mass and specific γ radiation, D = Δ A, where Δ = (Ω/m). Thus,

$$D \alpha\ A \alpha\ \ C \qquad (1)$$

Thus, as activity A and count rate C are proportional, any reduction in the error of the count rate will lead to the reduction in error in the activity as per law of error propagation [8]. In the gamma count rate measurements of ^{113m}In, several measurement systematic errors like back scattering from the room walls, geometrical factors are introduced during its formation and attenuation of the gamma ray and gamma ray intensity are introduced during its decay, requiring corrections. Applying the correction factors f to the count rate,

$$C = A\varepsilon \prod f\,(se) = N\Sigma\varphi\varepsilon \prod f\,(se), \text{ since } A = N\sum \varphi$$

Here, $f\,(se)$ is the correction factor for each of the (se) systematic errors. For generality, the above equation is rewritten for any element i as

$$C_i = N_i \Sigma_i \varphi\,\varepsilon_i \prod f_i\,(se) \qquad (2)$$

In Equation (2), both Σ and φ are unknown. An unknown cross-section is often determined by means of reaction rate ratio measurement relative to a well-known cross-section. The principle behind such relative measurements is the parallel irradiation of two different foils in the same neutron field of ^{252}Cf and subsequently counting their induced activities. Hence, by ratio measurement, unknown neutron flux density φ is eliminated,

$$R_{ij} = C_i / C_j$$

$$= \left[N_i / N_j \right]\left[\Sigma_i / \Sigma_j \right]\left[\varepsilon_i / \varepsilon_j \right]\left[\prod f_i\,(se) \right]/\left[\prod f_j\,(se) \right]$$

In our analysis, we consider formation of ^{113m}In by two sets of ratio measurement, i.e. ^{113}In (n, n') relative to ^{27}Al (n, α) and ^{115}In (n, n') respectively in the standard neutron field of ^{252}Cf.

Differentiating Equation (2) and writing dC/C = δC, one can obtain

$$\delta C_i = \delta N_i + \delta \sum_i + \delta \varepsilon_i + \sum \delta f_i\,(se) \qquad (3)$$

The relative covariance M_C between the count rate C_i and C_j with its various components as in Equation (2) with their respective correlation coefficient ρ_{ij} is obtained as follows,

$$M_C = <\delta C_i \delta C_j> = \rho_{ij}(N) <\delta N_i \delta N_j> + \rho_{ij}\!\left(\sum\right) <\delta \sum_i \delta \sum_j>$$
$$+ \rho_{ij}(\varepsilon) <\delta \varepsilon_i \delta \varepsilon_j> + \rho_{ij}(f) <\delta f_i\,(se)\delta f_j\,(se)>$$

The Determinant of M_C is designated as G. i.e. G = Det. M_C.

The complete list of all the error contributions with their magnitude and the correlation coefficients depicted in curly braces are given is **Table 1**. The two ^{113}In foils used for the relative measurements have same correction factors for mass N and hence the errors are fully correlated {P}. The correction for geometrical factor $f\,(gf)$ is the same for all measurements and so the corresponding errors show full correlation {Q}. ^{113m}In is the product nucleus in both the sets of relative measurement and hence the correction factors for half lives $f\,(hl)$ of ^{113m}In are fully correlated {R}. Similarly since ^{113m}In is a γ emitter, the correction factors for gamma ray attenuation $f\,(ga)$ and intensity $f\,(gi)$ are fully correlated {S, T}. The sources of back scattering are the room walls and the correction factors $f\,(bs)$ are fully correlated {U}. The details of constructing the covariance matrix given the various correlation coefficients for the above activation measurement is described in detail in [8].

3. Reduction of Systematic Error by Entropy Based Mutual Information

The fundamental property of systematic error is that they vary between a lower limit L and an upper limit U. If a probability distribution of the error over these limits is

Table 1. Errors and their magnitudes and their correlation in ^{113}In activation measurement.

Description of Error Quantities	Symbols of Errors	Ratio I		Ratio II	
		^{113}In (n,n') %	^{27}Al (n,α) %	^{113}In (n,n') %	^{115}In (n,n') %
Efficiency	ε	2.08	1.06	2.08	2.23
Mass Determination	N	0.1 {P}	0.1	0.1 {P}	0.1
Geometrical Factor	f(gf)	2.0 {Q}	2.0 {Q}	2.0 {Q}	2.0 {Q}
Half Life	f(hl)	0.11 {R}	0.13	0.14 {R}	0.01
Gamma Ray Attenuation	f(ga)	1 {S}	0.5	1 {S}	1
Gamma Ray Intensity	f(gi)	1 {T}	0.1	1 {T}	1
Back Scattering	f(bs)	1 {U}	0.7 {U}	1 {U}	1 {U}
Irradiation and Cooling Time	f(ic)	0.35	0.08	0.35	0.13
Cross-Section	Σ	4.77	2.77	4.77	4.08

{P} represents fully correlated with $\rho = 1.0$, as Indium foils are used in the two measurements; {Q} represents fully correlated with $\rho = 1.0$ as the correction factor is the same for the same geometry of foils and {R,S,T} represents fully correlated with $\rho = 1.0$, as ^{113m}In is the end product in the two measurements; {U} represents fully correlated with $\rho = 1.0$ as the source of back scattering is from the room walls.

known, then it can be used to describe the error. However, in most of the cases, such a distribution is not known, and it seems reasonable to choose a maximally uncertain density function. The appropriate measure of uncertainty for a distribution is a positive quantity called entropy H, given by,

$$H = \int a(x)\log(a(x))dx$$

The larger the entropy, the greater the uncertainty and hence we should choose a density function $a(x)$ that maximizes the entropy subject to the following constraints, $a(x) \geq 0$ for all values of x

$$L \leq x \leq U$$

$\int a(x)dx = 1$ between the Upper and lower limits.

When data on known standards are available, then, the standard techniques for estimating the end points of a uniform distribution can be used to estimate the limits L and U. But, where no such standards are available, the value of end points should be estimated using the knowledge of the measurement process. Further, in our case, systematic error in cross-section dominates over the statistical and hence conventional central limit theorem fails [9] making it necessary to use information theory approach [10]. A novel property of entropy principle is that it remains valid even if one of the sources of errors is highly correlated and dominant [10]. Further, entropy principle does not make assumptions about the distribution of data thereby belonging to the non-parametric family of statistics. According to information theory, MI is a measure of statistical correlation between the variables, A and C [11]. We show below, how in maximizing the MI, the systematic

errors are reduced.

The MI between A and C is expressed as [12]

$$MI(A; C) = H(C) + H(A) - H(C, A)$$

or

$$H(C, A) = H(C) + H(A) - MI(A; C)$$

where $H(C)$ and $H(A)$ are the entropies or the uncertainty of C and A respectively and $H(C, A)$ is the joint entropy of C and A.

1) When, $MI(A; C) = 0$, then, $H(C, A) = H(C) + H(A)$, i.e. joint entropy or uncertainty is the sum of individual uncertainty of C and A.

2) When, $MI(A; C) > 0$, then $H(C, A) < H(C) + H(A)$ i.e. joint entropy or uncertainty is less than the sum of individual uncertainty of C and A. Thus, maximizing, MI implies, minimizing the total uncertainty in both C and A.

When, C is having Gaussian distribution, MI is given by [12],

$$MI(A; C) = \log(Det. M_C) = \log(G) \qquad (4)$$

MI thus depends on the determinant of the covariance matrix M_C and is always positive and maximizing MI is equivalent maximizing the determinant of the covariance matrix M_C.

As an illustration, let us consider a simple case of just two count rates,

$$Det. M_C = \begin{pmatrix} c_{11} & c_{12} \\ c_{21} & c_{22} \end{pmatrix}$$

The elements of the covariance matrix can be written as the variances of C_1 and C_2

$c_{11} = \sigma_1^2, c_{22} = \sigma_2^2, c_{12} = c_{21} = \sigma_1 \sigma_2 \rho_{12}$ where ρ_{12} is the correlation coefficient between C_1 and C_2.

$$G = \text{Det. } M_C = \sigma_1^2 \sigma_2^2 \left(1 - \rho_{12}^2\right) \quad (5)$$

From Equation (5), the maximum value of G is $\sigma_1^2 \sigma_2^2$ when $\rho_{12} = 0$ and minimum value of G is 0, when $\rho_{12} = \pm 1$.

Thus, G depends upon, ρ_{12} in addition to σ_1^2 and σ_2^2. As mentioned earlier, MI quantifies the amount of correlation between the variables A and C and G can be maximized by minimizing ρ_{12}. Minimization of ρ_{12} by estimating its limits leads to minimization of correlated systematic errors. Hence, an index of minimization correlated systematic errors is the maximization of MI by the estimation of upper and lower limits for the correlated elements of M. The technique of determinant Inequalities (TDI) to obtain these upper and lower limits [13] and the algorithm [14] based on TDI is described elsewhere and would not be repeated here. The total error (TE) in the count rate, is given by the sum of Uncorrelated Statistical Error (UCSTE) and the Correlated Systematic Error (CSYE).

$$(TE) = \left[(UCSTE) + (CSYE)^2\right]^{0.5} \quad (6)$$

where $UCSTE = \left[\{\sigma_1^2\} + \{\sigma_2^2\}\right]$ and

$$CSYE = [\rho_{12}][\sigma_1 \sigma_2]^c \quad (7)$$

Hence lesser the value of ρ_{12}, the lesser is the CSYE and consequently decreases the TE in the count rates.

4. Robustness of the Analysis

According to Hadamard's inequality [11],

$$G = \det M \le \prod M_{ij} \quad (8)$$

The equality is achieved if and only if $\rho_{ij} = 0$. The maximum value of the determinant is the product of the diagonal elements and the least positive value is zero, when ρ_{ij} is either +1 or −1. Since, MI cannot be negative, the value of either the upper or the lower limit of ρ_{ij} which maximizes G is the robust value which maximizes the MI.

5. Results

The complete covariance matrix M_C generated with the contributions due to uncorrelated and correlated errors mentioned in **Table 1** is depicted in **Table 2**. As we have to minimize the correlated systematic error, we focused our attention on the correlated non-diagonal elements in the Matrix M_C. The values of the upper and lower limits for these elements are determined by TDI and are tabulated in **Table 3** along with the corresponding values of G.

6. Discussion

Systematic errors pervade in all types of physical measurements and affected by errors due to instrumentation, environment and personnel. The best investment is therefore to spend the maximum possible effort on identification and minimization of systematic errors. According to **Table 3**, the maximum value of G is 465303 for the ideal case of $\rho_{ij} = 0$. Only the values of lower limit of matrix M_C yield the second largest value of 424657 as compared to both the upper limits and the existing element values of matrix M_C. Further, it is apparent that the value of correlation coefficient is also less for the these lower limits and consequently, the corresponding correlated systematic errors is also less according to Equation (7). Since the systematic error is reduced by the lower limits, the total error in count rate is also reduces according to Equation (6). As activity and dose are proportional to the count rate as per Equation (1), any reduction in error in the count rate is reflected in the reduced errors of dose and activity. From **Table 3**, it is evident that in using our TDI, element wise processing to minimize the systematic error is feasible by using the concept of MI where the entire structure of the covariance matrix is taken and not by the

Table 2. Covariance Matrix M_C for the ^{113}In Activation Measurement.

$$\begin{pmatrix} 34.22 & 4.7 & 7.03 & 5.0 \\ 4.7 & 13.57 & 4.7 & 4.7 \\ 7.03 & 4.7 & 34.23 & 5.0 \\ 5.0 & 4.7 & 5.0 & 28.64 \end{pmatrix}$$

Table 3. Upper and lower limits of the non-diagonal elements of the Matrix M_C of Table 2 and the corresponding values of G where G = Det. M_C.

No.	Value of M_{ij} of matrix M_C of **Table 2**	Lower limit of M_{ij} of matrix M_C	Upper limit of M_{ij} of matrix M_C	Value of G with M_{ij} as in **Table 2**	Value of G with Upper limits of M_{ij}	Value of G with Lower limits of M_{ij}	Value of G by Hadamard Inequality with $\rho_{ij} = 0$
1	$M_{12} = 4.7$	−1.3	8.11				
2	$M_{13} = 7.03$	−3.61	11.47				
3	$M_{14} = 5.0$	−2.86	11.41	370401	236656	424657	465303
4	$M_{23} = 4.7$	−1.3	8.11				
5	$M_{24} = 4.7$	−3.17	8.25				
6	$M_{34} = 5.0$	−2.87	11.42				

Principal Component Analysis (PCA) and Factor analysis methods where the covariance matrix is only decorrelated and factored into dominant eigen values. The element wise processing of covariance matrix is a boon to experimentalists as it gives them enough maneuverability to improve the different factors causing systematic errors by way of improving either the quality of measurement or the associated instrumentation. Hence in our case, upper and lower limits have been given for all the correlated elements as an aid for the experimentalists to venture and such flexibility exists only by method of MI and not by PCA and other methods.

REFERENCES

[1]   L. Carole, C. Claude, E. Paul, F. Nuno, B. Bernard and T. Regine, "Optimization of Injected Dose Based on Noise Equivalent Count Rates for 2- and 3-Dimensional Whole-Body PET," *Journal of Nuclear Medicine*, Vol. 43, No. 9, 2002, pp. 1268-1278.

[2]   K. Weise and W. Woger, "A Bayesian Theory of Measurement Uncertainty," *Measurement Science and Technology*, Vol. 4, No. 1, 1993, pp. 1-11.

[3]   D. L. Smith, "Probability, Statistics and Data Uncertainties in Nuclear Science and Technology," American Nuclear Society, Inc., La Grange Park, Illinois, 1991.

[4]   H. J. Kim, B. Zeeberg and R. C. Reba, "Evaluation of Reconstruction Algorithms in SPECT Neuroimaging II: Computation of Deterministic and Statistical Error Components," *Physics in Medicine and Biology*, Vol. 38, No. 7, 1993, pp. 881-895.

[5]   D. J. Kadarmas, E. V. R. Di Bella, R. H. Huesman and G. T. Gullberg, "Analytical Propagation of Errors in Dy-

namic SPECT: Estimators, Degrading Factors, Bias and Noise," *Physics in Medicine and Biology*, Vol. 44, No. 8, 1999, p. 1997.

[6]   P. Talou, *et al.*, "Covariance Matrices for ENDF/B-VII^{235}, ^{238}U and ^{239}Pu Evaluated Files in the Fast Energy Range," *Proceedings of the International Conference on Nuclear Data for Science and Technology*, 22-27 April 2007, Nice, France.

[7]   M. Neves, A. Kling and A. Oliveira, "Radionuclides Used for Therapy and Suggestion for New Candidates," *Journal of Radioanalytical and Nuclear Chemistry*, Vol. 266, No. 3, 2005, pp. 377-384.

[8]   D. L. Smith, "Covariance Matrices and Applications to the Field of Nuclear Data," ANL/NDM-62, 1981.

[9]   P. Meyer, "Introductory Probability and Statistical Applications," Addison-Wesley, Reading, MA, 1970.

[10]   F. H. Frohner, "Evaluation of Data with Systematic Errors," *Nuclear Science and Engineering*, Vol. 145, No. 8, 2003, pp. 342-353.

[11]   G. Deco and D. Obradovic, "An Information—Theoretic Approach to Neural Computing," Springer, New York, 1996.

[12]   T. Cover and J. Thomas, "Elements of Information Theory," Wiley, New York, 1991.

[13]   P. T. K. Kumar, "Estimation of Bounds for Monte Carlo Uncertainty Analysis," *Annals of Nuclear Energy*, Vol. 17, No. 9, 1990, pp. 483-486.

[14]   P. T. Krishna Kumar, "Algorithm for Evaluation of Bounds for Covariance Based Uncertainty Analysis," *Annals of Nuclear Energy*, Vol. 18, No. 8, 1991, pp. 479-481.

4

Discussion on IAEA and China Safety Regulation for NPP Coastal Defense Infrastructures against Typhoon/Hurricane Attacks

Guilin Liu[1], Defu Liu[2]*, Huajun Li[1], Fengqing Wang[2], Tao Zou[2]
[1]College of Engineering, Ocean University of China, Qingdao, China
[2]Disaster Prevention Research Institute, Ocean University of China, Qingdao, China
*

ABSTRACT

The World Meteorological Organization estimates that about 90 percent of all natural disasters is extreme meteorological hazards like typhoon/hurricane and tropical cyclone triggered disasters. With the increasing tendency of natural hazards, the typhoon induced surge, wave, precipitation, flood and wind as extreme external loads menacing Nuclear Power Plants (NPP) in coastal and inland provinces of China. For all of the planned, designed and constructed NPP in China the National Nuclear Safety Administration of China and IAEA recommended Probable Maximum Hurricane/Typhoon/(PMH/T), Probable Maximum Storm Surge (PMSS), Probable Maximum Flood (PMF), Design Basis Flood (DBF) as safety regulations recommended for NPP defense infrastructures. This paper discusses the joint probability analysis of simultaneous occurrence typhoon induced extreme external hazards and compared with IAEA 2003-2011 recommended safety regulations for some NPP along China coast to make safety assessment based on the "As Low As Reasonable Practice" (ALARP) principle.

Keywords: IAEA Safety Regulation; Typhoon/Hurricane Attacks; Design Basis Flood; Multivariate Compound Extreme Value Distribution; Risk Assessment; ALARP

1. Introduction

In China, three NPP have been built along coasts in 1980, and more than 37 NPP along coast of South-East China Sea are in the stages of planning, design, or construction. In the "2007 China Long Term NPP Plan" estimated that before 2020 about 70 billion USD (450 billion RMB) will be invested in 6 coastal provinces.

China has a wide continental slope to decay tsunami energy. If M9 earthquake occurs at Manila trench or Rykyu trench, the wave produced by tsunami wave at south and southeast China coast would be no more than 5 - 6 m [1]. In 2006, typhoon disasters were especially serious in China. Five of the most severe typhoon disasters brought about 1600 deaths and disappearances, and affected 66.6 million people. The economic loss reached 80 billion RMB and influenced agriculture areas of totally more than 2800 thousand hectares. Among these disasters, typhoon Saomai induced 3.76 m surges and 7 m waves, causing 240 deaths, sinking 952 ships and damaging 1594 others in Shacheng harbor. If the typhoon Saomai had landed 2 hours later, then the simultaneous occurrence of the typhoon surge and high spring tide

with 7 m wave would have inundated most areas of the Zhejiang and Fujian provinces, where located several NPP. The results would be comparable with 2011 Japanese nuclear disaster.

With the global warming and sea level rising, the frequency and intensity of extreme external natural hazards would increase. All the coastal areas having NPP are menaced by possibility of future typhoon disasters. So calibration of typhoon disaster prevention criteria is necessary for existed and planning NPP. In China Nuclear Safety Regulations: "HAF101, HAD101/09~11" [2-5] and IAEA Engineering Safety Section: "Extreme External Events in the Design or Assessment of NPP" [6-10] there are appeared some vague definitions and they should be dissected and described with probability characteristics by using statistical analysis.

This paper discusses the joint probability analysis of simultaneous occurrence typhoon induced extreme external hazards and compare China and IAEA recommended safety regulation design criteria for some constructed NPP coastal defense infrastructures along China coasts.

2. Discussion on Design Basic Flood (DBF)

For coastal sites, IAEA, US NRC and China safety re-

*Corresponding author.

gulations [2-10] recommended evaluation of DBF, which should be the combination of following three parts: Probable maximum storm surge (PMSS) induced by Probable Maximum Typhoon (PMT), spring tide and simultaneous extreme wave height. Probable wind-wave effects considered independently or in combination.

IAEA recommends that deterministic and probabilistic methods for evaluating the design basis flood should be considered complementary. The estimated flood hazard should be compared to historical data to verify that the specified design basis exceeds the historical extreme by a substantial margin. For example, the characteristics of PMT and Probabilistic models for estimating design basis floods are generally based on approaches that characterize the extreme flood as a random event, describe the properties of this random phenomenon using probability distributions, and use these probability distributions to estimate extreme floods corresponding to a specified probability of exceedance.

The two key components of the probabilistic models are 1) the historical flood data and 2) the probability distribution used to describe the historical flood data. Typically, the historical record of annual maximum instantaneous peak discharge at the site of interest. Because the random variable of interest—the peak discharge—is represented as a continuous variable, a continuous probability distribution is appropriate. The design-basis flood at the site can be selected from the frequency distribution of extreme floods.

A suitable combination of flood causing events depends on the specific characteristics of the site and involves considerable engineering judgment. The following is an example of a set of combinations of events that cause floods for use in determining the design conditions for flood defense in coastal areas: the astronomical tide, storm surge and simultaneous wave. The design basis flood associated with an established probability of exceedance (e.g. 1×10^{-4}) for the combination of events should be determined.

The above definitions in safety regulation of coastal defenses against typhoon attacks for nuclear power plant are influenced by many uncertainty factors such as the differences in comprehensions and calculation methods of them. DBF as design criteria is used for all of planned, designed, uncompleted and constructed nuclear power plants in China but not included any joint probability consideration. There are some facts must be taken into account: First, prediction of typhoon induced extreme events instead of traditional annual maximum data sampling, the typhoon process maximum data sampling is used, it significantly improved description of the probability laws of extraordinary floods; Second, PMT in different sea area is related to annual occurring frequency of typhoon (λ) and six typhoon characteristics. It means

that different PMT can be derived from different combinations of typhoon characteristics; Third, according to the randomness of annual typhoon occurrence frequency along different sea areas, it can be considered as a discrete random variable. Typhoon characteristics or typhoon-induced extreme sea events are continuous random variables. The Multivariate Compound Extreme Value Distribution (MCEVD) can be then derived by compounding a discrete distribution and the extreme distribution for typhoon induced extreme events along coasts [11,12].

3. Theory of Multivariate Compound Extreme Value Distribution (MCEVD)

In 1972, Typhoon Rita attacked Dalian port in the North Bohai Bay of China, causing severe damage in this port. The authors found that, using traditional extrapolation (such as a Pearson type III model), it was difficult to determine the design return period for the extreme wave height induced by a typhoon. According to the randomness of annual typhoon occurrence frequency along different sea areas, it can be considered as a discrete random variable. Typhoon characteristics or typhoon-induced extreme sea events are continuous random variables. The Compound Extreme Value Distribution (CEVD) can then be derived by compounding a discrete distribution and the extreme distribution for typhoon-induced extreme events along China's coasts [13]. Then the CEVD is used to analyze long-term characteristics of hurricanes along the Gulf of Mexico and the Atlantic US coasts [14]. During the past few years, CEVD has been developed into MCEVD and applied to predict and prevent typhoon induced disasters for coastal areas, offshore structures, and estuarine cities [15-19]. Both CEVD and MCEVD have the following advantages: instead of traditional annual maximum data sampling, the typhoon process maximum data sampling is used; and the typhoon frequency is used in the models. Derivation of Poisson-Nested Logistic Trivariate Compound Extreme Distribution (PN-LTC-ED):

As mentioned above, frequency of occurrences of extreme events can be fitted to Poisson distribution, as

$$P_i = \frac{e^{-\lambda}\lambda^i}{i!} \tag{1}$$

and substitute nested-logistic trivariate distribution [20] for the continuous distribution $G(x_1, \cdots, x_n)$, the PNLT-CED can be obtained using:

$$F_0(x_1, x_2, x_3) = e^{-\lambda}\left(1 + \lambda \int_{-\infty}^{x_3}\int_{-\infty}^{x_2}\int_{-\infty}^{x_1} e^{\lambda \cdot G_1(u_1)} g(u_1, u_2, u_3)\, du_1 du_2 du_3\right) \tag{2}$$

Nested-logistic trivariate distribution is expressed as:

Discussion on IAEA and China Safety Regulation for NPP Coastal Defense Infrastructures against Typhoon/Hurricane Attacks

25

$$G\left(x_1, x_2, x_3\right) = \exp\left[-\left\{\left[\left(1+\xi_1\frac{x_1-\mu_1}{\sigma_1}\right)^{-\frac{1}{(\alpha\beta\xi_1)}}+\left(1+\xi_2\frac{x_2-\mu_1}{\sigma_2}\right)^{-\frac{1}{(\alpha\beta\xi_2)}}\right]^\beta+\left(1+\xi_3\frac{x_3-\mu_3}{\sigma_3}\right)^{-\frac{1}{(\alpha\xi_3)}}\right\}^\alpha\right] \quad (3)$$

In which ξ_j, μ_j, σ_j are the shape, location and scale parameters of the marginal distributions $G\left(x_j\right)$ to x_j $(j = 1, 2, 3)$, respectively. And dependent parameters α, β can be obtained by moment estimation

$$\hat{\alpha} = \frac{\sqrt{1-r_{13}}+\sqrt{1-r_{23}}}{2} \quad (4)$$

$$\hat{\beta} = \frac{\sqrt{1-r_{12}}}{\hat{\alpha}} \quad (5)$$

where $r_{i,j}$ is the correlation coefficient between x_i and x_j for $i < j$, $i, j = 1, 2, 3$.

Let

$$s_j = \left(1+\xi_j\frac{x_j-\mu_j}{\sigma_j}\right)^{-\frac{1}{\xi_j}} \quad j = 1, 2, 3. \quad (6)$$

Then Equation (3) can be written as

$$G\left(x_1, x_2, x_3\right) = \exp\left\{-\left[\left(s_1^{1/(\alpha\beta)}+s_2^{1/(\alpha\beta)}\right)^\beta+s_3^{1/\alpha}\right]^\alpha\right\} \quad (7)$$

Its corresponding probability density function is

$$g\left(x_1, x_2, x_3\right) = \frac{\partial^3 G}{\partial x_1 \partial x_2 \partial x_3}$$
$$= \frac{1}{\sigma_1\sigma_2\sigma_3}e^{-u}u^{1-2/\alpha}v^{1/\alpha-2/\alpha\beta}s_1^{1/(\alpha\beta)-\xi_1}s_2^{1/(\alpha\beta)-\xi_2}s_3^{1/\alpha-\xi_3}Q \quad (8)$$

In which

$$v = \left(s_1^{1/(\alpha\beta)}+s_2^{1/(\alpha\beta)}\right)^{\alpha\beta} \quad (9)$$

$$u = \left[\left(s_1^{1/(\alpha\beta)}+s_2^{1/(\alpha\beta)}\right)^\beta+s_3^{1/\alpha}\right] = \left(v^{1/\alpha}+s_3^{1/\alpha}\right)^\alpha \quad (10)$$

$$Q = \left(\frac{v}{u}\right)^{1/\alpha}Q_3\left(u;\alpha\right)+\frac{1-\beta}{\alpha\beta}Q_2\left(u;\alpha\right) \quad (11)$$

$$Q_2\left(u;\alpha\right) = u+\frac{1}{\alpha}-1 \quad (12)$$

$$Q_3\left(u;\alpha\right) = u^2+3\left(\frac{1}{\alpha}-1\right)u+\left(\frac{1}{\alpha}-1\right)\left(\frac{2}{\alpha}-1\right) \quad (13)$$

Trivariate layer structure (α-outside, β-inside layer) shows that the correlation between x_1 and x_2 is better than those among x_1, x_3 and x_2, x_3.

As shown above, the PNLTCED can be obtained through the estimation of parameters of the marginal distributions and their dependent parameters. The PNLT-CED considers the extreme event occurring frequency

and combination of trivariate variable and it has a simple structure. Consequently, it is easy for use in engineering applications.

Many application of MCEVD in engineering design and risk analysis show the scientific and reasonable of its predicted results in China and abroad [21-24]. As mentioned in "Summary of flood frequency analysis in the United States" [25]: "The combination of the event-based and joint probability approaches promises to yield significantly improved descriptions of the probability laws of extraordinary floods". MCEVD is the model which follows the development direction of the extraordinary floods prediction hoped for by Kirby and Moss. Since 2005 hurricane Katrina and Rita disasters proved accuracy of 1982 predicted hurricane characteristics and after disaster calculated results. It stands to reason that MCEVD is a practicable model for prediction of typhoon/hurricane/tropical cyclone induced extreme events. Our proposed methods in [13,14,24,25] are used as design criteria of wind-structure interaction experimentation for mitigating hurricane-induced coastal disasters [26].

4. Design Code Calibration of Coastal Defense against Typhoon Attacks from Lesson Hurricane Katrina Disaster

In 1979, American National Oceanic and Atmospheric Administration (NOAA) divided Gulf of Mexico and Atlantic coasts into 7 areas according to hurricane intensity, in which corresponding Standard Project Hurricane (SPH) and Probable Maximum Hurricane (PMH) were proposed as hurricane disaster prevention criteria [27]. Using Compound Extreme Value Distribution (CEVD) [14], the predicted hurricane central pressures with return period of 50 yr and 1000 yr were close to SPH and PMH, respectively, except that for the sea area nearby New Orleans (Zone A) and East Florida (Zone1) coasts, hurricane intensities predicted using CEVD were obviously severer than NOAA proposed values. SPH and PMH are only corresponding to CEVD predicted 30 - 40 yr and 120 yr return values, respectively. In 2005, hurricane Katrina and Rita attacked coastal area of the USA, which caused deaths of about 1400 people and economical loss of $400 billion in the city of New Orleans and destroyed more than 110 platforms in the Gulf of Mexico. The disaster certified that using SPH as flood-protective standard was a main reason of the catastrophic results [28-30]. **Figure 1**, **Tables 1** and **2** indicate that both

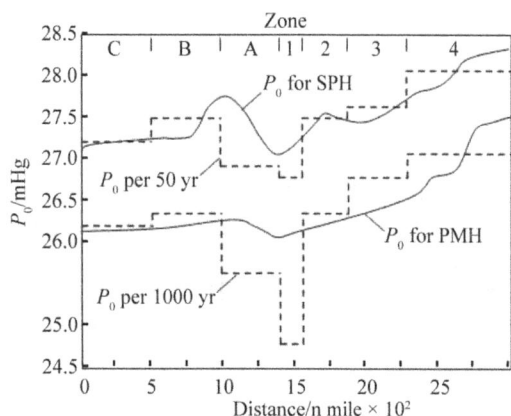

Figure 1. Comparison between the results of CEVD and NOAA (see Figure 6 in [14]).

Table 1. Comparison between NOAA and CEVD.

Zone		NOAA (hPa)	CEVD (hPa)		Hurricane (hPa)
A	SPH	941.0	50-yr	910.8	Katrina
	PMH	890.5	1000-yr	866.8	902.0
1	SPH	919.3	50-yr	904.0	Rita
	PMH	885.4	1000-yr	832.9	894.9

Table 2. Comparison between MCEVD and other method.

Methods	MCEVD [24]	Coles et al. [31]	Casson & Coles [32]	Georgion et al. [33]
100 yr Wind speed (m/s)	70.6	46.0	38.0	39.0

CEVD and MCEVD (see next part) predicted and hindcast results are more reasonable than NOAA or other methods.

According to safety regulations for NPP in China, USA and IAEA [2-10]: DBF in coastal areas should be taken as combinations of spring tide, PMSS and simultaneous 100 years return period wave height.

The above definitions in safety regulation of coastal defenses against typhoon attacks for NPP are influenced by many uncertainty factors such as the differences in comprehensions and calculation methods of them.

The spring tide, maximum wave height and PMSS can be seen as non-Gaussian random variables with different correlation. The PMT and PMSS must involve the joint probability characters, and then DBF can be actually obtained by multivariate joint probability prediction. For example, the characteristics of PMT and PMSS in different sea area is related to annual occurring frequency of typhoon (λ), maximum central pressure difference (ΔP), radius of maximum wind speed (R_{max}), moving speed of typhoon center (s), minimum distance between typhoon center and target site (δ), typhoon moving angle (θ) and typhoon duration (t). It means that different PMT and

PMSS can be derived from different combinations of typhoon characteristics. For this reason, the characteristics of PMT and PMSS inevitably involve a selection of discrete distribution (λ) and multivariate continuous distribution of other typhoon characteristic factors (ΔP, R_{max}, s, δ, θ, t), which can be described by Multivariate Compound Extreme Value Distribution (MCEVD) [23]. The calculation of PMT and PMSS by a numerical simulation method can remove the uncertainties of typhoon characteristics and may be led to different results, while the PMSS obtained on basis of them may has some arbitrary and cause wrong decision making. The lesson from 2005 hurricane Katrina showed that unreasonable calculation of the Probable Maximum Hurricane (PMH) and Standard Project Hurricane (SPH) is one of the most important reasons of New Orleans catastrophe [2,17,19].

According to IAEA and China safety regulations, PMSS should be obtained based on PMT. So aiming at PMT with different combinations of typhoon characteristics, some sensitive factors should be selected as control factors and substituted into procedure of Global Uncertainty Analysis (GUA) and Global Sensitivity Analysis (GSA) [34]. The PMSS corresponding to PMT of different sea areas can be derived by repeated forward-feedback calculations.

Based on MCEVD (analytical solution and stochastic simulation), Multi-Objective Nested Probability Model (MONPM) can be established for long term probability prediction of typhoon characteristics and corresponding disaster factors.

As shown in **Figure 2**, GUA and GSA are introduced into DLNMPM. In the model, typhoon characteristics in the first layer need to be varied repeatedly, and then their sensitivities to storm surge can be calculated. The PMSS corresponding to PMT of different typhoon characteristic combinations in certain sea area can be calculated by numerical simulation of repeated forward-feedback calculations of GUA, GSA in input-output procedure. The most sensitivity combination of typhoon characteristics and their induced storm surge can be selected as PMT and PMSS. PMSS with corresponding spring tide and 100 years return period wave height with joint return period calculated by MCEVD will be determined the probabilistic definition of DBF. This model also can be used for dominated by wave case. The ALARP（As Low As Reasonable Practice）[35] can be used to estimate failure probability of NPP coastal defense infrastructures against extreme external hazards.

5. Joint Probability Safety Assessment for DYW-NPP Defense Infrastructure in South China Sea Coast

Nuclear power plant DYW is located at coast f South

Discussion on IAEA and China Safety Regulation for NPP Coastal Defense Infrastructures against
Typhoon/Hurricane Attacks

27

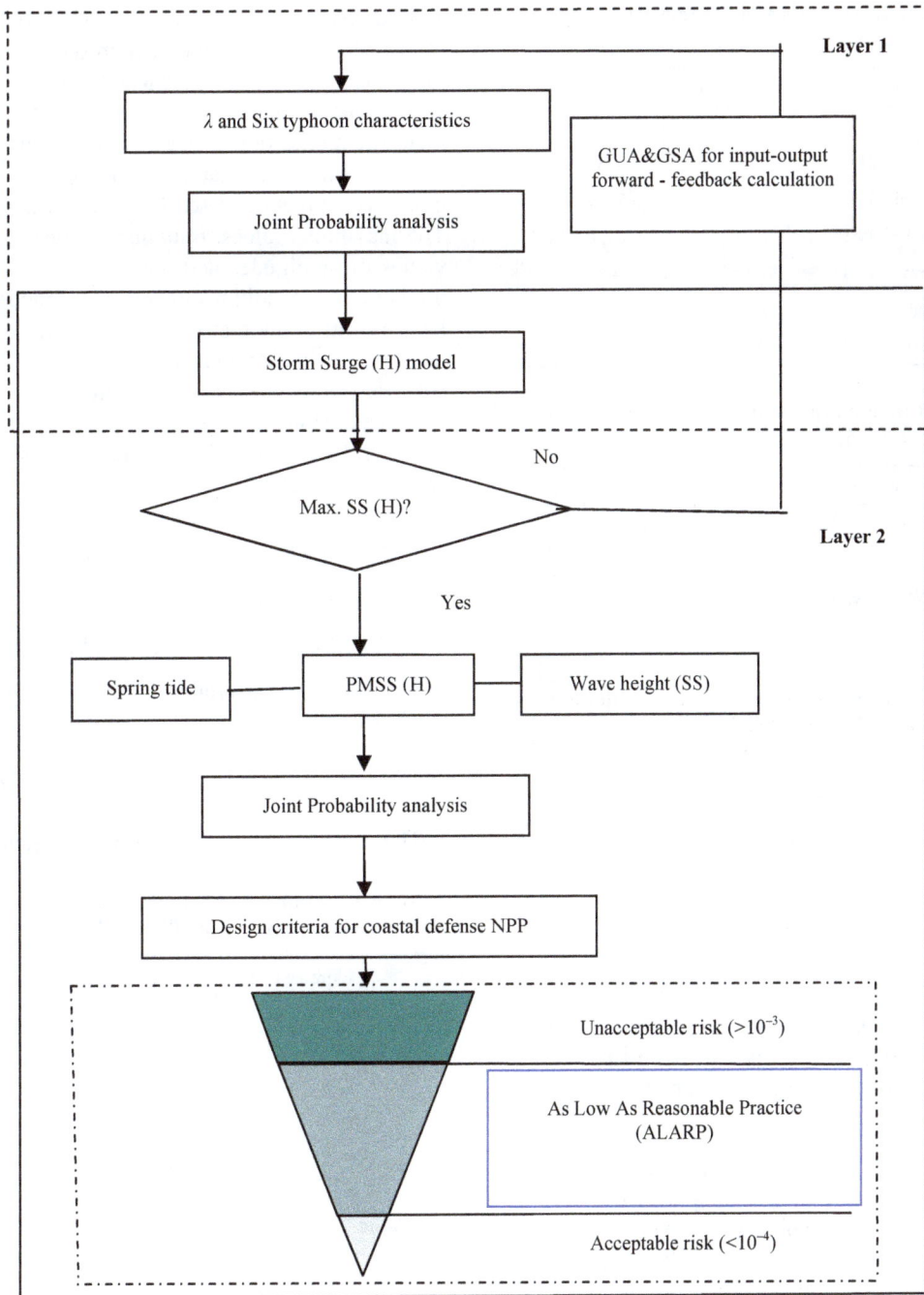

Figure 2. Application of MONPM with GUA, GSA to safety regulation calibration for NPP coastal defense infrastructures by ALARP.

China Sea, where the combined extreme external events are dominated by waves. The typhoon characteristics and design criteria of coastal engineering is listed in **Tables 3** and **4**.

For discussion on joint return period of wave height with storm surge and corresponding spring tide, the PNLTCED can be used for analytical solution. Different combinations of typhoon characteristics in first layer of PNLCED (see **Table 3**) can be induced different combi-

nations of storm surge and waves. The diagnostic checks of spring tide, surge and wave in **Figures 3(a)-(c)** show that PNLTCED is applicable model.

The **Tables 4** and **5** show the present design criteria by China design code and by PNLTCED predicted extreme external events with different joint return period. The DYW NPP constructed breakwater is 14 - 16 m height which is lower than 1000 years return period combined extreme events.

Table 3. Marginal distribution and parameters of typhoon characteristics.

	Distributions	Mean	Standard variance	Parameters
λ	Poisson		$\lambda = 6.19$	
ΔP (hPa)	Gumbel	21.89	14.96	$a = 0.073, b = 14.45$
R_{max} (km)	Lognormal	45.79	25.22	$\mu = 3.71, \sigma = 0.5$
s (m/s)	Gumbel	30.19	15.95	$a = 0.07, b = 22.4$
δ (km)	Uniform	44.37	169.63	$a = 294.6, b = 333.8$
θ (°)	Normal	15	37.36	$\mu = 15, \sigma = 37.36$
t (h)	Gumbel	12.95	5.56	$a = 0.20, b = 10.29$

Table 4. Present design criteria for coastal defense of DYW nuclear power plant [36].

Design water level	Design value (m)
DBF	6.35
PMSS	5.30
Extreme Wave Height	6.6
Design low water level	−1.93

Table 5. Combined extreme external events with joint return period for DYW NPP by PNLTCED.

Joint Probability \ Extreme Event	Spring Tide (m)	Surge (m)	Wave (m)
100	2.14	2.79	6.6
500	2.19	3.49	7.3
1000	2.75	3.85	7.9
10000	3.15	4.50	9.7

6. Joint Probability Safety Assessment for QS-NPP Defense Infrastructure in Qiantang River Estuarine Area, East China Sea

The combination of typhoon induced storm surge with the strongest spring tide in Qiantang river estuarine always lead to disasters. The observed maximum surge and spring tide more than 9 m. The QS NPP locates in south coast of estuarine Qiantang River and face to East China Sea where always occurred the severest spring tide in China.

The height of constructed breakwater is 9.76 m. So the joint probability safety assessment of combined extreme external events for coastal defense infrastructure dominated by spring tide should be taken into account.

As the severest extreme external events for QS NPP are combined effect of spring tide and surge, two dimensional joint probability model can be used to calculate corresponding joint probability density function and cumulative distribution function (**Figures 4(a)** and **(b)**). Joint probability distribution of spring tide, storm surge

and corresponding extreme wave with 1000 years joint return period much severe than present design criteria by [37], it can be seen in **Table 6** and **Figure 5**.

For mentioned above joint probability safety assessment the spring tide is used as one of the random variables, because the statistical analysis shows that harmonic constituent and sea level vary from year to year. The maximum values, minimum values and some other values of amplitudes and mean sea level are chosen as the boundary conditions of numerical model. The calculated results show that the sea level caused by semi-diurnal tide of major lunar tidal constituent M_2 and major solar tidal constituent S_2 (see **Table 7**) [37].

Calculated mean value μ, variance σ and coefficient of variation Cov of the harmonic constituents and sea level are shown as follows:

For harmonic constituents:

$$\text{Cov1} = \sigma/\mu = 0.0053; \quad \mu = 166.65; \quad \sigma = 0.88;$$

For mean sea level:

$$\text{Cov2} = \sigma/\mu = 0.8256; \quad \mu = 4.80; \quad \sigma = 3.96.$$

The resulting uncertainty for spring tide can be obtained as:

$$\text{Cov} = \left[\left(\text{cov}1\right)^2 + \left(\text{cov}2\right)^2\right]^{1/2} = 0.825$$

The confidence intervals of predicted joint return value

Table 6. Combined extreme external events with joint return period for QS NPP by PNLTCED.

Joint Probability \ Extreme Event	Spring Tide (m)	Surge (m)	Wave (m)
100	4.2	3.0	2.5
500	5.0	3.5	3.0
1000	5.5	4.0	3.5
10000	6.5	4.8	4.0

Table 7. Calculated spring tide from different inputs.

Input			output
A_n (cm)	$A(M_2)$ (cm)	$A(S_2)$ (cm)	Maximum sea level (cm)
1[*]	2[*]	3[*]	
4.08	127.54	41.60	156.12
2.08	123.47	41.53	151.11
−14.92	127.05	40.14	139.62
4.08	126.41	41.39	155.32
20.8	125.22	41.25	152.45
14.92	124.69	41.39	146.41
2.08	124.44	41.22	151.64

Note: 1[*]: Annual mean sea level; 2[*]: Amplitude of constituent M_2; 3[*]: Amplitude of constituent S_2.

Discussion on IAEA and China Safety Regulation for NPP Coastal Defense Infrastructures against
Typhoon/Hurricane Attacks

29

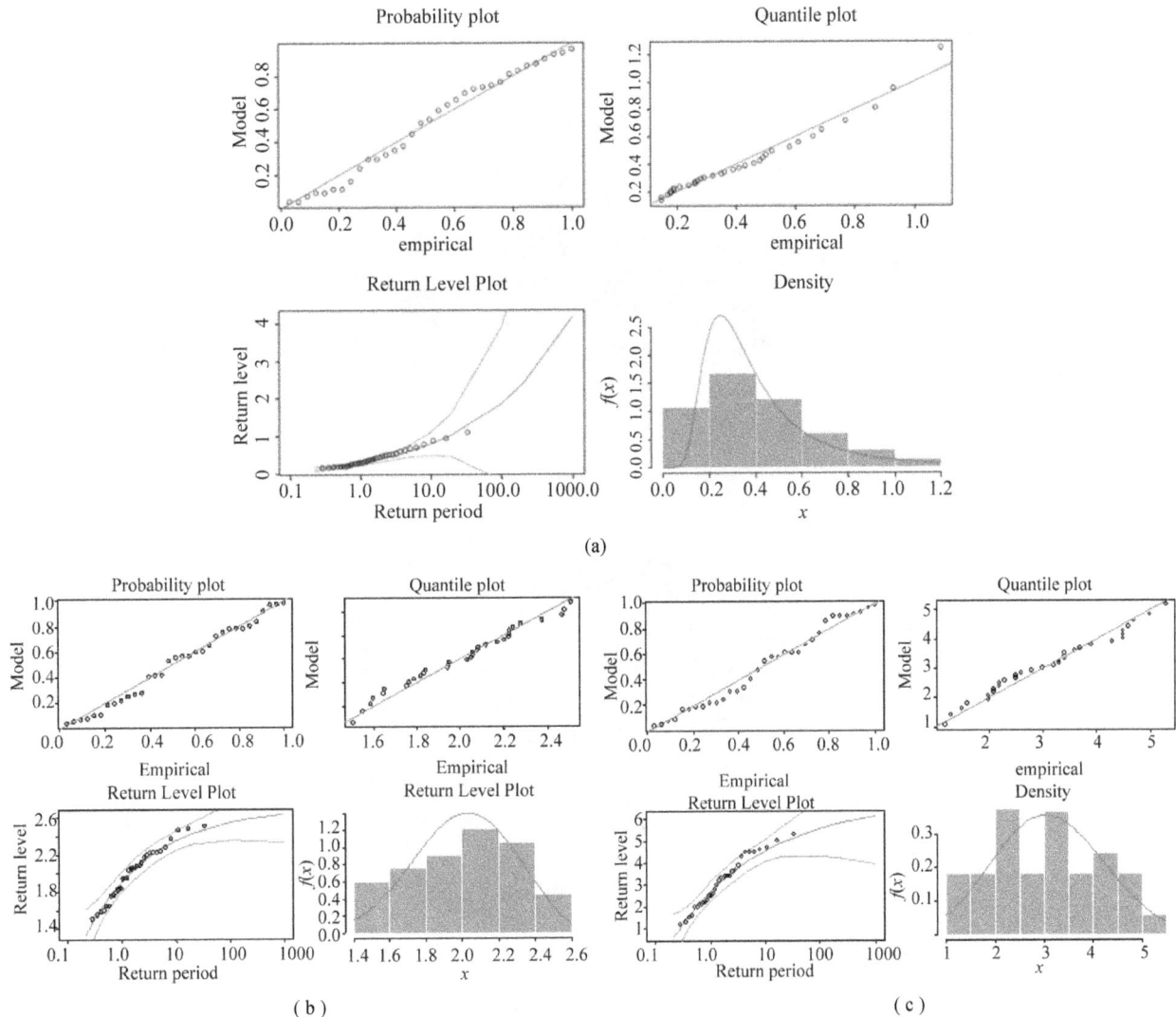

Figure 3. (a) Distribution diagnostic testing of storm surge; (b) Distribution diagnostic testing of spring tide; (c) Distribution diagnostic testing of wave height.

are estimated by authors proposed formula as follows [38]:

$$\Delta H = H_T + \frac{\sigma}{\sqrt{N}}\sqrt{1+\frac{\left(X_{N,P}\right)^2}{2}} \qquad (14)$$

In formula

ΔH—confidence interval

H_T—design value with T years return period

σ—standard deviation

N—total data number

$X_{N,P}$—coefficient dependent on data number N and probability P, it can be calculated by formula:

$$P = \left[\int_{-\infty}^{x_{np}} \frac{1}{\sqrt{2\pi}} e^{\frac{-t^2}{2}}\, dt\right]^N \qquad (15)$$

The confidence intervals of predicted 500 years joint return spring tide can be calculated by mentioned above formulas as 1.7 m, the 500 years joint return value of spring tide, surge (5.0 + 1.7 + 3.5) = 10.2 m with corresponding wave height 3.0 m should be over constructed breakwater height 9.76 m. Joint probability safety assessment shows that constructed coastal defense infrastructure for QS NPP can not against 500 years return period extreme hazards.

7. Conclusions

Joint probability safety assessment for NPP coastal defense infrastructure against extreme external hazards shows that China and IAEA recommended safety regulations appear to have some vague definitions and different kinds of uncertainties. Both of two constructed NPP located along South China Sea and East China Sea where dominated external events are wave and spring tide, and

(a)

(b)

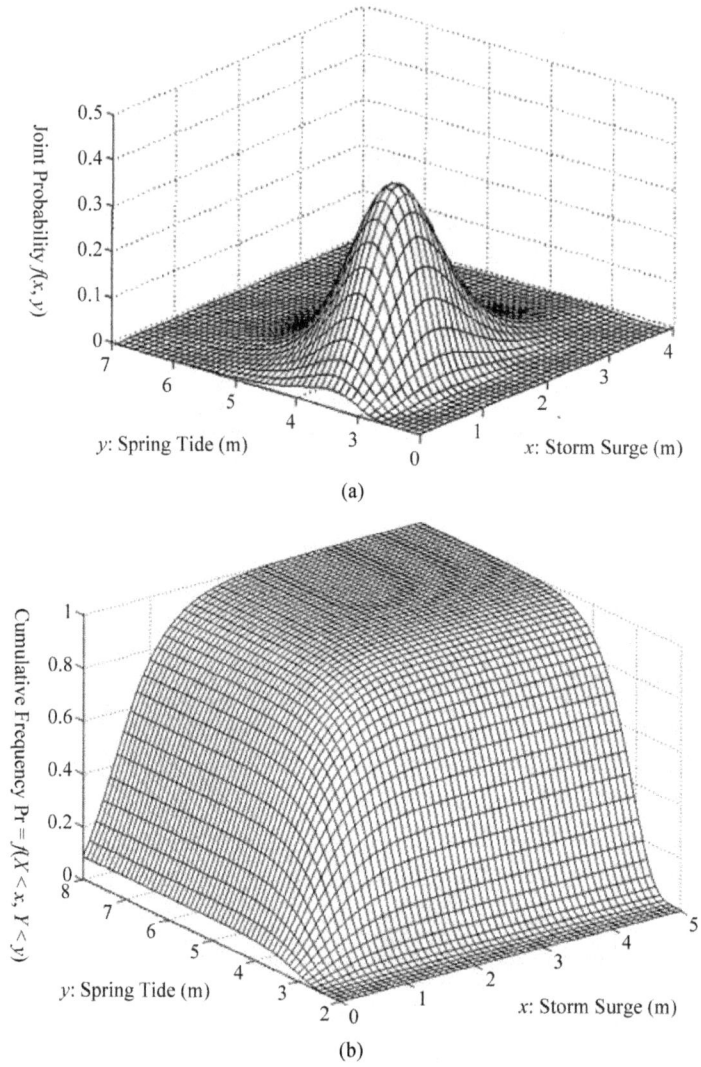

Figure 4. (a) Probability density distribution of spring tide and storm surge; (b) Cumulative probability distribution of spring tide and storm surge.

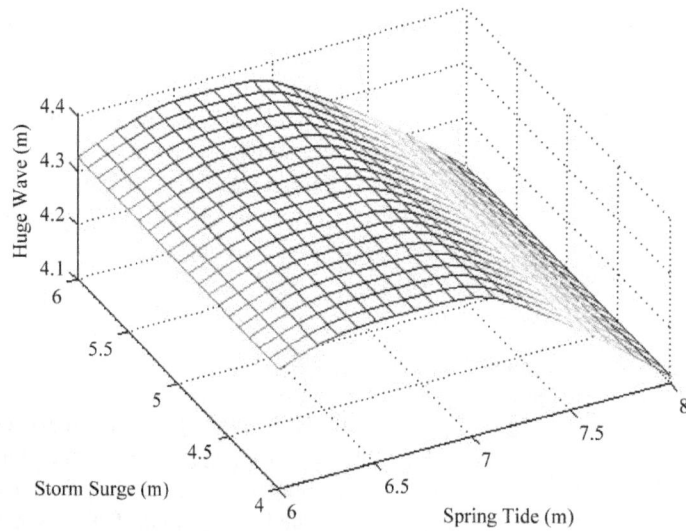

Figure 5. Joint probability distribution of spring tide, storm surge and extreme wave with 1000 years joint return period.

Discussion on IAEA and China Safety Regulation for NPP Coastal Defense Infrastructures against
Typhoon/Hurricane Attacks

31

the China and IAEA recommended safety regulation are much lower than 1000 years return period typhoon induced sea hazards predicted by DLNMPM, that means by ALARP principle [39] the risk level of mentioned above two NPP constructed coastal defense infrastructures is unacceptable (**Figure 2**).

Face up to the frequently occurrence of typhoon hazards and disasters with 1000 years and higher return periods in the world during the past few years, we have to worry about consequence of typhoon disasters for nuclear power plant.

8. Acknowledgements

This work is supported by the National Natural Foundation of China (No. 5101000).

REFERENCES

[1] Y. Liu, A. Santos, *et al.*, "Tsunami Hazards along Chinese Coast from Potential Earthquakes in South China Sea," *Physics of the Earth and Planetary Interiors*, Vol. 163, No. 1-4, 2007, pp. 233-244.

[2] National Nuclear Safety Administration of China, "Determination of Design Basis Flood Level on Coastal Nuclear Power Plant Site," HAD101/09, 1990.

[3] National Nuclear Safety Administration of China, "Design Tropical Cyclone for Nuclear Power Plant," HAD-101/11, 1991.

[4] National Nuclear Safety Administration of China, "Extreme Meteorological Condition for Selection of Nuclear Power Plant Site," HAD101/10, 1991.

[5] National Nuclear Safety Administration of China, "Regulations for Selection of Nuclear Power Plant Site," HAF-101, 1991.

[6] IAEA, "Extreme External Events in the Design or Assessment of Nuclear Power Plants," TECDOC-1341, Vienna, 2003.

[7] IAEA, "Flood Hazard for Nuclear Power Plants on Coastal and River Sites," Safety Standards Series No. NS-G-3.5, 2003.

[8] IAEA, "Advanced Nuclear Plant Design Options to Cope with External Events," 2006.

[9] IAEA, "Criteria for Use in Preparedness and Response for a Nuclear or Radiological Emergency, General Safety Guide," Safety Standards Series No. GSG-2, 2011.

[10] NRC, "Design-Basis Flood Estimation for Site Chracterization at NPP in the United States of America," 2011.

[11] D. F. Liu, L. P. Wang, *et al.*, "Theory of Multivariate Compound Extreme Value Distribution and Its Application to Extreme Sea State Prediction," *Chinese Science Bulletin*, Vol. 51, No. 23, 2006, pp. 2926-2930.

[12] D. F. Liu, L. Pang, *et al.*, "Typhoon Disaster in China: Prediction, Prevention and Mitigation," *Natural Hazards*, Vol. 49, No. 3, 2009, pp. 421-436.

[13] T. F. Liu and F. S. Ma, "Prediction of Extreme Wave Heights and Wind Velocities," *Journal of the Waterway Port Coastal and Ocean Division*, Vol. 106, No. 4, 1980, pp. 469-479.

[14] T. F. Liu, "Long Term Distribution of Hurricane Characteristics," *Proceedings of Offshore Technology Conference*, Houston, 3-6 May 1982, pp. 305-313.

[15] D. F. Liu, H. D. Shi, *et al.*, "Disaster Prevention Design Criteria for the Estuarine Cities: New Orleans and Shanghai—The Lesson from Hurricane Katrina," *Acta Oceanologica Sinica*, Vol. 25, No. 4, 2006, pp. 131-142.

[16] D. F. Liu, Y. Song, *et al.*, "Combined Environmental Design Loads Criteria for Marine Structures," *Offshore Technology Conference*, Houston, 6-9 May 2002, pp. 1-4.

[17] D. F. Liu, H. J. Li, *et al.*, "Prediction of Extreme Significant Wave Height from Daily Maxima," *China Ocean Engineering*, Vol. 15, No. 1, pp. 97-106.

[18] D. F. Liu, H. J. Li, G. L. Liu and F. Q. Wang, "Design Code Calibration of Offshore, Coastal and Hydraulic Energy Development Infrastructures," *World Science and Engineering Academy Society (WSEAS) International Journal Energy and Environment*, Vol. 5, No. 6, 2011, pp. 733-747.

[19] L. Pang, D. F. Liu, *et al.*, "Improved Stochastic Simulation Technique and Its Application to the Multivariate Probability Analysis of Typhoon Disaster, Lisbon," *Proceedings of the 16th International Offshore and Polar Engineering Conference*, Lisbon, 1-6 July 2007, pp. 1800-1805.

[20] D. J. Shi and S. S. Zhou, "Moment Estimation for Multivariate Extreme Value Distribution in a Nested Logistic Model," *Annals of the Institute of Statistical Mathematics*, Vol. 51, No. 2, 1999, pp. 253-264.

[21] M. G. Naffa, A. M. Fanos and M. A. Elganainy, "Characteristics of Waves off the Mediterranean Coast of Egypt," *Journal of Coastal Research*, Vol. 7, No. 3, 1991, pp. 665-676.

[22] M. C. Ochi, "Stochastic Analysis and Probabilistic Prediction of Random Seas," *Advanced Hydro Science*, Vol. 13, 1982, pp. 218-375.

[23] S. T. Quek and H. F. Cheong, "Prediction of Extreme 3-Sec. Gusts Accounting for Seasonal Effects," *Structural Safety*, Vol. 11, No. 2, 1992, pp. 121-129.

[24] L. R. Muir and A. H. EL-Shaarawi, "On the Calculation of Extreme Wave Height: A Review," *Ocean Engineering*, Vol. 13, No. 1, 1986, pp. 93-118.

[25] W. H. Kirby and M. E. Moss, "Summary of Flood-Frequency Analysis in the United States," *Journal of Hydrology*, Vol. 96, No. 1-4, 1987, pp. 5-14.

[26] A. G. Chowdhury and P. Huang E., "Novel Full-Scale Wind-Structure Interaction Experimentation for Mitigating Hurricane Induced Coastal Disasters," *Far East*

Journal of Ocean Research, Vol. 2, 2009, pp. 1-27.

[27] R. W. Schwerdt, F. P. Ho and R. R. Wakins, "Meteorological Criteria for Standard Project Hurricane and Probable Maximum Hurricane Wind Fields, Gulf and East Coasts of the United States," *NOAA Technical Report NWS*-23, 1979, pp. 309-316.

[28] R. Bea, "Reliability Assessment and Management Lessons from Hurricane Katrina," *ASME* 2007 *26th International Conference on Offshore Mechanics and Arctic Engineering*, San Diego, 10-15 June 2007, pp. 467-478.

[29] D. T. Resio, *et al.*, "White Paper on Estimation Hurricane Inundation Probabilities," *US Army Corps of Engineering Report*, 2007, pp. 1-10.

[30] Army Corps of Engineers, "History of Lake Pontchartrain and Vicinity Hurricane Protection Project," Report of US Government Accountability Office GAO-06-244T, 2005, pp. 1-4.

[31] S. Coles and E. Simiu, "Estimating Uncertainty in the Extreme Value Analysis of Data Generated by a Hurricane Simulation Model," *Journal of Engineering Mechanics*, Vol. 129, No. 11, 2003, pp. 1288-1294.

[32] E. Casson and S. Coles, "Simulation and Extremenal Analysis of Hurricane Events," *Applied Statistics*, Vol. 49, No. 2, 2000, pp. 227-245.

[33] P. N. Georgiou, A. G. Davenport and P. J. Vickery, "Design Wind Speeds in Regions Dominated by Tropical Cyclones," *Journal of Wind Engineering & Industrial Aero-*

dynamics, Vol. 13, No. 1-3, 1983, pp. 139-152.

[34] S. Tarantola, N. Giglioli, N. J. Jesinghaus and A. Saltelli, "Can Global Sensitivity Analysis Steer the Implementation of Models for Environmental Assessments and Decision-Making?" *Stochastic Environmental Research and Risk Assessment*, Vol. 16, No. 1, 2002, pp. 63-76.

[35] S. L. Xie, "Design Criteria for Coastal Engineering Works for Nuclear Power Plant," *China Harbour Engineering*, Vol. 1, 2000, pp. 6-9.

[36] L. M. Wang and J. L. Liu, "Research into PMSS of Coastal Nuclear Power Station," *Electric Power Survey*, Vol. 2, 1999, pp. 49-53.

[37] D. F. Liu, G. L. Liu, H. J. Li and F. G. Wang, "Risk Assessment of Coastal Defense against Typhoon Attacks for Nuclear Power Plant in China," *Proceedings of ICAPP* 2011, Nice, 2-5 May 2011, pp. 2484-2492.

[38] D. F. Liu and L. S. Kong, "Stochastic-Numerical Model of Tidal Current Field for Jiaozhou Bay of Yellow Sea," *Proceedings of the 7th International Offshore and Polar Engineering Conference*, Stavanger, Vol. 3, 17-22 June 2001, pp. 682-686.

[39] D. N. Veritas, "Marine Risk Assessment, for the Health and Safety Executive," *Offshore Technology Report*, 2001, p. 3.

5

Gamma Radiation Measurements in Tunisian Marbles

Kais Manai[1], Chiraz Ferchichi[1], Mansour Oueslati[2], Adel Trabelsi[1,3]
[1]Research Unit of Nuclear and High Energy Physics, Faculty of Sciences of Tunis,
Tunis El-Manar University, Tunis, Tunisia
[2]National Center for Nuclear Sciences and Technologies, Sidi-Thabet, Tunisia
[3]Physics Department, College of Science, King Faisal University, Hofuf, KSA

ABSTRACT

The radioactivity of 15 kinds of different granites collected in Tunisia was determined by gamma-ray spectrometry using hyper-pure germanium (HPGe) detector. The average activity concentrations for primordial radionuclides ^{238}U, ^{232}Th and ^{40}K were respectively 33.24, 8.01 and 116.98 Bq/kg. The activity concentrations ranged from 3.59 to 87.37 Bq/kg for ^{238}U, from 0.45 to 25.34 Bq/kg for ^{232}Th and from 24.06 to 380.23 Bq/kg for ^{40}K. The measured activity concentrations were used to assess of the radium equivalent activity ranged from 22.2 to 995.8 Bq/kg, the absorbed dose rate in air from 7 to 1209 nGy/h and the internal (0.1 to 2.8) and external (0.1 to 2.7) hazard indices. The data obtained in this study may be useful for natural radioactivity mapping.

Keywords: Radioactivity; Marble; Radium Equivalent; External and Internal Hazard Indices

1. Introduction

Natural radioactive materials (rocks, soil, sediment, water and aliment) contain low-level radioactivity. Marble used in building and decoration are also possible sources of radioactivity. Radioactivity in the marbles consists of the ^{238}U series [1-3] with a relative abundance of 99.284% and a half-life of 4.468×10^9 years, the thorium series ^{232}Th with a relative abundance of 100% and a half-life of 14.05×10^9 years and the actinium series ^{235}U with a relative abundance of 0.716% and a half-life of 9.47×10^{10} years. There are also several singly occurring radionuclides, the most important one is ^{40}K with a relative abundance of 0.0117% and a half-life of 1.248×10^9 years. The emanation of radon (^{222}Rn) is associated with the presence of ^{226}Ra and its ultimate precursor ^{238}U in the ground. The inhalation of its short-lived daughter products is a major contributor to the total radiation dose.

The objective of the work is to estimate the radioactivity concentrations of primordial radionuclides; ^{238}U, ^{232}Th and ^{40}K in marble samples collected in Tunisia. The absorbed and effective dose rate, the external and internal hazard indices in marble samples have been determined. The measurements have been carried out using a HPGe spectrometer.

2. Experimental Setup

A total of 15 different kinds of marble in Tunisia have been collected. **Figure 1** shows the Tunisian states where the commercial marble samples were collected. The samples were pulverised, dry-weighed [4,5], sealed in 1 l plastic Marinelli beakers and stored for 4 weeks in order to allow the reaching of equilibrium between ^{226}Ra and ^{222}Rn and its decay products [6,7].

Spectra for different samples were measured with a hyper-pure germanium (HPGe) detector of high-resolution gamma-ray spectrometer. **Figure 2** show a spectrum of Tunisian marble sample 1. The detector has a photopeak efficiency of 30% and an energy resolution of 1.85 keV (*FWHM*) for the 1.332 MeV reference transition of ^{60}Co. The measurements have been carried out at the Center for Nuclear Sciences and Technology, Sidi Thabet, Tunisia (*CNSTN*).

To obtain the gamma spectrum with good statistics, the accumulation time for each sample is set at 20 hours. Marble samples container were placed inside a shield, in order to minimize the background radiation. The background distribution in the environment around the detector was counted in the same manner and in the same geometry as the samples. It was determined using an empty sealed beaker. The background spectra were used to correct the net peak area of gamma rays of measured isotopes.

3. Results and Discussion

3.1. Calculation of Activity Concentrations

The specific activity A_{E_i} (Bq/kg) of a nuclide i and for

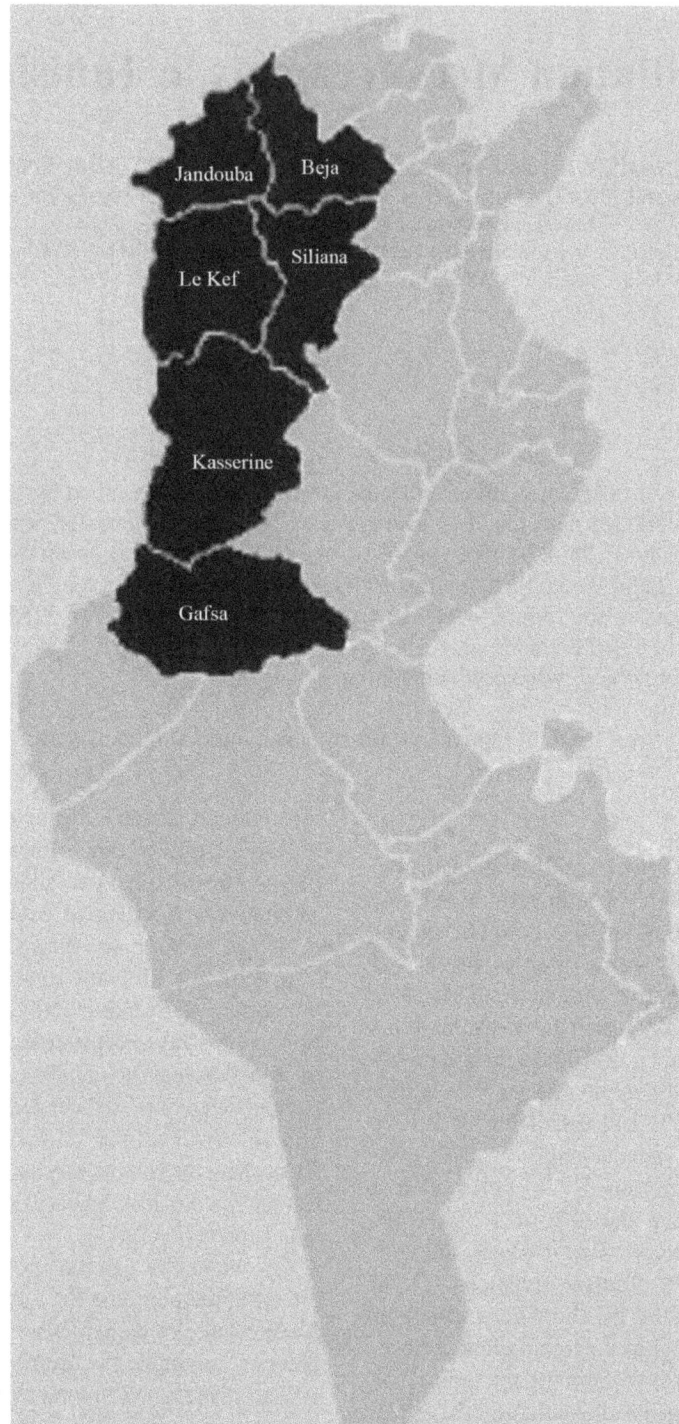

Figure 1. Localization of the Tunisian states where the commercial marble samples were collected.

a peak at energy E, is given by:

$$A_{E_i} = N_{E_i} / \left(\varepsilon_E \times \gamma_D \times t \times M_s \right) \qquad (1)$$

where N_{E_i} is the net peak area, ε_E the detection efficiency at energy E, γ_D the gamma ray yield per disintegration of the specific nuclide for the transition at energy E, M_s the mass of the measured sample (kg)

and t the counting live-time (s). If there is more than one peak in the energy range of analysis, then the result is the weighted average nuclide activity.

The gamma-ray transitions of ^{214}Pb, ^{214}Bi were used to determine the concentration of the ^{238}U series. The gamma-ray transitions ^{228}Ac, ^{208}Tl ^{228}Ac were used to calculate the concentration of the ^{232}Th series. The gama-ray

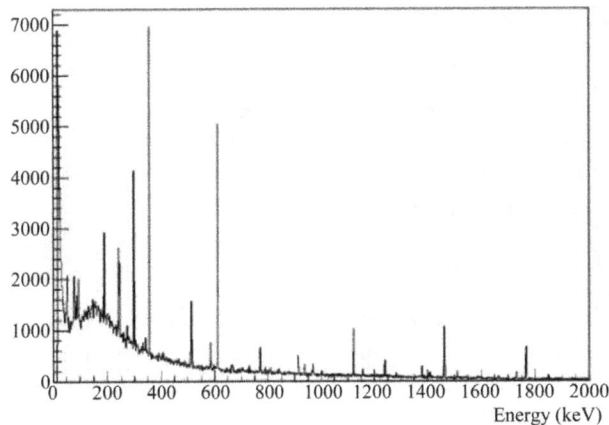

Figure 2. A spectrum of Tunisian marble sample 1.

transition of ^{40}K was used to determine the concentration of ^{40}K in different samples.

The activity concentrations of the natural radionuclides of the 15 kinds of marbles are shown in **Figure 3**. The activity concentrations ranged from 3.59 to 87.37 Bq/kg for ^{238}U, from 0.45 to 25.34 Bq/kg for ^{232}Th and from 24.06 to 380.23 Bq/kg for ^{40}K. The average activity concentrations are 33.24 Bq/kg for ^{238}U, 8.01 Bq/kg for ^{232}Th, and 116.98 Bq/kg for ^{40}K.

From the 15 samples measured in this study, Sample (2) appears to present the highest concentrations for ^{238}U, Sample (11) presents the highest concentration for ^{232}Th and Sample (7) presents the highest concentration for ^{40}K.

3.2. Absorbed and Effective Dose Rate

Absorbed dose rate D (in nGy/h) can be calculated using the following formula [8,9]:

$$D = A_{E_i} \times C_F \qquad (2)$$

where A_{E_i} is the activity concentration (in Bq/kg), and C_F is the dose conversion factor (absorbed dose rate in air per unit activity per unit of soil mass in nGy/h per Bq/kg).

To estimate the annual effective dose, one has to take into account the conversion coefficient from absorbed dose in air to effective dose and the indoor occupancy factor [10]. A value of 0.7 Sv/year was used for the conversion coefficient from absorbed dose in air to effective dose received by adults, and 0.8 for the indoor occupancy factor, implying that 20% of time is spent outdoors. The effective dose rate $\left(D_{eff}\right)$ indoors (in mSv/year), is calculated by the following formula:

$$D_{eff} = D \times 8760\left(\text{h/year}\right) \times 0.2 \times 0.7 \times 10^{-6} \qquad (3)$$

where D (nGy/h) is the calculated dose rate, 0.7 (Sv/Gy) is the conversion coefficient from the absorbed dose rate in air to the effective dose rate and 0.2 is the outdoor occupancy factor [11].

Table 1 shows the absorbed and effective dose rate. The average absorbed and effective dose rates are 25.7 nGy/h and 0.23 mSv/year respectively. The highest value of the dose rate is for Sample (15), 51.20 nGy/h and the minimum value is for Sample (1), 4.31 nGy/h. The maximum effective dose rate is 0.45 mSv/year and the minimum is 0.04 mSv/year.

3.3. Radium Equivalent Activity

The distribution of ^{238}U, ^{232}Th and ^{40}K in marbles is not uniform. Uniformity in respect of exposure to radiation

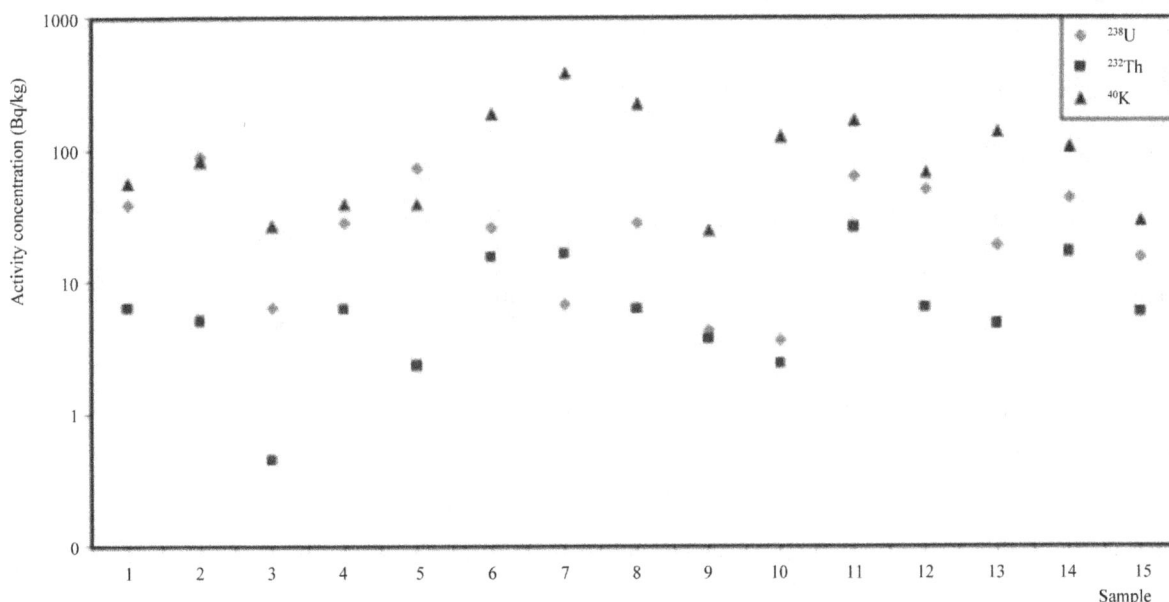

Figure 3. The activity concentrations for primordial radionuclides ^{238}U, ^{232}Th and ^{40}K.

Table 1. Absorbed dose rate (n Gy/h), effective dose rate (mSv/an), H_{ex}, H_{in} and Ra_{eq} (Bq/kg) for the marble samples.

No.	Commercial name	D	D_{eff}	H_{ex}	H_{in}	Ra_{eq}
1	Thala beige	4.31	0.04	0.05	0.04	9.04
2	Thala royal	18.27	0.16	0.09	0.18	39.63
3	Thala imperial	36.70	0.32	0.30	0.41	79.30
4	Thala noir	29.18	0.26	0.21	0.24	62.15
5	Thala gris	28.93	0.25	0.09	0.18	59.17
6	Gris foussana	25.81	0.23	0.09	0.22	53.67
7	Chamtou	5.24	0.05	0.05	0.04	11.35
8	Kadhel gris	8.33	0.07	0.14	0.05	16.65
9	Kadhel rose	51.20	0.45	0.21	0.47	110.9
10	Jaune moutard	29.83	0.26	0.18	0.31	64.36
11	Courteau	17.16	0.15	0.14	0.15	35.88
12	Noisette beige	34.86	0.31	0.15	0.32	75.61
13	Gris Marmis	11.95	0.11	0.10	0.11	26.01
14	Noir Aziza	25.70	0.23	0.15	0.24	54.88
15	Marron Kesra	51.20	0.45	0.30	0.51	110.9

index is calculated by the following formula [15-17]:

$$H_{in} = A_U/185 + A_{Th}/259 + A_K/4810 \qquad (6)$$

The calculated external and internal hazard indices values (**Table 1**) are ranged from 0.05 to 0.3 and from 0.04 to 0.51 respectively with an average value of 0.15 and 0.24 respectively.

4. Conclusions

Natural radioactivity of 15 samples of marbles collected in Tunisia was determined by gamma-ray spectrometry using hyper-pure germanium (HPGe) detector. In the case of uranium the highest level was found in the "Thala royal" marble sample and the lowest level in the "Gris Marmis" marble sample. For thorium the highest level was found in "Courteau" marble sample and the lowest level in "Thala imperial" marble sample, while for potassium the highest level was found in "Chamtou" marble and the lowest level in the "Kadhel rose" marble sample.

Based on the results obtained in this study for absorbed and effective dose rate, radium equivalent activity, external and internal hazard indices, we conclude that the radioactivity levels of marbles are within the international recommended values from world-wide areas due to terrestrial gamma radiation [11]. According to the recommended values and the calculated external and internal hazard indices values, the marble samples are acceptable for use as building materials and decoration.

has been defined in terms of radium equivalent activity (Ra_{eq}) in Bq/kg to compare the specific activity of marbles containing different amounts of ^{238}U, ^{232}Th and ^{40}K. It is assumed that 370 Bq/kg of ^{238}U, 259 Bq/kg of ^{232}Th and 4810 Bq/kg of ^{40}K produce the same gamma dose rate [12]. A radium equivalent of 370 Bq/kg in marbles will produce an exposure of about 1.5 mSv/year to the inhabitants [11]. The radium equivalent activity which can be calculated by the following formulae [13,14]:

$$Ra_{eq} = A_U + A_{Th} \times 1.43 + A_K \times 0.077 \qquad (4)$$

where A_U is the ^{238}U activity concentration (Bq/kg), A_{Th} is the ^{232}Th activity concentration (Bq/kg) and A_K is the ^{40}K activity concentration (Bq/kg).

The average calculated (Ra_{eq}) values for marble samples were ranging from 9.04 to 110.9 Bq/kg with the average value of 54.88 Bq/kg.

3.4. External and Internal Hazard Indices

To estimate the radiation dose expected to be delivered externally if a building is constructed using marbles, the external hazard index (H_{ex}), due to the emitted γ-rays, can be calculated using the following equation [8,15]:

$$H_{ex} = A_U/370 + A_{Th}/259 + A_K/4810 \qquad (5)$$

In addition to the external irradiation, radon and its daughter products are also hazardous to the respiratory organs. The internal exposure to radon and its short-lived products is proscribed by the internal hazard index (H_{in}). For the safe use of a marbles in the construction of dwellings H_{in} should be less than unity. The Internal hazard

REFERENCES

[1] M. A. Misdaq and A. Amghar, "Radon and Thoron Emanation from Various Marble Materials Impact on the Workers," *Radiation Measurements*, Vol. 39, No. 4, 2005, pp. 421-430.

[2] A. J. A. H. Khatibeh, N. Ahmad, Matiullah and M. A. Kenawy, "Natural Radioactivity in Marble Stones, Jordan," *Radiation Measurements*, Vol. 28, No. 1-6, 1997, pp. 345-348.

[3] E. S. Larsen and D. Gottfried, "Uranium and Thorium in Selected Sites of Igneous Rocks," *American Journal of Science*, Bradley Volume, Vol. 258A, 1960, pp. 151-169.

[4] M. Brai, S. Basile, S. Bellia, S. Hauser, P. Puccio, S. Rizzo, A. Bartolotta and A. Licciardello, "Environmental Radioactivity at Stromboli (Aeolian Islands)," *Applied Radiation and Isotopes*, Vol. 57, No. 1, 2002, pp. 99-107.

[5] IAEA, "Measurement of Radionuclides in Food and the Environment," *Technical Report*, Series No. 295, 1989, Vienna.

[6] R. R. Benke and K. J. Kearfott, "Soil Sample Moisture Content as a Function of Time during Oven Drying for Gamma-Ray Spectroscopic Measurements," *Nuclear Instruments and Methods in Physics Research Section A: Accelerators, Spectrometers, Detectors and Associated*

Equipment, Vol. 422, No. 1-3, 1999, pp. 817-819.

[7]   A. S. Mollah, G. U. Ahmed, S. R. Hussain and M. M. Rahman, "The Natural Radioactivity of Some Building Materials Used in Bangladesh," *Health Physics*, Vol. 50, No. 6, 1986, pp. 849-851.

[8]   A. M. El-Arabi, "Gamma Activity in Some Environmental Samples in South Egypt," *Indian Journal of Pure and Applied Physics*, Vol. 43, No. 6, 2005, pp. 422-426.

[9]   C. Kohshi, I. Takao and S. Hideo, "Terrestrial Gamma Radiation in Koshi Prefecture, Japan," *Journal of Health Science*, Vol. 47, No. 4, 2001, pp. 362-372.

[10] M. Tzortzis, H. Tsertos, S. Christo and G. Christodoulides, "Gamma-Ray Measurements of Naturally Occurring Radioactive Samples from Cyprus Characteristic Geological Rocks," *Radiation Measurements*, Vol. 37, No. 3, 2003, pp. 221-229.

[11] UNSCEAR, "Sources, Effects and Risks of Ionizing Radiation," United Nations Scientific Committee on the Effects of Atomic Radiation, New York, 2000.

[12] E. Stranden, "Some Aspects on Radioactivity of Building Materials," *Pyhsica Norvegica*, Vol. 8, 1976, pp. 167-177.

[13] J. Beretka and P. J. Mathew, "Natural Radioactivity of Australian Building Materials, Industrial Wastes and by Products," *Health Physics*, Vol. 48, No. 1, 1985, pp. 87-95.

[14] M. Tufail, N. Ahmad, S. M. Mirza, N. M. Mirza and H. A. Khan, "Natural Radioactivity from the Building Materials Used in Islamabad and Rawalpindi, Pakistan," *Science of the Total Environment*, Vol. 121, 1992, pp. 283-291.

[15] R. Krieger, "Radioactivity of Construction Materials," *Betonwerk Fertigteil Technik*, Vol. 47, 1981, pp. 468-473.

[16] L. S. Quindos, P. L. Fernandez and J. Soto, "Building Materials as Source of Exposure in Houses," In: B. Seifert and H. Esdorn, Eds., *Indoor Air '87*, Institute for Water, Soil and Air Hygiene, Berlin, 1987, p. 365.

[17] E. Cottens, "Actions against Radon at the International Level," *Proceedings of the Symposium on SRBII, Journee Radon*, Royal Society of Engineers and Industrials of Belgium, Brussels, 17 January 1990.

6

Molybdenum Carbide Nano-Powder for Production of Mo-99 Radionuclides

Vladimir D. Risovany, Konstantin V. Rotmanov, Genady I. Maslakov, Yury D. Goncharenko, Grigory A. Shimansky, Aleksandr I. Zvir, Irina M. Smirnova, Irina N. Kuchkina

Joint Stock Company, State Scientific Center—Research Institute of Atomic Reactors, Dimitrovgrad, Russia

ABSTRACT

At present, there are two ways to produce ^{99}Mo in a reactor: 1) fission process—from U fission product by reaction ^{235}U (n, f) ^{99}Mo and 2) activation process—by radiation capture reaction ^{98}Mo (n, γ) ^{99}Mo. This paper presents the results of experiments performed with molybdenum carbide nano-powder to produce ^{99}Mo. These results show the implementation of the above idea in practice.

Keywords: Carbide Nano-Powder; Molybdenum-99; Activation Process; Fission Process

1. Introduction

At present, there are two ways to produce ^{99}Mo in a reactor: 1) fission process—from U fission product by reaction ^{235}U (n, f) ^{99}Mo and 2) activation process—by radiation capture reaction ^{98}Mo (n, γ) ^{99}Mo.

The main advantage of ^{99}Mo production from U fission products is a possibility to have large amounts of the final radionuclide ^{99}Mo of a high specific activity (more than ten thousand Ci/g) per one production cycle with practically no carrier. In U fission, the total ^{99}Mo yield makes up 6% in average. At a neutron flux density of about 10^{14} n/(cm^{2}s), the maximum ^{99}Mo accumulation is achieved in 5 - 7 days from the beginning of the target irradiation. There is only one significant drawback in this way of ^{99}Mo production that is a large amount of radwaste that contains more than 97% of the initial U in the target. The remaining ^{235}U isotopes can be extracted from radwaste for further treatment but this is rather a costly process.

The activation process of ^{99}Mo production by reaction ^{98}Mo (n, γ) ^{99}Mo has a great advantage that is practically no radwaste. In addition, when producing molybdenum by the above process, there is no problem with the purification of the final ^{99}Mo from high-active ^{235}U fission product impurities as compared to the fission one. The cost of a ^{99}Mo activity unit produced by the activation process is an order of magnitude lower as compared to the ^{235}U fission product reprocessing. The main disadvantage of this process is the low specific activity of the final ^{99}Mo radionuclide that is several Ci/g. It makes the activation process unsuitable for a commercial produc-

tion of sorption-type ^{99m}Tc generators.

When producing ^{99}Mo by the activation process, molybdenum oxide (MoO$_3$) is used as a target, as a rule [1]. The irradiated target is reprocessed by dissolving molybdenum oxide in alkali with further correction of the solution composition depending on the way of the generator production [2]. The solution, along with the final ^{99}Mo, contains a significant amount of a carrier (non-activated ^{99}Mo and other Mo isotopes if a target was made from natural Mo). In this case, the specific activity of produced ^{99}Mo is about 1-5 Ci/g [3]. The purpose of this paper is to search for a way of selective extraction of final ^{99}Mo from the irradiated target.

Molybdenum carbide nano-power was proposed as a target since it has a well-developed surface that can achieve from tens to hundreds square meters per on gram of powder. The particle surface can be considered as a grain boundary that is an extended defect able to accumulate inter-crystalline segregations. Under irradiation, atoms capture neutrons and turn into excited state. Further, during the decay, energy releases that is enough to destroy the chemical bonding of atoms in a molecule. Atoms go out of the crystalline lattice and may come to the grain boundaries. The surface is like a trap for any non-uniform elements of the material base structure. Atoms of non-uniform elements achieved the surface have an advantage from the energy viewpoint since they can concentrate on grain surfaces. In this case, they can be easily dissolved chemically in alkali or acid solutions.

For the first time the idea of using nano-powders for radio nuclides accumulation was published in 2006 and patented [4]. This paper presents the results of experiments performed with molybdenum carbide nano-powder to produce ^{99}Mo. These results show the implementation of the above idea in practice.

2. Investigation of Molybdenum Carbide Powder Structure

Molybdenum carbide with the natural Mo isotopes content was used, **Table 1**. Molybdenum carbide powder was examined at the super-high resolution field-emission scanning electron microscope Zeiss SUPRA55VP equipped with energy dispersive spectrometer Inca Energy 350, wave dispersive spectrometer Inca Wave 500 and HKL Electron Back Scatter Diffraction Premium registration and analysis system.

We applied some amount of powder (2 - 5 mg) on an Agar electro-conductive band and put it into the microscope vacuum chamber.

Figure 1 presents high magnification secondary electron image of nano-powder particles at accelerating voltage 20 kV.

Then, we examined separate areas by energy dispersive spectrometer Inca Energy 350, registered X-ray characteristic spectra and analyzed the chemical element composition of these areas.

Powder particles were of a globular shape and their sizes varied from 10 to 50 nm. However, the majority of particles was about 20 - 40 nm in size.

It should be mentioned that the examination of the chemical element composition of separate nano-powder particles by SEM is impossible since the typical dimension of the X-ray excitation area (about 1 μm) comprising data on the material elementary composition were certainly larger than the physical size of a particle. This area may consist of not only numerous powder particles but also of carbon-containing substrate that makes difficult the evaluation of the real carbon content in the powder. Besides, the evaluation of the carbon content may be affected by the thickness of a carbon-containing film (the same refers to oxygen) that appears on any surface being in contact with air for some time.

Table 2 presents the results of the chemical composition examination. The ultimate composition of the nano-powder consists, mainly, of molybdenum and carbon. There can be insignificant impurities of calcium and platinum.

3. Irradiation in Research Reactor RBT-6

There were fabricated four quartz capsules with molybdenum carbide powder, **Figure 2** and **Table 3** present the powder mass per each capsule. The capsules were placed into an irradiation rig (**Figure 3**) that was loaded into an FA cell. Irradiation was performed in RBT-6 channel 4 for 10 effective days until the calculated amount of ^{99}Mo

Table 1. Mo isotopes.

Nuclide	Content, %
^{92}Mo	15.5
^{94}Mo	9.4
^{95}Mo	16.1
^{96}Mo	16.7
^{97}Mo	9.4
^{98}Mo	23.6
^{100}Mo	9.2

Table 2. Chemical composition of the nano-powder.

Spectrum	C	O	Ca	Mo	Pt
Spectrum 1	57.72	10.85	0.26	30.40	0.77
Spectrum 2	60.48	10.03	0.21	28.49	0.79
Spectrum 3	60.98	10.34	0.25	27.68	0.75
Spectrum 4	60.02	9.38		29.75	0.84
Spectrum 5	65.01	7.03	0.21	27.76	
Spectrum 6	64.12	8.97	0.34	26.56	

Table 3. Amount of molybdenum carbide powder in capsules.

Capsule No	Weighted amount, mg	Mo content in weighted amount, mg
1	354.75	71.66
2	354.05	71.52
3	354.45	71.60
4	395.05	79.80

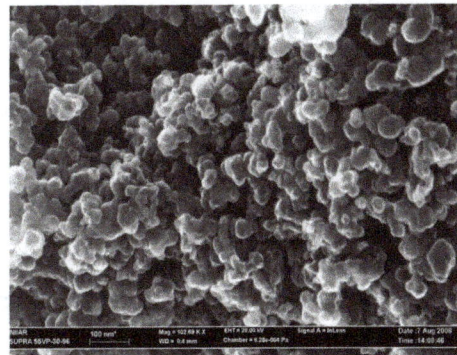

Figure 1. Mo$_2$C nano-powder particles.

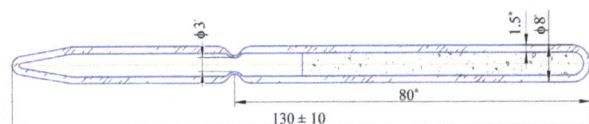

Figure 2. Capsule (target) design with Mo$_2$C powder.

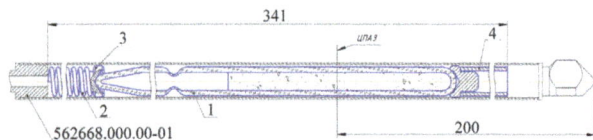

Figure 3. Irradiation rig design.

with a specific activity of about 1Ci/kg was accumulated.

The Mo activation was calculated at a neutron flux density of $1.16 \times 10^{14}\,\mathrm{cm}^{-2} \times \mathrm{s}^{-1}$ (E > 0), $5.17 \times 10^{13}\,\mathrm{cm}^{-2} \times \mathrm{s}^{-1}$ (E > 0.1 MeV) and reactor power 6 MW. To calculate the rate of inelastic interaction reactions between neutrons and nuclei, the ADL-3 cross-section library was used. The radionuclide constants were taken from the FENDL-2 data file. The kinetics of nuclide concentrations under irradiation was calculated by the TRANS_MU code. All the calculation results are in Bq/g unit that is the activity of radionuclide in 1 g of molybdenum at the end of irradiation.

Figures 4 and **5** present the ^{99}Mo decay scheme and its specific activity vs. irradiation time.

It can be seen that the ^{99}Mo specific activity becomes practically stationary after irradiation for 15 effective days (97.4% from the stationary value). After irradiation for 10 effective days, this value is 91.7% from the stationary one and it is 71.5% after irradiation for 5 effective days. Since the specific activity of a long-lived ^{93}Mo nuclide increases almost linearly, then, from the viewpoint of a purity of ^{99}Mo generating nuclide in irradiated molybdenum, the minimal irradiation period is more preferable. All radio nuclides involved into the calculations have half-lives less than 1 minute and activity no less than 1000 Bq/g. Their activities are given in **Table 4**. The relative decrease of ^{99}Mo activity observed during its cooling for 60 days is shown in **Figure 6**.

4. Extraction of ^{99}Mo from Irradiated Material

^{99}Mo was extracted as soon as capsules were removed from the reactor. The capsules preserved their integrity. Powder Mo_2C did not differ from its initial state. It freely ran inside capsules when we overturned them. Close to the capsules, the gamma-emission dose rate did not exceed 1000 μSv/h. Gamma-spectra from the capsules were measured and they showed the dominant presence of ^{99}Mo, **Figure 7**.

The capsules were placed into a glove box and opened. SEM sowed that the size of molybdenum carbide particles remained the same, **Figure 8**.

We added 2 cm^3 of 1×10^{-4} mol/l KOH into a test tube with molybdenum carbide and agitated the mixture for 5 minutes. Then, the suspension was filtrated through a double "blue ribbon" filter into a 25 cm^3 volumetric tube. The deposit remained on the filter was washed 5 - 6

Table 4. Activities of radio nuclides in irradiated molybdenum, Bq/g.

Radionuclide	$T_{1/2}$	Irradiation time, days		
		5	10	15
^{89}Zr	3.27 days	1.20×10^6	1.61×10^6	1.76×10^6
^{89m}Zr	4.18 min	5.74×10^5	5.74×10^5	5.74×10^5
^{91m}Nb	62 days	72784	1.42×10^5	2.07×10^5
^{92m}Nb	10.1 days	6.86×10^7	1.17×10^8	1.52×10^8
^{93}Mo	3×10^3 years	16,500	33,000	49,500
^{99}Mo	2.75 days	2.64×10^{10}	3.39×10^{10}	3.61×10^{10}
^{101}Mo	14.6 min	1.08×10^{10}	1.08×10^{10}	1.08×10^{10}
^{102}Mo	11.2 min	3560	3560	3560
^{99}Tc	2.11×10^5 years	642	2010	3580
^{99m}Tc	6.01 h	2.24×10^{10}	2.96×10^{10}	3.17×10^{10}
^{101}Tc	14.2 min	1.08×10^{10}	1.08×10^{10}	1.08×10^{10}

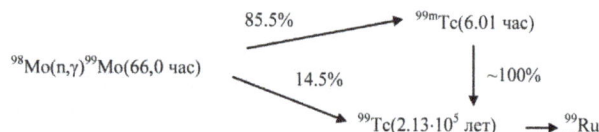

Figure 4. ^{99}Mo decay scheme.

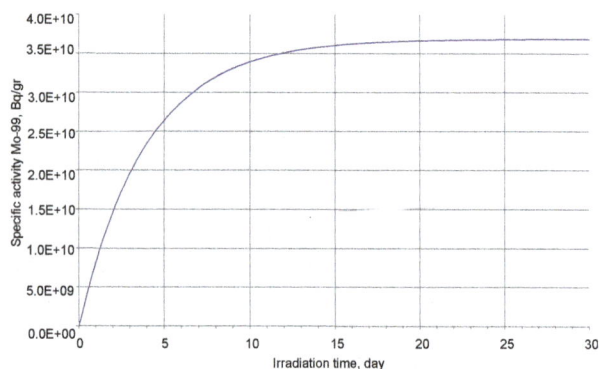

Figure 5. ^{99}Mo specific activity vs. irradiation time.

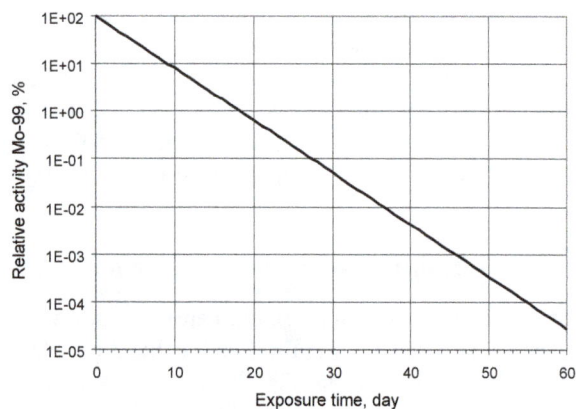

Figure 6. Decrease of ^{99}Mo relative activity after irradiation.

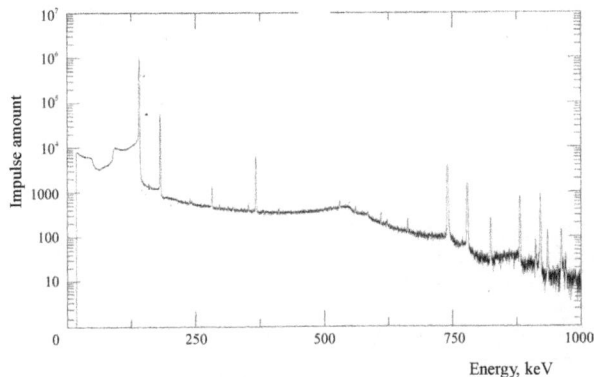

Figure 7. Gamma-spectrum of an irradiated Mo sample.

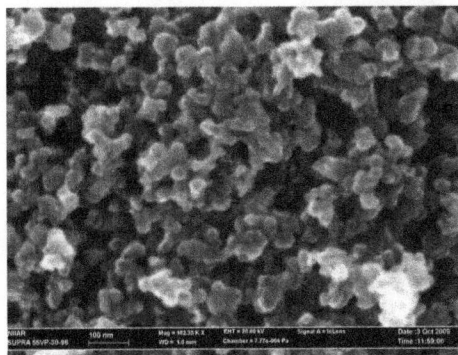

Figure 8. Mo₂C nano-powder particles after irradiation in RBT-6.

Table 5. Measured activities of filter solutions and filtrates samples.

Capsule No.	Operation	Filtrate activity, Bq	Weighted amount activity, Bq	^{99}Mo specific activity, Ci/g Mo
3	Powder washing 0.0001 mol/l KOH	3.3×10^8	-	-
4	Powder washing 0.1 mol/l KOH	7.0×10^8	3.5×10^9	1.2
	Dissolution of filter in H₂SO₄ + H₂O₂	2.8×10^9		
2	Powder washing 0.1 mol/l KOH	8.9×10^8	2.3×10^9	0.9
	Dissolution of filter in H₂SO₄ + H₂O₂	1.4×10^9		
1	Powder washing 0.0001 mol/l KOH	5.6×10^8	1.8×10^9	0.7
	Dissolution of filter in H₂SO₄ + H₂O₂	1.2×10^9		

times with de-ionized water, the filtrate color being dark-green. Then, there was one more filtration though the "blue ribbon" filter. The final filtrate volume made up 20.5 cm³. Then, 1 cm³ of filtrate was diluted with de-ionized water by 10000 times and 50 cm³ of solution was placed into a screw-top plastic glass (∅40 mm, H = 60 mm) and sent for radiometry. It took ~1.5 hours to ex-tract ^{99}Mo. Table 5 presents the measured filtrate sam-ples activities.

Quartz capsule No.4 was opened the next day, on 09. 09.2009. Molybdenum carbide powder was poured into a 5 cm³ test tube through a fennel (visually, almost all powder was removed from the capsule). The gamma emission rate from the powder made up ~350 μSv/h. We added 2 cm³ of 0.1 mol/l KOH into a test tube with mo-lybdenum carbide and agitated the mixture for 5 minutes. Then, the suspension was filtrated through a double "blue ribbon" filter into a 25 cm³ volumetric tube. The deposit remained on the filter was washed 5 - 6 times with de-ionized water, the filtrate color being yellow. Then, there was one more filtration though the "blue ribbon" filter. The final filtrate volume made up 14.5 cm³.

We sampled 1 cm³ to measure the ^{99}Mo and ^{99m}Tc ac-tivity. Then, 1 cm³ of filtrate was diluted with deionized water by 10,000 times and 50 cm³ of solution was placed into a screw-top plastic glass (∅40 mm, H = 60 mm) and

sent for radiometry. All operations took ~2 hours. The results of the filtrate sample activity measurement are given in Table 5.

The activity of filtrates and filter solutions was mea-sured at gamma-spectrometer with an HPGe-detector and DSPEC^{PLUS} analyzer (EG&G ORTEC). All the filtrates and filter solutions were measured in the same plastic glasses with covers ∅40 × 55 mm; the volume of each solution to be measured was 50 ml.

To have a high recording accuracy, the spectrometer was calibrated by means of a certified standard source ^{152}Eu (ρ = 1.01 r/cm³) for volume of 50 ml.

^{99}Mo characteristics are given in Table 6.

The filtrates were cooled for about 1.5 months. During this time, practically all Mo-99 decayed that allowed us to subject these solutions to chemical analysis to evaluate the content of the carrier-stable molybdenum. The analy-sis was done at the "Spectroflame Module S" ICP field-emission spectrometer and the results are given in Table 7. The data on the content of stable molybdenum allowed us to calculate the specific activity of Mo-99 (in filtrates) as well as to evaluate the level of filtrates "en- richment" in Mo-99, i.e. to see how much the filtrate activity was higher than the specific of the whole weighted amount. As it can be seen from Table 6, the "enrichment" of fil-trates No. 1 and No. 2 made up 1.5 and 1.6, respectively. The "enrichment" of filtrate No. 4 was different that could be explained by the insufficient amount of the washing solution (14.5 ml) as compared to filtrates No. 1 and No. 2 (20 and 25 ml, respectively). So, some amount of Mo-99 could remain in the filtrate.

5. Discussion

Figure 9 presents the crystalline structure of molybde-num carbide. Atoms of molybdenum and carbon are lo-

Table 6. ^{99}Mo characteristics.

Reaction	$T_{1/2}$, day	E_γ, keV	yield E_γ, %
^{98}Mo (n, γ) ^{99}Mo	2.75	140.5 (+ ^{99m}Tc)	4.52 (+88.8)
		181.1	6.08
		366.5	1.16
		739.4	12.1
		778.8	4.36

Table 7. Characteristics of Mo$_2$C powder treatment for capsules No. 1, 2, 4.

Capsule No.	Concentration KOH mol/l	Mo content in weighted amount, mg	Mo content in filtrate, mg	Activity of weighted amount, Bq	Activity of filter, Bq	Specific activity of weighted amount, Ci/g Mo	Specific activity of filtrate, Ci/g Mo	Enrichment
1	1×10^{-4}	71.66	13.39	1.8×10^9	5.6×10^8	0.7	1.13	1.6
2	0.1	71.52	17.84	2.3×10^9	8.9×10^8	0.9	1.35	1.5
4	0.1	79.80	14.998	3.5×10^9	7.0×10^8	1.2	1.28	1.07

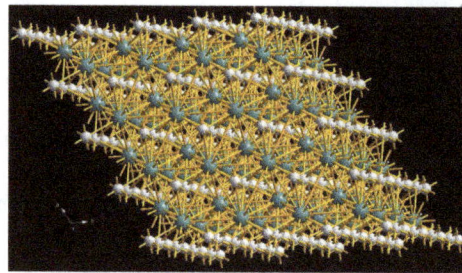

Figure 9. Mo$_2$C crystalline structure.

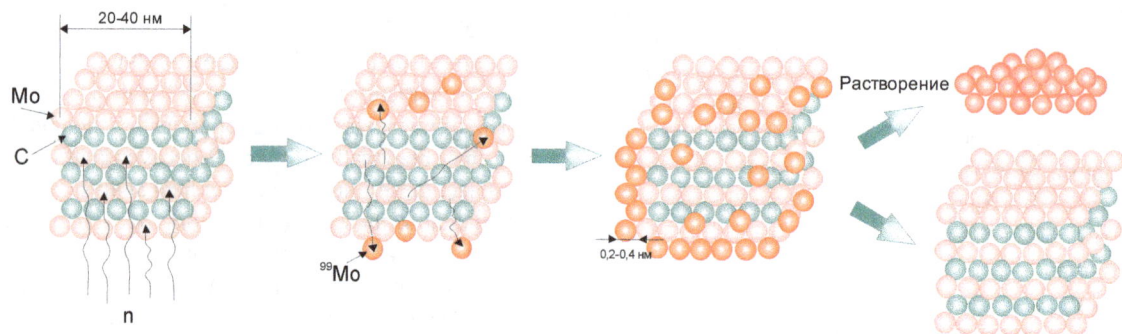

Figure 10. Production of ^{99}Mo from molybdenum carbide nano-powder.

cated in layers and come out to the particles surfaces. Under irradiation, the following mechanisms could be considered. As a neutron is captured, a molybdenum nucleus becomes excited and, as a gamma-quantum is emitted, energy starts releasing that exceeds significantly the chemical bound energy of atoms in the lattice. It allows the generated radionuclide to move out of the lattice at a rather long distance. The surface of powder particles is a good barrier (trap), on which radio nuclides escaped from the lattice accumulate. The second mechanism is when near-surface molybdenum atoms capture neutrons. For these atoms, due to self-shielding, the capture rates could be higher as compared to those located at a dis-

tance from the surface.

The examined molybdenum carbide nano-powder has a very developed surface that achieves tens and hundreds of square meters per one gram of powder. In this case, after irradiation, ^{99}Mo accumulated on the surface can be converted into solution by dissolving a thin near-surface layer about several atom layers thick. The above mechanism is schematically shown in **Figure 10**.

The performed examinations showed that this scheme can be principally used for the ^{99}Mo production. It has several advantages that are, first of all, minimization of waste, simple production operations and low cost. The whole process from the completion of irradiation to the production of solution with ^{99}Mo takes several hours.

It is quite evident that the ^{99}Mo specific activity of more than 50 Ci/g can be achieved if we produce Mo$_2$C nano-powder with Mo enrichment from 24.1% to 99% in ^{99}Mo and irradiate it in higher neutron flux (5 - 10 times higher), in the SM reactor, for instance. Such specific activity can be achieved only under irradiation in a high-flux research reactor. It is expected that if the size of molybdenum carbide nano-powder particles and chemical agent is optimized, the extraction of ^{99}Mo from a target by simple washing may achieve more than 40%, the target being preserved and reused.

It is expedient to use molybdenum enriched in ^{98}Mo since more pure solution with ^{99}Mo, *i.e.* less contaminated with other radio nuclides, can be produced. It is known that the following impurities may generate under irradiation of natural molybdenum: ^{93}Mo (T$_{1/2}$ = 3.5 $\times$ 10^3 years) from ^{92}Mo converting into ^{93m}Nb (T$_{1/2}$ = 13.6 years) and then into ^{93}Nb; ^{101}Mo (T$_{1/2}$ = 14.6 min) from ^{100}Mo generating ^{101}Tc (T$_{1/2}$ = 14.2 min) when decaying

and then ^{101}Ru [3]. Besides, ^{188}W (T$_{1/2}$ = 75 days) and daughter ^{188}Re (T$_{1/2}$ = 167 h) generating ^{188}Os when decaying may be observed in irradiated samples in the form of impurity radio nuclides because of insufficient purification of molybdenum from foreign-metal impurities.

REFERENCES

[1] E. L. R. Hetherington and R. E. Boyd, "Targets for the Production of Neutron Activated Molybdenium-99," *Proceedings of the IAEA Consultants Meeting*, Faure Island, 10-12 April 1997, pp. 19-24.

[2] G. E. Kodina and V. N. Korsunsky, "Status and Progress of Application of Tc-99m Radiopharmaceuticals in Russia," *Radiochemistry*, Vol. 39, No. 5, 1997, pp. 385-388.

[3] M. P. Zykovand and G. E. Kodina, "Methods for ^{99}Mo Production," *Radiochemistry*, Vol. 41, No. 3, 1999, pp. 193-204.

[4] G. I. Maslakov, V. G. Maslakov and L. G. Babikov, "Production of ^{99m}Tc and ^{188}Re," RF Patent No. 2268516, 2006.

7

U and Th Determination in Natural Samples Using CR-39 and LR-115 Track Detectors

S. A. Eman, S. H. Nageeb, A. R. El-Sersy

Department of Ionizing Radiation, National Institute for Standards, Giza, Egypt

ABSTRACT

In this work, a simple and accurate method for U and Th determination in natural samples is proposed. This method is based on simplified calculations of the efficiency factor of the solid state track detector using a thin source approach. Samples were firstly saluted using a concentrated H_2SO_4 acid and then distributed on a glassy slide where the thickness of the sample was about 7 μm. CR-39 and LR-115 track detectors were exposed to the thin layer of the natural samples for few days and then the track densities were obtained. By the mean of originated track densities in CR-39 and LR-115 as a function of exposure time and sample weight, the concentration of U and Th in Bq/kg were obtained by the thin source approach of SSNTD.

Keywords: U; Th; CR-39 and LR-115

1. Introduction

Solid State Nuclear Track Detectors, SSNTD are widely used in different applications related to α-track registration [1-4]. One of the most applications is the U and Th determination in natural materials. Th in natural samples is usually more abundant than U by a factor of about 4, so it is necessary to calculate the ratio of Th/U in order to perform a quantitative analysis of uranium and thorium in natural samples [5,6] but this ratio is related to the geological structure of samples.

Different tools have been used for Th content as isotope dilation mass spectroscopy chromatography. Inductive coupled plasma mass spectrometry, neutron activation analysis, α-spectrometry using SSNTD. The use of SSNTD has many advantages than the others due to the long term exposure property and the alpha track counting is performed without any electronic attachment.

Using SSNTDs, the general relation used for concentration determination of U or Th in CR-39 or LR-115 are given in Equations (1) through (4) [5].

$$\rho_U^{CR} = k_U^* C_U \quad (1)$$

$$\rho_{Th}^{CR} = k_{Th}^* C_{Th} \quad (2)$$

where ρ_{Th}^{CR} and ρ_{Th}^{CR} are the registered track densities of U and Th, series respectively. C_U and C_{Th} are the concentrations of U and Th in the natural sample. k_U and k_{Th} are the α registration efficiency from U and Th series, respectively and so for LR-115 C_U and C_{Th} are:

$$\rho_U^{LR} = k_U C_U \quad (3)$$

$$\rho_{Th}^{LR} = k_{Th} C_{Th} \quad (4)$$

Then the total track density of CR-39 is

$$\rho_t^{CR} = k_U^* C_U \left[1 + \frac{k_{Th}^* C_{Th}}{k_U^* C_U} \right] \quad (5)$$

And for LR

$$\rho_t^{LR} = k_U C_U \left[1 + \frac{k_{Th} C_{Th}}{k_U C_U} \right] \quad (6)$$

From Equations (5) and (6), the track density ratio of CR-39 to LR-115 is

$$\bar{\rho} = \frac{\rho_t^{CR}}{\rho_t^{LR}} = \frac{k_U^*}{k_U} \left[\frac{1 + \frac{k_{Th}^* C_{Th}}{k_U^* C_U}}{1 + \frac{k_{Th} C_{Th}}{k_U C_U}} \right] \quad (7)$$

The efficiency factor for thick source is given by Equations (8) and (9):

$$k_U = \frac{1}{4} \sum_{i=1}^{8} R_i \cos^2 \theta_{ci} \quad (8)$$

$$k_{Th} = \frac{1}{4} \sum_{i=1}^{6} R_i \cos^2 \theta_{c_i} \quad (9)$$

where R_i and θ_{ci} are the range and the critical angle of isotopes in U and Th series, respectively [7].

2. Method of Calculations

From Equations (8) and (9), both of α range (R) and the critical angle of etching (θ_c) should be determined for efficiency calculation. For accurate calculation the α range from U and Th series, chemical analysis should be performed for the natural containing radioactive material. Also θ_c for 14 different alpha energies should be obtained, which sometimes very sophisticated.

In this work, a thin source approach was proposed, where the efficiency factor in Equations (8) and (9) could be written as for U and Th:

$$k_U = \sum_{i=1}^{8}\left(1-\sin\theta_{c_i}\right) \qquad (10)$$

$$k_{Th} = \sum_{i=1}^{6}\left(1-\sin\theta_{c_i}\right) \qquad (11)$$

From Equations (10) and (11), one can notice that efficiency depends only on the critical angle θ_c which is easily known as a function of α energy [7].

Using the thin source technique, with CR-39 Equation (5) can be rewritten as:

$$\rho_{CR\,thin} = k_{U\,thin}C_U + k_{Th\,thin}C_{Th} \qquad (12)$$

where $k_{U\,thin}$ and $k_{Th\,thin}$ can be calculated from Equations (10) and (11) which yield to the determination of $C_U + C_{Th}$. So other relations are needed to get both C_U and C_{Th}, which can be obtained from LR-115 data.

By considering Th series [9-10], one can notice that the series contains 6 alpha particle emitters with energies higher than 4.5 (window of LR-115 detector [8]) only ^{232}Th has α energy of value 4.08 MeV [9,10]. So when thin layer of sample is used with LR-115 it will only register the 4 α from U-series with energy smaller than 4.5 MeV, then

$$\rho_t^{LR} = \sum_{i=1}^{4}\left(1-\sin\theta_c\right)C_U \qquad (13)$$

From Equations (12) and (13) both of C_U and C_{Th} are easily obtained.

3. Experimental

Natural samples are collected from Oil fields at the beach of the Meditraian Sea near Domiat and Kafer El-Skeekh north east of Cairo, Egypt. Samples were grinded to very small grain size. A very small mass of each sample was added to concentrated solution of H_2SO_4 acid. The resulted solution was distributed on a glass slide where the thickness of the sample on the slide is in the range of 7 μm.

CR-39 and LR-115 detectors were exposed, in closed contact, to the samples for few days. Detectors were collected and etched at the optimum etching conditions. Track density originated in CR-39 and LR-115 was counted using an image analyzer model ELBEK SAMICA, Germany.

4. Results and Discussion

Colleted samples were coded with numbers from 1 to 10 and about 300 mg of sample was used and distributed on a glassy slide. The originated tracks ($cm^{-2} \cdot g^{-1} \cdot sec^{-1}$) in CR-39 and LR-115 within 15 day exposure were counted and illustrated in **Figure 1.**

From the track densities in CR-39 and LR-115, two methods of calculation were used. These methods were method 1 where the contribution of track density in LR-115 from Th is neglected and Method 2 where track density from Th in LR is considered.

4.1. First Method

Values of alpha energies in U and Th series are given in references 9 and 10.

In this approximation neglecting the contribution in track density from Th series in LR-115, the U activity concentration can be determined from Equation (13) and represented in **Figure 2**.

From the obtained U concentration, and by substituting in Equation (12), the Th concentration is obtained and illustrated in **Figure 3**.

4.2. Second Method

In this method, all alpha energies closed to 4.5 MeV or below are considered in LR-115 registration from both U and Th series. In this method Equation (13) can be written as:

$$\rho_t^{LR} = \sum_{i=1}^{4}\left(1-\sin\theta_c\right)C_U + \left(1-\sin\theta_c\right)C_{Th} \qquad (14)$$

The obtained U concentration from this method is

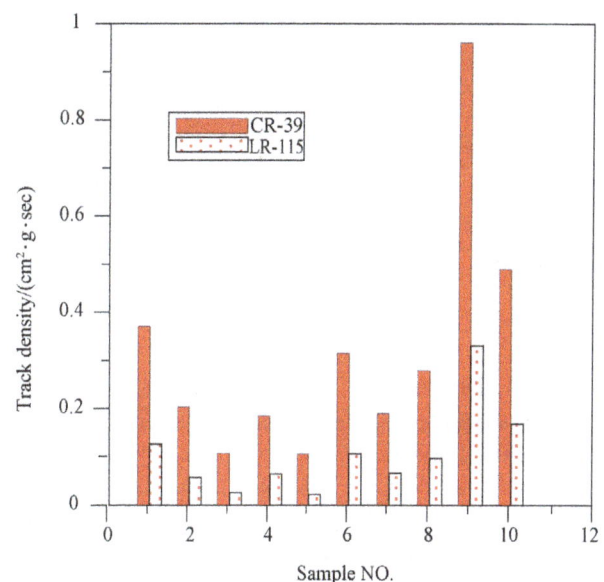

Figure 1. Originated track density in CR-39 and LR-115 exposed to natural samples.

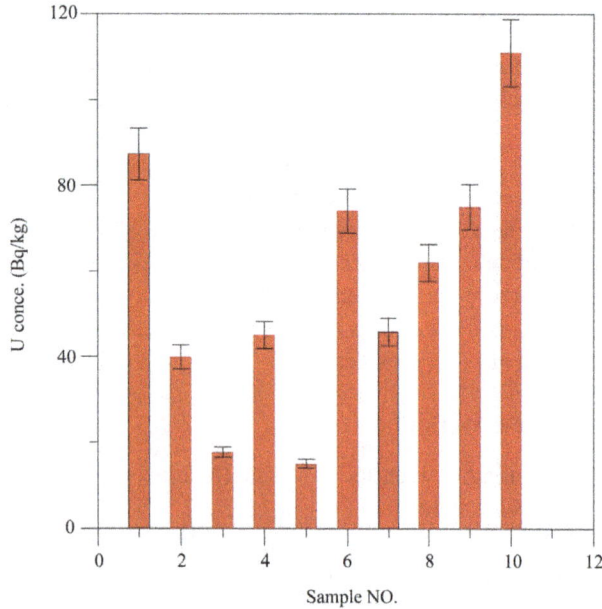

Figure 2. Activity concentration of U (Bq/kg) calculated from Equation (13).

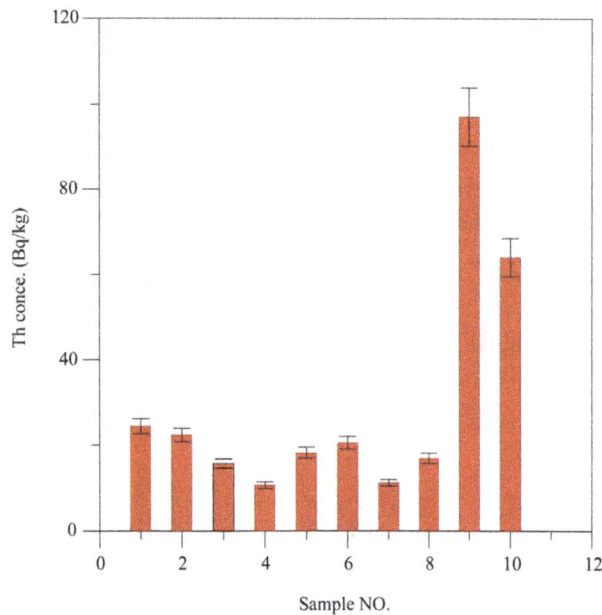

Figure 3. Th activity concentration (Bq/kg) for the studied samples.

illustrated in **Figure 4**.

The Th concentration is calculated from Equations (13) and (14) using the values of C_U and represented in **Figure 5**.

To cheek the variation between these two methods, a comparison between U concentrations calculated from the two approximations is represented in **Figure 6**. From **Figure 6**, one can notice that, the U concentration calculated by Method 1 is slightly grater that obtained from Method 2. By the same procedure a comparison between Th concentrations calculated by the two methods is represented

in **Figure 7** where nearly same value of Th concentration is noticed.

From **Figures 6** and **7**, one can notice that, there is a good agreement between theses two methods of U and Th calculations where a correlation factor of 86% and 89% for U and Th, respectively. Also one can use Method 1 as an easy method for U and Th calculation using SSNTDs with reasonable accuracy.

To assure the U and Th concentrations obtained from this study, one has two directions. The first one is the confirmation of the results by other technique used for U and

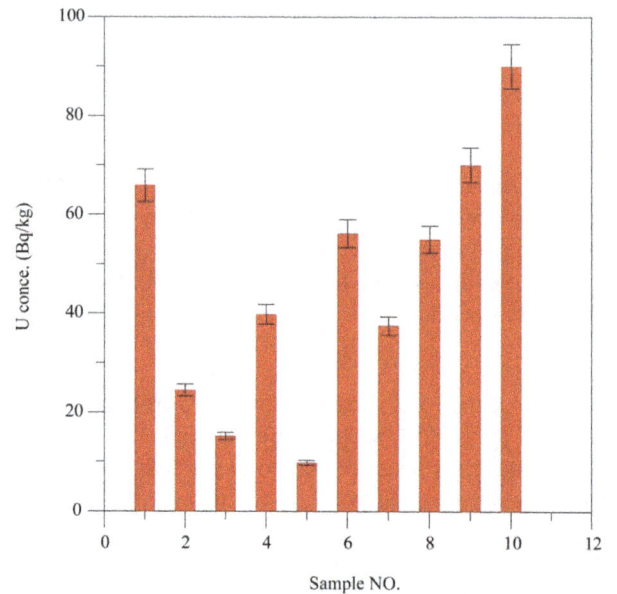

Figure 4. Activity concentration of U (Bq /kg) calculated by Method 2.

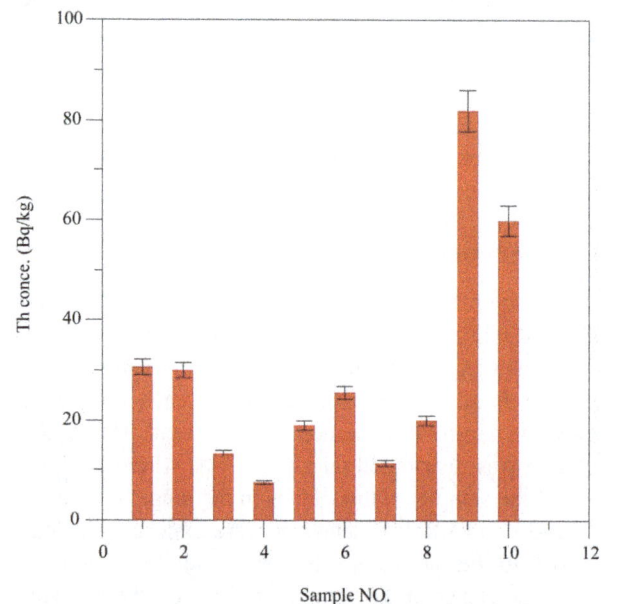

Figure 5. Th activity concentration (Bq/kg) calculated by method 2.

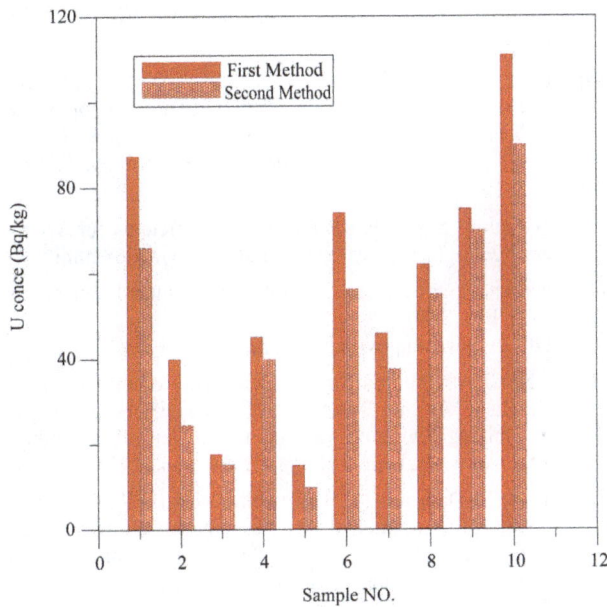

Figure 6. Comparison between U concentrations as calculated by the two methods.

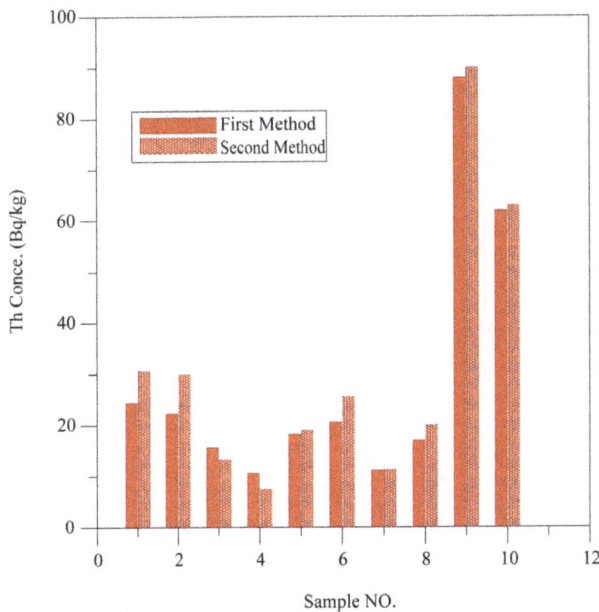

Figure 7. Comparison between Th concentrations as calculated by the two methods.

Th detection. Three samples were selected and then the content U and Th were measured by gamma spectroscopic technique calibrated by a standard multi-radionuclide source traceable to the National Institute for Standard and Technology (NIST) at USA. The obtained correlation coefficient is 87% and 84% for U and Th concentrations that reflects a good agreement between the two techniques.

The second direction of confirmation is the comparison of the obtained data with that published in the literature. The world average of U and Th concentrations as re-

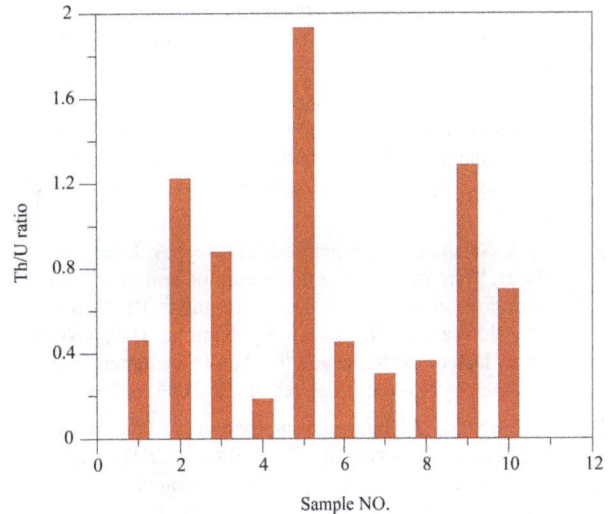

Figure 8. Values of Th/U ratio for the studied samples.

ported in the UNSCEAR, 1988 [11] is 25 Bq/kg. The average U and Th concentrations in the studied samples as obtained from Method 2 are 46.38 and 31.02 Bq/kg, respectively. From this data, one can notice that the obtained results are in the same order of that obtained in the UNSEAR and El-Daly et al., 2008 [12].

Ratio of Th/U is calculated and represented in **Figure 8** where the average ratio for all samples is 0.80 with stander deviation of 0.55. The variation of this ratio may be attributed to the collection of the studied samples was from one zone and different depth that reflected the difference of the geological stricture for each depth.

5. Conclusion

From this work, one can conclude that, SSNTDs can be used for U and Th determination in natural samples. The suggested thin source used is applicable for U and Th determination without sophisticated calculations. Also, the first approximation is accurate and easy for U and Th determination. The correlation between the two methods of calculations is 88% and 86% for U and Th values that reflected the applicability of method one. Also the obtained U and Th concentrations were in a good agreement with the gamma spectroscopic technique.

REFERENCES

[1] T. EL-Zakla, H. A. Abdel-Ghny and A. M. Hassan, "Natural Radioactivity of Some Local Fertilizers," *Romanian Journal of Physics*, Vol. 52, No. 5-7, 2007, p. 731.

[2] H. Florou, K. Kehagia, C. H. Chaloulou, V. Koukouliou and C. H. Lykomitrou, "Determination of Radionuclide in Mystiques Galloprovicialis by Alpha and Gamma-Spectroscopy," *Mediterranean Marine Science*, Vol. 5, No. 1, 2004, p. 117.

[3] C. A. Simion, C. Cimpeanu, C. Barna and E. Duta, "The

Uranium Determination in Commercial Iodine Salt," *Romanian Journal of Physics*, Vol. 51, No. 7-8, 2006, p. 845.

[4] N. W. El-Dine, A. El-Shershaby, F. Ahmed and A. S. Abdel-Haleem, "Measurement of Radioactivity and Radon Exhalation Rate in Different Kinds of Marbles and Granites," *Applied Radiation and Isotopes*, Vol. 55, 2001, p. 853.

[5] T. A. Salama, U. Seddik, T. M. Dsoky, A. Morsy and R. Asser, "The Influence of the Nature of Soil and Plant and Pollution on the ^{238}U, ^{232}Th, ^{222}Rn and ^{220}Rn Concentrations in Various Natural Honey Samples Using Nuclear Track Detectors: Impact on the Adult Consumers," *Indian Academy of Sciences*, Vol. 62, No. 2, 2006, p. 269.

[6] H. Surbeck, "Alpha Spectrometry Sample Preparation Using Selectively Adsorbing Thin Films," *ICRM Conference on Low Level Radioactivity Measurement Techniques*, Mol, Belgium, 18-22 October, 1999, pp. 97-100.

[7] M. M. El-Hawary, M. M. Mansy, A. Hussein, A. A. Ammar and A. R. El-Sersy, "Construction of a Charged Particle Irradiation Chamber for the Use with Plastic Detectors," *Radiation Physics and Chemistry*, Vol. 54, 1999, p. 547.

[8] H. El-Samman, M. Mansy, A. Hussein, M. El-Hawary and A. R. El Sersy, "Registration Efficiency of Some SSNTDs with Source Area Included," *Nuclear Instruments and Methods in Physics Research Section B*, Vol. 155, 1999, p. 426.

[9] "Uranium Series ICRP 68 Dose Coefficients for Workers," 2003. http://www.wise-uranium.org/rdfi68.html

[10] National Service Center for Environmental Publications (NSCEP), 2007. http://www.epa.gov/radiation/heast/docs/heast2

[11] UNSCEAR (United Nations Scientific Committee on the Effects of Atomic Radiation), "Sources Effects and Risks of Ionizing Radiation," Report to the General Assembly, With Annexes, United Nations, New York, 1988.

[12] T. A. El-Daly and A. S. Hussein, "Natural Radioactivity Level in Environmental Samples in North Western Desert of Egypt," *Proceedings of the 3rd Environmental Physics Conference*, Aswan, Egypt, 19-23 February 2008, pp. 79-88.

8

A Closed-Form Formulation for the Build-Up Factor and Absorbed Energy for Photons and Electrons in the Compton Energy Range in Cartesian Geometry

Volnei Borges, Julio Cesar Lombaldo Fernandes, Bardo Ernest Bodmann, Marco Túllio Vilhena, Bárbara Denicol do Amaral Rodriguez

Department of Applied Math, Universidade Federal do Rio Grande do Sul, Porto Alegre, Brazil

ABSTRACT

In this work, we report on a closed-form formulation for the build-up factor and absorbed energy, in one and two dimensional Cartesian geometry for photons and electrons, in the Compton energy range. For the one-dimensional case we use the LTS_N method, assuming the Klein-Nishina scattering kernel for the determination of the angular radiation intensity for photons. We apply the two-dimensional LTS_N nodal solution for the averaged angular radiation evaluation for the two-dimensional case, using the Klein-Nishina kernel for photons and the Compton kernel for electrons. From the angular radiation intensity we construct a closed-form solution for the build-up factor and evaluate the absorbed energy. We present numerical simulations and comparisons against results from the literature.

Keywords: Build-Up Factor; Compton Energy; Cartesian Geometry; Fokker-Plank Equation

1. Introduction

In radiological protection the effectiveness of a material as a biological shield is related to its cross-section for scattering and absorption which is cast into physical parameters. The determination of those parameters require the solution of a systems of linear transport equations, *i.e.* the fluence for photons and electrons. Established methods that solve the transport equations are the PN approximation [1], the discrete ordinate method and their variants [2]. The S_N method has been used successfully in photon transport calculation, whereas recently the P_N approximation was applied to electron transport. In this work we present the solution of a couple system of linear transport equations, using the LTS_N method for a rectangular domain considering the Klein-Nishina scattering kernel and a multi-group model for photons. The electron contribution to energy deposition induced by incident photons is quantified solving the two-dimensional Fokker-Planck equation for electron transport [3,4] by the P_N approximation in the angular variable followed by applying the Laplace Transform to one of the spatial variables (here x). This procedure leads to a closed-form formulation for the build-up factor and absorbed energy, in one and two dimensional Cartesian geometry for photons and electrons, in the energy range where Compton scattering is dominant [5].

2. The LTS_N Nodal Solution in Two Dimensional for Photons

The two-dimensional S_N nodal problem for photons, assuming Klein-Nishina scattering kernel and a multi-group model is

$$
\begin{aligned}
&\mu_n \frac{\partial}{\partial x} I_{jn}(x,y) + \eta_n \frac{\partial}{\partial y} I_{jn}(x,y) + \mu_{1j} I_{jn}(x,y) \\
&= \frac{\Delta}{3} \sum_{l=0}^{L} \frac{2l+1}{2} \sum_{r=1}^{G} c_r \alpha k_{rj} P_l (1 + \lambda_r - \lambda_j) P_l(\mu_n) \\
&\times \sum_{i=1}^{N} P_l(\mu_i) I_{ri}(x,y) w_i
\end{aligned}
\quad (1)
$$

subject to vacuum boundary conditions in a rectangle $0 \le x \le a$ and $0 \le y \le b$. Where $j = 1, \cdots, G; n = 1, \cdots, N;$ $N = M(M+2)/2$ is the cardinality of the discrete ordinate set, M represents the order of the angular quadrature, G is the number of energy groups in units of wavelengths, μ_{lj} is the linear attenuation coefficient, $I_{jn}(x,y) = I(x,y,\lambda_j,\Omega_n)$ is the angular flux into the discrete direction $\Omega_n = (\mu_n, \eta_n)$ for the j-th energy group, w_i are the Lewis-Miller quadrature weights and $k_{rj} = k(\lambda_r, \lambda_j)$ is the Klein-Nishina scattering kernel, defined as

$$
k_{rj} = \frac{3}{8} \frac{\lambda_r}{\lambda_j} \left(\frac{\lambda_r}{\lambda_j} + \frac{\lambda_j}{\lambda_r} - \sin^2 \sin^2 \theta \right)
\quad (2)
$$

The LTS_N nodal approach for photons yields the following S_N equation system

$$\eta_n \frac{\partial}{\partial y} I_{jny}(y) + \frac{\mu_n}{a}\left[I_{jn}(a,y) - I_{jn}(0,y)\right] + \mu_{lj} I_{jny}$$

$$= \frac{\Delta}{3}\sum_{l=0}^{L}\frac{2l+1}{2}\sum_{r=1}^{G}c_r\alpha k_{rj}P_l\left(1+\lambda_r-\lambda_j\right)P_l(\mu_n) \qquad (3)$$

$$\times \sum_{i=1}^{N}P_l(\mu_i)I_{riy}(x,y)w_i$$

For $j = 1,\cdots,G$ and $= 1,\cdots,N$. Here $I_{jn}(a,y)$ and $I_{jn}(0,y)$ are the angular fluxes at the boundary edges with $x = 0$, $x = a$ and the average angular flux is

$$I_{jny}(y) = \frac{1}{a}\int_0^a I_{jn}(x,y)\,\mathrm{d}x \qquad (4)$$

Further,

$$\mu_n \frac{\partial}{\partial x} I_{jnx}(x) + \frac{\eta_n}{b}\left[I_{jn}(x,0) - I_{jn}(x,b)\right] + \mu_{lj}I_{jnx}$$

$$= \frac{\Delta}{3}\sum_{l=0}^{L}\frac{2l+1}{2}\sum_{r=1}^{G}c_r\alpha k_{rj}P_l\left(1+\lambda_r-\lambda_j\right)P_l(\mu_n) \qquad (5)$$

$$\times \sum_{i=1}^{N}P_l(\mu_i)I_{rix}(x)w_i$$

For $j = 1,\cdots,G$ and $= 1,\cdots, N$. In Equation (5), $I_{jn}(x,b)$ and $I_{jn}(x,0)$ are the angular fluxes at the boundary edges with $y = 0$, $y = b$ and the average angular flux is

$$I_{jnx}(x) = \frac{1}{b}\int_0^b I_{jn}(x,y)\,\mathrm{d}y \qquad (6)$$

Application of the LTS_N method consists in Laplace transform of the set of S_N equations and solving the resulting algebraic equations by matrix diagonalization and subsequent inversion of the transformed angular flux by standard procedures of integral transform theory. The Laplace transform of Equation (3) is

$$s\overline{I}_{jny}(s) + \frac{\mu_{lj}}{\eta_n}\overline{I}_{jny}(s) - \frac{\Delta}{3\eta_n}\sum_{l=1}^{L}$$

$$\times \frac{2l+1}{2}\sum_{r=1}^{G}c_r\alpha k_{rj}P_l(\mu_n)\sum_{i=1}^{N}P_l(\mu_i)\overline{I}_{riy}(s)w_i \qquad (7)$$

$$= I_{jny}(0) - \frac{\mu_n}{a\eta_n}[\overline{I}_{jn}(a,s) - \overline{I}_{jn}(0,s)]$$

For $j = 1,\cdots,G$ and $n = 1,\cdots,N$ which can be cast in matrix form,

$$\left(sI - B_{jny}\right)\overline{I}_{jny}(s) = I_{jny}(0) + \overline{Z}_{(j-1)y}(s) + \overline{S}_{jny}(s) \qquad (8)$$

Here, $\overline{I}_{jny}(s)$ are the N componentes of the Laplace transformed angular flux in the y variable and $\overline{I}_{jny}(0)$ are the respective components of the angular flux at the edge $y = 0$. The components have the following forms,

$$\overline{I}_{jny}(s) = \left[\overline{I}_{j1y}(s), \overline{I}_{j2y}(s), \cdots, \overline{I}_{jNy}(s)\right]^T \qquad (9)$$

and

$$\overline{I}_{jny}(0) = \left[\overline{I}_{j1y}(0), \overline{I}_{j2y}(0), \cdots, \overline{I}_{jNy}(0)\right]^T \qquad (10)$$

The components of the matrix B_{jny} are given by,

$$b_y(p,q)$$
$$= \begin{cases} -\dfrac{\mu_{lj}}{\eta_p} + \dfrac{\Delta}{3\eta_p}\sum_{l=1}^{L}\dfrac{2l+1}{2}c_j\alpha k_{jj}P_l(\mu_p)P_l(\mu_p)w_p, & p = q \\[2ex] \dfrac{\Delta}{3\eta_p}\sum_{l=0}^{L}\dfrac{2l+1}{2}c_j\alpha k_{jj}P_l(\mu_p)P_l(\mu_q)w_p, & p \neq q \end{cases}$$

The scattering term is,

$$\overline{Z}_{(j-1)y}(s) = \sum_{i=1}^{j-1}H_{ij}\overline{I}_{iny}(s) \qquad (11)$$

with the components of matrix H_{ij} (see the equation below). The vector $\overline{S}_{jny}(s)$ has the components,

$$\overline{S}_{jiy}(s) = -\frac{\mu_{lj}}{a\eta_i}\left[\overline{I}_{ji}(a,s) - \overline{I}_{ji}(0,s)\right] \qquad (12)$$

A similar methodology in the x variable leads to the linear algebric system which can be matrix form,

$$\left(sI - A_{jnx}\right)\overline{I}_{jnx}(s) = I_{jnx}(0) + \overline{Z}_{(j-1)x}(s) + \overline{S}_{jnx}(s) \qquad (13)$$

where $\overline{I}_{jnx}(s)$ has the following form,

$$\overline{I}_{jnx}(s) = \left[\overline{I}_{j1x}(s), \overline{I}_{j2x}(s), \cdots, \overline{I}_{jNx}(s)\right]^T \qquad (14)$$

The matrix elements of A_{jnx} are,

$$a_x(p,q)$$
$$= \begin{cases} -\dfrac{\mu_{lj}}{\eta_p} + \dfrac{\Delta}{3\eta_p}\sum_{l=0}^{L}\dfrac{2l+1}{2}c_j\alpha k_{jj}P_l(\mu_p)P_l(\mu_p)w_p, & p = q \\[2ex] \dfrac{\Delta}{3\eta_p}\sum_{l=0}^{L}\dfrac{2l+1}{2}c_j\alpha k_{jj}P_l(\mu_p)P_l(\mu_q)w_p, & p \neq q \end{cases}$$

and the scattering term is

$$\overline{Z}_{(j-1)x}(s) = \sum_{i=1}^{j-1}H_{ix}\overline{I}_{inx}(s) \qquad (15)$$

where the matrix elements of H_{ix} are given

$$h_x(p,q)$$
$$= \begin{cases} \dfrac{\Delta}{3\eta_p}\sum_{l=0}^{L}\dfrac{2l+1}{2}c_j\alpha k_{ij}P_l(1+\lambda_i-\lambda_j)P_l(\mu_p)P_l(\mu_p)w_p \\[2ex] -\dfrac{\Delta}{3\eta_p}\sum_{l=0}^{L}\dfrac{2l+1}{2}c_j\alpha k_{jj}P_l(1+\lambda_i-\lambda_j)P_l(\mu_p)P_l(\mu_q)w_p \end{cases}$$

$$h_y(p,q) = \begin{cases} \dfrac{\Delta}{3\eta_p}\sum_{l=0}^{L}\dfrac{2l+1}{2}c_j\alpha k_{ij}P_l\left(1+\lambda_i-\lambda_j\right)P_l(\mu_p)P_l(\mu_p)w_p, & p = q \\[2ex] -\dfrac{\Delta}{3\eta_p}\sum_{l=0}^{L}\dfrac{2l+1}{2}c_j\alpha k_{jj}P_l\left(1+\lambda_i-\lambda_j\right)P_l(\mu_p)P_l(\mu_q)w_p, & p \neq q \end{cases}$$

A Closed-Form Formulation for the Build-Up Factor and Absorbed Energy for Photons and Electrons in the
Compton Energy Range in Cartesian Geometry

51

and the vector $\bar{S}_{jnx}(x)$ is

$$\bar{S}_{jix}(s) = -\frac{\eta_i}{b\mu_i}\left[\bar{I}_{ji}(s,b) - \bar{I}_{ji}(s,0)\right] \quad (16)$$

The LTS_N solution for Equations (8) and (16) are given by

$$\bar{I}_{jny}(s) = \left(sI - B_{jny}\right)^{(-1)}\left[I_{jny}(0) + \bar{Z}_{(j-1)y}(s) + \bar{S}_{jny}(s)\right] \quad (17)$$

and

$$\bar{I}_{jnx}(0) = \left(sI - A_{jnx}\right)^{-1}\left[I_{jnx}(0) + \bar{Z}_{(j-1)x}(s) + \bar{S}_{jnx}(s)\right]^{T} \quad (18)$$

Taking the Laplace inversion by applying the Heaviside expansion yields

$$I_{jny}(y) = \sum_{k=1}^{jn}\beta_k e^{s_k y}I_{jny}(0) + Z_{(j-1)y}(y) * \mathcal{L}^{-1}\left\{\left(sI - B_{jny}\right)^{-1}\right\}$$
$$+ S_{jny}(y) * \mathcal{L}^{-1}\left\{\left(sI - B_{jny}\right)^{-1}\right\} \quad (19)$$

and

$$I_{jnx}(x) = \sum_{k=1}^{jn}\beta_k e^{s_k x}I_{jny}(0) + Z_{(j-1)x}(x) * \mathcal{L}^{-1}\left\{\left(sI - A_{jnx}\right)^{-1}\right\}$$
$$+ S_{jnx}(x) * \mathcal{L}^{-1}\left\{\left(sI - A_{jny}\right)^{-1}\right\} \quad (20)$$

For the fluxes at boundary one may use reasonable approximation [6]

$$I_{jn}(x,0) = F_{jn}e^{-sign(\mu_n)\Lambda x} \quad (21)$$

$$I_{jn}(0,y) = G_{jn}e^{-sign(\eta_n)\Lambda y} \quad (22)$$

$$I_{jn}(x,b) = O_{jn}e^{-sign(\mu_n)\Lambda x} \quad (23)$$

$$I_{jn}(a,y) = P_{jn}e^{-sign(\eta_n)\Lambda y} \quad (24)$$

Here the signal function follows the usual definition $sign(\mu) = 1$ for $\mu > 0$ and $sign(\mu) = -1$ for $\mu < 0$ and Λ represents an attenuation parameter, here the macroscopic absorption crosssection.

The generalization of the LTS_N nodal solution for a heterogeneous rectangular geometry assuming the Klein-Nishina scattering kernel and a multi-group model may be implemented in a completely analogue procedure. In such a problem the LTS_N solution is determined for each subdomain and the integration constants are evaluated upon applying the boundary and interface.

3. The Two-Dimensional Fokker-Planck Equation Solution for Electrons

The purpose of radiation transport is to determine how particles move through materials and what effects their propagation have on the material through the mechanisms of deposited energy and charge deposition. The angular flux of electrons in a rectangular domain can be deter-

mined by solving the following two-dimensional, time independent electron transport equation, in a rectangle $0 \leq x \leq a$ and $0 \leq y \leq b$, subject to vacuum boundary conditions.

$$\mu\frac{\partial\psi(x,y,\bar{\Omega},E)}{\partial x}$$
$$+ \eta\frac{\partial\psi(x,y,\bar{\Omega},E)}{\partial y} + \sigma_t(E)\psi(x,y,\bar{\Omega},E) \quad (25)$$
$$= \int_{4\pi}d\bar{\Omega}\sigma_s\left(E' \to E,\bar{\Omega}'\cdot\bar{\Omega}\right)\psi(x,y,\bar{\Omega}',E')$$

Here, the angular flux (x,y,E,Ω), represents the flux of particles at position (x,y), with energy E travelling in direction $\Omega = (\mu,\eta)$. The quantity σ_s in Equation (25) is the differential scattering cross-section, in this work, we focus on screened Rutherford scattering which is

$$\sigma_s(E,\mu_0) = \frac{\sigma_t(E)\eta^*(\eta^* + 1)}{\pi\left(1 + 2\eta^* - \mu_0^2\right)} \quad (26)$$

where $\eta^* > 0$ is typically a small constant called the screening parameter parameter is given by

$$\eta^* = \frac{h^2 Z^{2/3}}{4\left(a_H\right)^2\left(m_e v\right)^2} \quad (27)$$

with Z the atomic number of the nucleus, mv the relativistic momentum of the scattered electron and

$$C = \frac{\bar{h}^2}{4a_H^2} \quad (28)$$

which $\bar{h}$ is Planck's constant and a_H is the Bohr radius.

We assume that the Fokker-Planck (FP) scattering description is appropriate, so that the transport problem (25) is given by,

$$\mu\frac{\partial\psi^{FP}(x,y,\bar{\Omega}',E)}{\partial x} + \eta\frac{\partial\psi^{FP}(x,y,\bar{\Omega}',E)}{\partial y}$$
$$= \frac{\sigma_{tr}}{2}\frac{\partial}{\partial\mu}\left[\left(1 - \mu^2\right)\frac{\partial}{\partial\mu}\right]\psi^{FP}(x,y,\bar{\Omega}',E) \quad (29)$$

Here the coefficient σ_{tr} is he transport cross-sectionand defined as,

$$\sigma_{tr} = 2\pi\int_{-1}^{1}\int_0^1\sigma_s(E,\mu_0)(1 - \mu_0)d\mu_0 d\eta \quad (30)$$

Upon multiplying Equation (29) by $P_N(\mu)$, integrating over μ and using the recursion formula, as well as general properties of Legendre polynomials leads to the following P_N equations

$$\frac{n+1}{2n+1}\frac{\partial}{\partial x}\psi_{n+1}^{FP}(x,y,E) + \frac{n}{2n+1}\frac{\partial}{\partial x}\psi_{n-1}^{FP}(x,y,E)$$
$$+ \frac{2n+1}{n}\frac{\partial}{\partial y}\psi_n^{FP}(x,y,E)T_n = \frac{\sigma_{tr}}{2}\left[-n(n+1)\right]\psi_n^{FP}(x,y,E)$$

$$(31)$$

The angular flux moments in discrete ordinates may be expressed using a quadrature approximation,

$$\psi^{FP}\left(x,y,\bar{\Omega}'E\right)=\sum_{l=0}^{L}\frac{2l+1}{2}\psi_{n}^{FP}\left(x,y,E\right)P_{n}\left(\mu\right)\quad(32)$$

For $n=0,\cdots,N$ with $\psi_{N+1}^{FP}\left(x,y,E\right)=0$ in the P_{N} approximation and T_{n} represented by an integral term which can be solved analytically,

$$T_{n}=\int_{(-1)}^{1}\sqrt{1-\mu^{2}}\,P_{n}\left(\mu\right)P_{(n+1)}\left(\mu\right)\mathrm{d}\mu\quad(33)$$

After applying the Laplace transform in Equation (31) in the spatial variable x, we came out with the linear algebric system,

$$\frac{n+1}{2n+1}\left[s\overline{\psi_{n+1}^{FP}}\left(s,y,E\right)-\overline{\psi_{n+1}^{FP}}\left(0,y,E\right)\right]$$
$$+\frac{n}{2n+1}\left[\overline{\psi_{n-1}^{FP}}\left(s,y,E\right)-\overline{\psi_{n-1}^{FP}}\left(0,y,E\right)\right]\quad(34)$$
$$+\frac{2n+1}{2}\frac{\partial}{\partial y}\overline{\psi_{n}^{FP}}\left(s,y,E\right)T_{n}=\frac{\sigma_{tr}}{2}\left[-n(n+1)\right]\overline{\psi_{n}^{FP}}$$

For $n=0,\cdots,N$ and where $\overline{\psi_{n-1}^{FP}}\left(x,y,E\right),\overline{\psi_{n}^{FP}}\left(x,y,E\right)$ and $\psi_{n+1}^{FP}\left(x,y,E\right)$ are the transformed angular fluxes in the spatial x variable. Equation (36) can be cast in matrix form,

$$A_{n}\overline{\psi_{n}^{FP'}}\left(s,y,E\right)$$
$$+B_{n}\left(s\right)\overline{\psi_{n}^{FP}}\left\{\left(s,y,E\right)-C_{n}\overline{\psi_{n}^{FP}}\left(0,y,E\right)=0\right.\quad(35)$$

where $\overline{\psi_{n}^{FP'}}$ are the N components of the derivative of the Laplace transformed angular flux in the x variable.

$$\overline{\psi_{n}^{FP'}}\left(s,y,E\right)=\left[\overline{\psi_{n}^{FP'}}\left(s,y,E\right),\cdots,\overline{\psi_{n}^{FP'}}\left(s,y,E\right)\right]^{T}\quad(36)$$

The components of matrices $A_{N},B_{N}\left(s\right)$ and C_{N} are for $i\in\{1,2,\cdots,N\}$, respectively

$$a_{ij}=\begin{cases}\left(2i+1\right)^{2}T_{i},\ i=j\\0,\ i\neq j\end{cases}$$

$$b_{ij}=\begin{cases}i\left(i+1\right)\left(2i+1\right)\sigma_{tr},i=j\\2is,\ i=\left|j-1\right|\\0,\ \text{otherwise}\end{cases}$$

$$c_{ij}=\begin{cases}2i,\ i=\left|j-1\right|\\0,\ \text{otherwise}\end{cases}$$

The solution of Equation (35) is

$$\overline{\psi_{n}^{FP}}\left(s,y,E\right)=c_{1}\left(s\right)e^{-\left(B_{n}(s)A_{n}^{-1}\right)}$$
$$+C_{n}\left(B_{n}\left(s\right)\right)^{-1}\psi_{n}^{FP}\left(0,y,E\right)\quad(37)$$

where $c_{1}\left(s\right)$ is an arbitrary constant determined by ap-

plying the boundary and interface conditions. Due to the linear character of the inverse Laplace transform operator, the solution is composed by

$$\psi_{n}^{FP}\left(x,y,E\right)=\mathcal{L}^{-1}\left\{c_{1}\left(s\right)e^{-\left(B_{n}(s)A_{n}^{-1}\right)}\right\}$$
$$+C_{n}\cdot\mathcal{L}^{-1}\left\{\left(B_{n}\left(s\right)\right)^{-1}\right\}\cdot\psi_{n}^{FP}\left(0,y,E\right)\quad(38)$$

4. Numerical Results

In the following we apply the closed-form formulation to the build-up factor and absorbed energy for photons and electrons in the Compton energy range and in Cartesian geometry. Moreover, we the one-dimensional, two-dimensional problem for photons and two-dimensional problem for photons and electrons. To this end we evaluate the exposure build-up factor, considering a scalar flux of the photons with energy 1 MeV, incident in a multi-layered slab with two regions, composed of water and lead, water and iron, lead and iron. Assuming that the kernel is described by the Klein-Nishina differential scattering cross-section and the energy variables may be simplified in form of a multi-group model in the wavelength (energy) variable.

Here the exposure build-up factor is defined as the sum of the product of the attenuation coefficient of the air with the scalar flux for all photons, including the incident flux, divided by the attenuation coefficient of the air for the incident flux multiplied by the incident scalar flux. The numerical results for three problems are shown in **Table 1** for water and lead, in **Table 2** for water and iron and in **Table 3** for lead and iron. In **Table 1** we present the LTS_{N} numerical simulations for the exposure build-up factor and comparisons with results from the EGS_{4} code [7] generated for the one-group model. We consider a multi-layered slab with two regions, composed of water ($\mu_{ij}=0.0707$ cm^{2}/g and 1.0 mpf) and lead ($\mu_{ij}=0.06848$ cm^{2}/g and depth in multiples of the mean free path, 4.0, 5.0, 10.0, 20.0, 30.0 and 40.0 mfp) together with the afore mentioned vacuum boundary conditions.

Table 1. Numerical exposure buildup factor simulations for a multilayered slab composed of water (1.0 mfp) and lead.

Iron (mfp)	LTS_{16}	EGS_{4}
4.0	2.30	2.31
5.0	2.07	2.08
10.0	3.57	3.59
20.0	5.29	5.31
30.0	6.77	6.79
40.0	8.26	8.27

A Closed-Form Formulation for the Build-Up Factor and Absorbed Energy for Photons and Electrons in the
Compton Energy Range in Cartesian Geometry

53

**Table 2. Numerical exposure buildup factor simulations for
a multilayered slab composed of water (1.0 mfp) and iron.**

Iron (mfp)	LTS_{16}	EGS_4
4.0	4.99	5.01
5.0	6.21	6.23
10.0	13.9	13.9
20.0	36.3	36.3
30.0	67.6	67.5
40.0	101.0	101.0

**Table 3. Numerical exposure buildup factor simulations for
a multilayered slab composed of lead (1.0 mfp) and iron.**

Iron (mfp)	LTS_{16}	EGS_4
4.0	4.87	4.86
5.0	6.34	6.28
10.0	15.4	15.3
20.0	41.5	41.4
30.0	78.4	78.3
40.0	118.0	117.0

In **Table 2** we present the LTS_N numerical simulations
for the exposure build-up factor and comparisons with
the results from EGS_4 generated for the one-group model.
We consider a multi-layered slab with two regions,
composed of water (μ_{ij} = 0.0707 cm²/g with depth of 1.0
mpf) and iron (μ_{ij} = 0.0596 cm²/g with depth 4.0, 5.0,
10.0, 20.0, 30.0 and 40.0 mfp) and vacuum boundary conditions.

From the analysis of the results encountered in **Tables
1-3**, one realizes a fairly good agreement between the
LTS_{16} and EGS_4 results. The numerical convergence of
the LTS_N results showed for increasing N a coincidence
of six significant digits for N = 14 and N = 16. For
two-dimensional problems for photons and electrons, we
applied the LTS_N method, for N = 8, in the transport
equation for photons and used N = 9 in the P_N approximation for the angular variable of the Fokker-Planck equation for electrons. We considered homogeneous rectangular geometries composed of water liquid (Z/A = 0.55508,
ρ = 1 g/cm²), soft tissue (ICRU44, Z/A = 0.54996, ρ =
1.06 g/cm²) and cortical bone (ICRU44, Z/A = 0.51478, ρ
= 1.92 g/cm²). Further, we assumed a mono-energetic (E
= 1.25 MeV) and monodirectional photon beam incident
on the edge of a rectangle with dimension 20 cm × 20 cm
and vacuum boundary conditions. The incoming photons

were tracked until their whole energy was deposited and/
or they left the domain of interest. In this problem also
the energy deposited by the secondary electrons, generated by the Compton Effect, were considered. Other possible effects, however with small or spurious contributions
were not taken into account. The numerical results encountered for absorbed energy in the domain were compared with simulations obtained with the program Geant4
v8 [8], using the low energy libraries and are presented in
Tables 4 and **5**.

The numerical convergence of the LTS_N results showed
for increasing N a coincidence of three significant digits
for N = 4, 6 and 8. Notice, the coincidence of four significant digits for the P_N approximation with N = 7 and 9.
These results were obtained in the homogeneous domain
with dimension 20 cm × 20 cm that was composed of
water. In **Table 4** we present the LTS_8 nodal numerical
simulations for the absorbed energy induced by photons
incident on a homogeneous rectangular domain, that is
composed of a variety of different materials. In **Table 5**
we present numerical simulations for absorbed energy by
the P_9 approximation, due to free electrons, arising from
Compton scattering in a homogeneous rectangular domain composed of different materials. These results were
compared with simulations obtained by Geant4.

In spite of the fact, that two different methods were
used to simulate energy deposition, the Monte Carlo method with Geant4 and our closed form solution the results
in **Tables 4** and **5** show a fairly good agreement. From
the results, we notice that the maximum discrepancy
found is 3.4% in the simulations for photons and 8.3%
for electrons. The difference of our numerical results in
comparison to the Geant simulations, that contain a cata-
logue of processes, demonstrate that other effects shall be
taken into account. As the material density increases, the

**Table 4. Absorbed energy (KeV/photon emitted from the
source) by the photons incident in a homogeneous domain
dimension 20 cm × 20 cm, composed of materials different.**

Domain composition	LTS_N	Geant4	Discrepancy (%)
Water liquid	0.00457	0.00468	2.3
Soft tissue	0.00531	0.00542	2.0
Cortical bone	0.0987	0.09487	3.4

**Table 5. Absorbed energy (KeV/photon emitted from the
source) by the free electrons in a homogeneous domain dimension 20 cm × 20 cm, composed of materials different.**

Domain composition	LTS_N	Geant4	Discrepancy (%)
Water liquid	0.03379	0.03609	6.4
Soft tissue	0.03317	0.03542	6.4
Cortical bone	0.79284	0.86380	8.3

number of interactions increases as well as the possibility of other production processes involving secondary electrons, responsible for more than 86% of the total energy absorbed in domain.

5. Conclusion

Concluding, we were successful in determinig the LTS_N solution in closed form forenergy deposition induced by photons in Cartesian geometry. From the solution we obtained the buildup factor and absorbed energy, for photons and electrons, in the Compton energy range. It is worth mentioning, that the LTS_N procedure maintains an analytical character of the solution and the unique approximation made was in the leakage angular flux at the boundary. The P_N solution of the Fokker-Planck equation remains analytical in the sense that no approximation is made along its derivation from P_N equations, except for the truncation. Finally, a variety of additional numerical experiments have shown us that the presented method is robust for problems of the considered transport equation type.

6. Acknowledgements

The authors are gratefully to CNPq (Conselho Nacional de Desenvolvimento Científico e Tecnológico) for the partial financial support of this work. A special acknowledgment for the project INCT (Instituto Nacional de Ciencia e Tecnologia—Reatores Nucleares Inovadores) for financial support.

REFERENCES

[1] C. F. Segatto, M. T. Vilhena and R. P. Pazos, "On the Convergence of the Spherical Harmonics Approximations," *Nuclear Science and Engineering*, Vol. 134, No. 1, 2000, pp. 114-119.

[2] M. T. Vilhena, C. F. Segatto and L. B. Barichello, "A Particular Solution for the SN Radiative Transfer Problems," *Journal of Quantitative Spectroscopy and Radiative Transfer*, Vol. 53, No. 4, 1995, pp. 467-469.

[3] C. Borges and E. W. Larsen, "The Transversely Integrated Scalar Flux of a Narrowly Focused Particle Beam," *SIAM Journal on Applied Mathematics*, Vol. 55, No. 1, 1995, pp. 1-22.

[4] C. Borges and E. W. Larsen, "On the Accuracy of the Fokker-Planck and Fermi Pencil Beam Equation for Charged Particle Transport," *Medical Physics*, Vol. 23, No. 10, 1996, pp. 1749-1759.

[5] B. D. A. Rodriguez, M. T. Vilhena, V. Borges and G. Hoff, "A Closed Form Solution for the Two-Dimensional Fokker-Planck Equation for Electron Transport in the Range of Compton Effect," *Annals of Nuclear Energy*, Vol. 35, No. 5, 2008, pp. 958-962.

[6] B. D. A. Rodriguez, M. T. Vilhena and V. Borges, "A Solution for the Two-Dimensional Transport Equation for Photons and Electrons in a Rectangular Domain by the Laplace Transform Technique," *International Journal of Nuclear Energy Science and Technology*, Vol. 5, No. 1, 2010, pp. 25-40.

[7] H. Hirayama and K. Shin, "Application of the EGS$_4$ Monte Carlo Code to a Study of Multilayer Gamma-Ray Exposure Buildup Factors," *Journal of Nuclear Science and Technology*, Vol. 35, No. 11, p. 816.

[8] D. H. Wright, "Physics Reference Manual," 2001.

[9] G. C. Pomraning, "Flux-Limited Diffusion and Fokker-Planck Equations," *Nuclear Science and Engineering*, Vol. 85, No. 2, 1983, p. 116.

[10] J. Wood, "Computational Methods in Reactor Shielding," Pergamon Press, Oxford, 1982.

[11] S. Agostinelli, *et al.*, "Geant4: A Simulation Toolkit," *Nuclear Instruments and Methods in Physics Research A*, Vol. 506, No. 3, 2003, pp. 250-303.

[12] DOORS 3.1, "One, Two and Three Dimensional Discrete Ordinates Neutron Photon Transport Code System," Radiation Safety Information Computational Center (RSICC), Code Package CCC-650, Oak Ridge, Tennessee, 1996.

[13] B. D. A. Rodriguez, M. T. Vilhena and V. Borges, "The Determination of the Exposure Buildup Factor Formulation in a Slab Using the LTS$_N$ Method," *Kerntechnik*, Vol. 71, No. 4, 2006, pp. 182-184.

9

Experimental Base of an Innovation S-Channel to Reduce MHD Pressure Drop[*]

Zengyu Xu, Chuanjie Pan, Xiujie Zhang, Xuru Duan, Yong Liu
Southwestern Institute of Physics, Chengdu, China

ABSTRACT

Consequent on MHD geometry sensibility phenomena was measured in an accident case; the more detail experiments have been conducted at the liquid metal experimental loop upgrade facility (LMEL-U). The experimental results indicate that MHD pressure drop can be greatly reduced in the special designed ducts. Base on the experimental data, an innovation channel concept (tentatively called as the secondary flow channel, short in "S-channel") is addressed as a reducing MHD pressure drop channel for the application of a liquid metal blanket system in fusion reactor. It may be a dawn for solving MHD pressure drop key issue of liquid metal blanket system.

Keywords: Liquid Metal Blanket; MHD Effect; Pressure Drop; S-Channel

1. Introduction

As known, liquid metal blanket concepts are still attractive ITER (International Thermonuclear Experimental Reactor) and DEMO (Demonstration fusion plant) blanket candidates as they have low operating pressure, simplicity, and a convenient tritium breeding cycle. While conducting an experimental investigation of MHD effect of the flow channel inserts (FCI) for liquid metal blanket system in ITER or DEMO potential application, some MHD geometry sensibility phenomena were measured in accident case [1], and the MHD geometry sensibility phenomena (or called the secondary flow MHD effect, short in "S-MHD" effect) is successfully used to understand FCI duct flow MHD behaviors [2,3], and the experimental results of FCI flow indicate that the problem how to reduce the MHD pressure drop and meet the heat transferred requirement still remain key issues. As well as, up to now, it is very little about the secondary flow knowledge. To understand the secondary flow MHD effect and to try to use the effect to reduce MHD pressure drop, four special design experiments are performed at the liquid metal experimental loop upgrade facility (see **Figure 1**) in southwestern institute of physics, China.

2. Experimental Description

Before we state the detail of the special design experiments, we review the MHD geometry sensibility phenomena (see

Figure 1. Picture of the liquid metal experimental loop up gradate facility (LMEL-U).

Figure 2, ref [1,2]): A rectangular duct with two small weld slots, *i.e.*, "Sc-duct", was fabricated with two simple U-type 304 stainless steel channels welded at the middle line of the lengthwise sides(designed to reduce the manufacture cost) while conducting experimental investigations on the MHD effect of the FCI flow. Yet, a dramatic influence on the flow pattern, pressure gradient and electric potential distribution had been measured [1]. It can be found that the small weld slots make the center velocity and the MHD pressure drop to increase, and the velocity distribution on the center-plane (y = 0) of the cross section of the duct deviates from the classical M-type profile. We make a supposition that the secondary flow (due to small weld slots) results in the center velocity and the MHD pressure drop raise and the velocity distribution de-

---
[*]China National Nature Science Foundation grants 10775042 and Nuclear Energy Development Project in Special Program, H6603100, China, support the work.

viated from the classical M-type profile (in this case, the classical M-type velocity profile strengthen the secondary flow effect). To proof the supposition, the special design experiments are conducted as the following:

1) Case 1, the normal duct (short in "N-duct") was designed: 304 SS duct of 68, 60 and 2 mm in 2a, 2b and t_w, respectively. This is as a non-secondary flow duct to be got the benchmark MHD experimental data.

2) Case 2, it is above stated MHD geometry sensibility phenomena (see **Figure 2**), the slots in the middle of 2a side walls is named "Sc-duct": 304 SS duct of 68, 60, 2, ~5 and ~10 mm in 2a, 2b, t_w, d_s (in weld zone produced small slots in depth) and $2w_s$(in width), respectively. It is the cause of the investigation of the secondary flow MHD effects.

3) Case 3, it is designed a protruding duct in 2a side walls (see **Figure 3**, short in "Pr-duct"): 304 SS duct of 68, 58, 2, 5, 5.5 and 2.54 mm in 2a, 2b t_w, $2w_n$ (the protruding twig in width), h_n (in height) and t_{we} (the equivalent wall thickness of the duct), respectively. If the supposition of the secondary flow MHD is correction, the MHD pressure drop will be reduced in the Pr-duct.

4) Case 4, it is designed as a slot in verges in 2a side walls (see **Figure 3**, short in "Sv-duct"): 304 SS duct of 68, 52, 6, 42, 6, 3 and 3.94 mm in 2a, 2b t_w, 2s (the verges of 2a side walls with distance), $2w_{sv}$ (in width), d_{sv} (in depth) and t_{we} (the equivalent wall thickness of the duct), respectively. If the supposition of the secondary flow MHD is correction, the MHD pressure drop will be also reduced in the Sv-duct.

Figure 3 shows the schematic of cross sections of Pr-duct and Sv-duct and the test units in the uniform magnetic field. The four test units, its rectangular duct made of stainless steel is 1500 mm long. The pressure drop is measured over a section of 500 mm long in center zone, which is well distanced from both the edges of the test section (500 mm apart) and the two fringing field zones at the entry to and the exit from the magnet (120 mm apart). So that the data measured is from the full developed flow zone.

3. Experimental Results and Discussion

The experimental results are shown in **Figures 4-6**. MHD pressure drop have evidently reduced for the Pr-duct and the Sv-duct flows. Even for N-duct flow, the velocity distribution in the cross section is deviating from classical M-type profile [4,5], it will be specially discussed in next section.

Figure 5 and **Figure 6** show that MHD pressure drop is evidently reduced (~65% lower than that in N-duct) in both cases of the protruding twigs at the middle of 2a side walls and slots in verge of 2a side walls (see **Figure 2**), and that the velocity distribution in cross section of the Pr-duct and the Sv-duct is similar with two peaks. It is difficult to explain clearly by current theory why the velocity

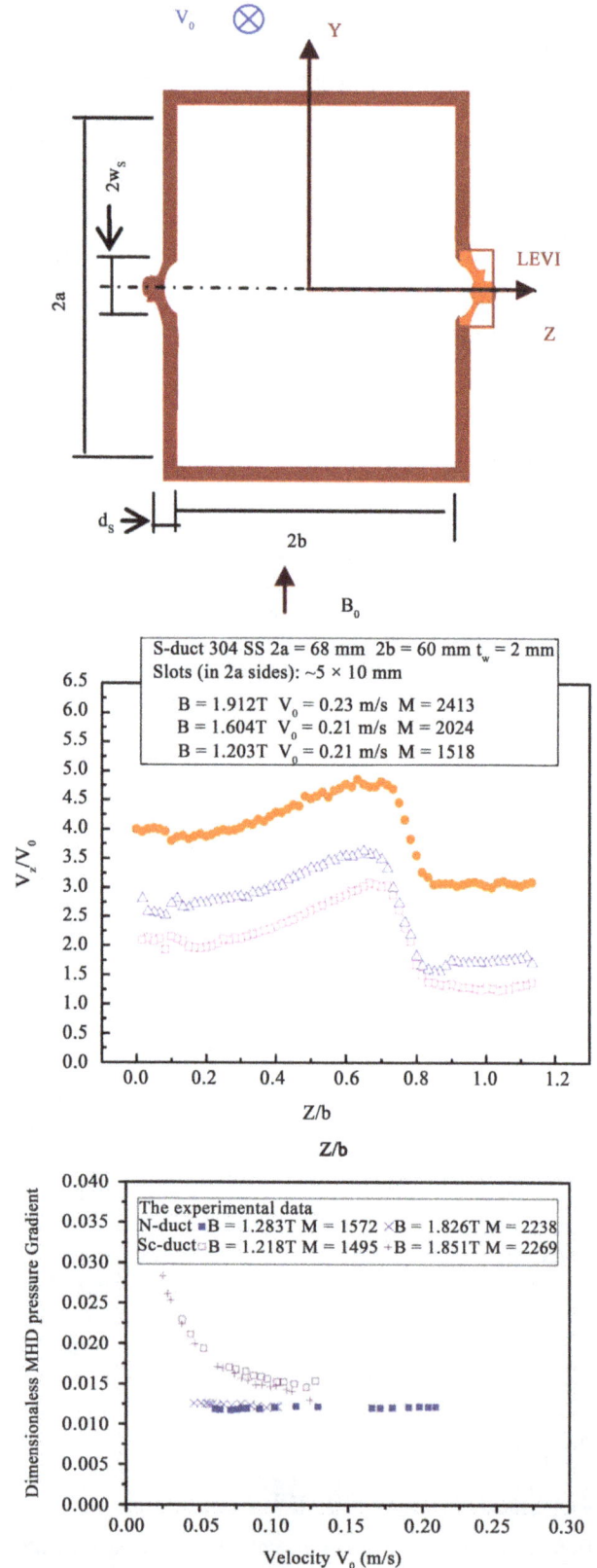

Figure 2. MHD experimental results of the velocity distribution in cross section (left upper) and MHD pressure drop (left downer) of MHD geometry sensibility phenomena (right upper, Sc-duct) re-built from Ref [1].

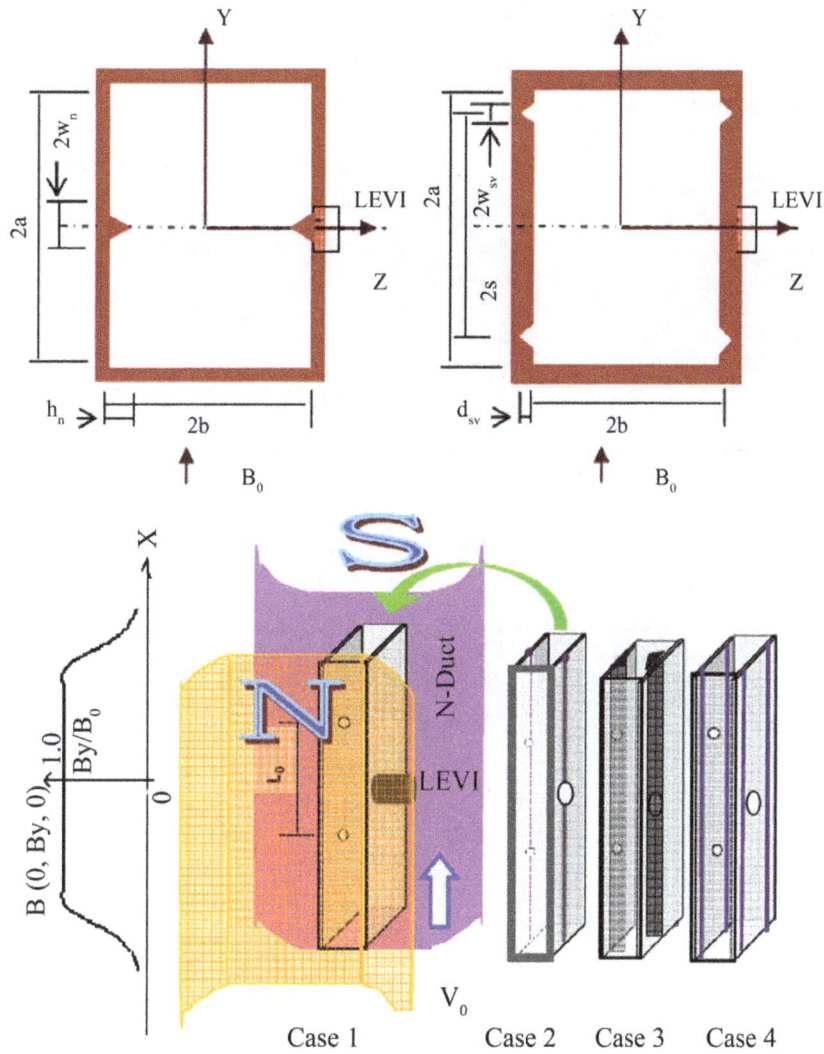

Figure 3. Schematic of cross section of Pr-duct and Sv-duct (upper part) and the test units in the uniform magnetic field (downer part).

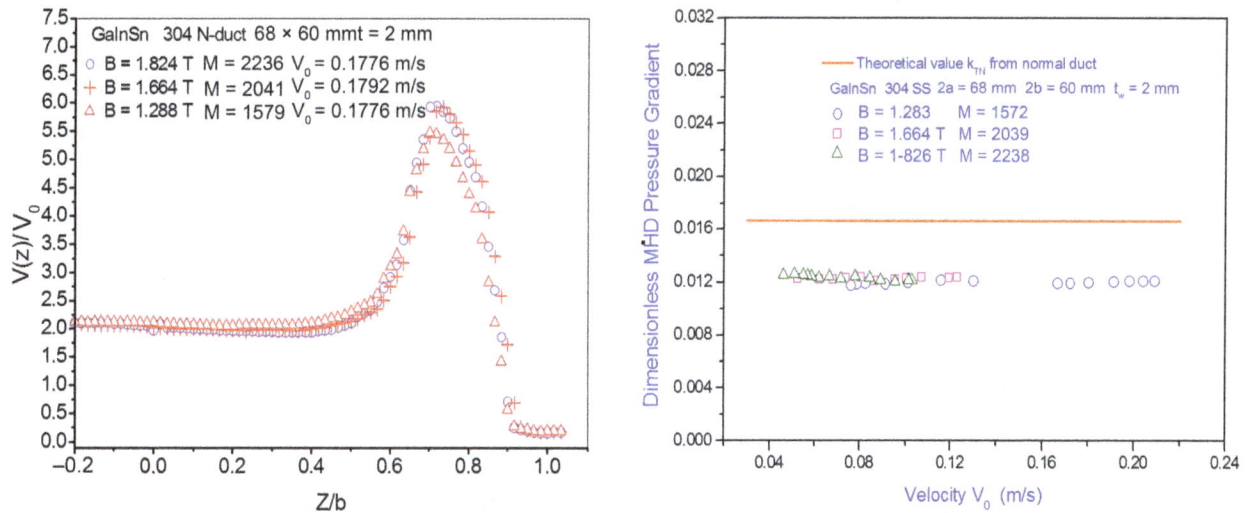

Figure 4. MHD experimental results of the velocity distribution in cross section (left) and MHD pressure drop (right) of the N-duct.

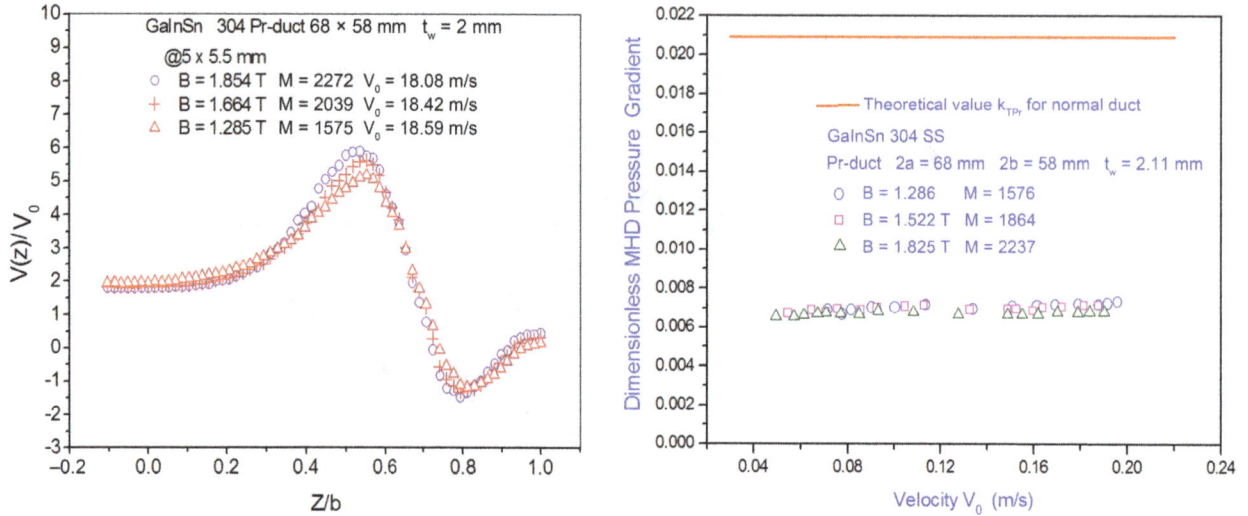

Figure 5. MHD experimental results of the velocity distribution in cross section (left) and MHD pressure drop (right) of the Pr-duct.

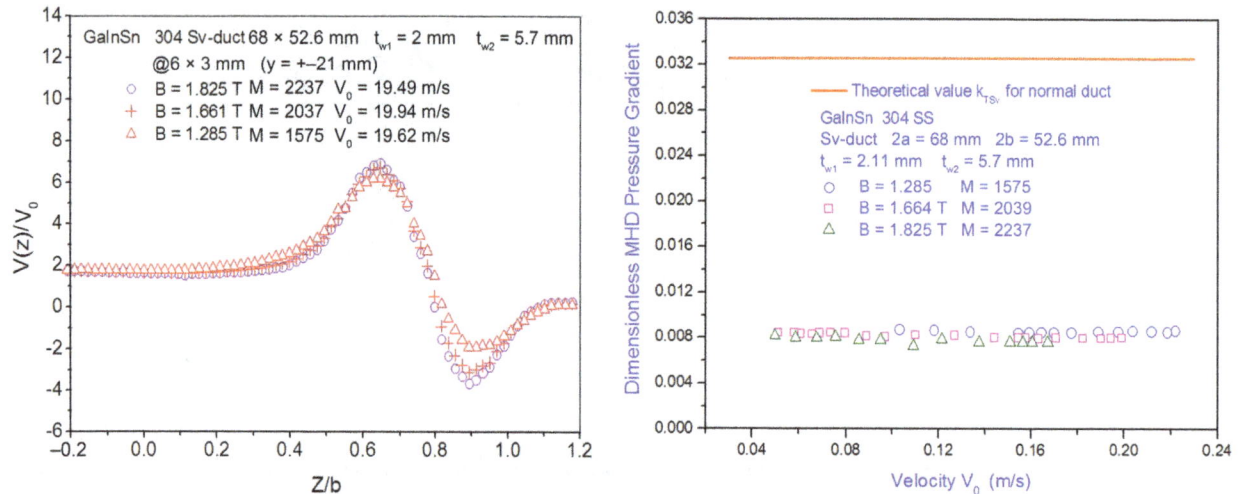

Figure 6. MHD experimental results of the velocity distribution in cross section (left) and MHD pressure drop (right) of the Sv-duct.

distributions like this? But we can understand why the peak is of difference in two cases (Pr-duct and Sv-duct cases). The diagram of the secondary flow due to the protruding twigs and slots (see **Figure 7**, and see ref [1]) can help to understand the difference of the velocity distribution. It can be seen that the secondary flow due to one pair of protruding twig in the Pr-duct, and that due to by two pair of slots in the Sv-duct. So it results in at y = 0 plane, the secondary flow effect is in double in the Sv-duct, and the two peaks of velocity in cross section of the Sv-duct are higher than that in the Pr-duct.

Above experimental data show that the slots or protruding twigs caused the secondary flow and results in MHD pressure drop increased (in Case 2) or decreased (in Cases 3 & 4), and indicated that it is correction of the supposition of the secondary flow MHD effect, and that the secondary flow MHD effect can be used to reduce

MHD Pressure drop.

For well understanding the experimental results, all experimental data are divided by a normalization factor to the N-duct case. As we know, according to classic magneto-hydrodynamics theory the MHD pressure drop in rectangular thin wall conducting duct is (Ref [5]):

$$\Delta P = k_t \sigma_f V_0 B_0^2 L_0 \qquad (1)$$

Here ΔP is the MHD pressure drop; k_t is the characteristic of the duct (called as the dimensionless MHD pressure drop/gradient), V_0 is the average velocity of the duct flow, B_0 is transverse magnetic field, L_0 is the distance of measured MHD pressure drop. The dimensionless MHD pressure drop from theory is:

$$k_t \approx \frac{\varphi}{1 + a/3b + \varphi} \qquad (2)$$

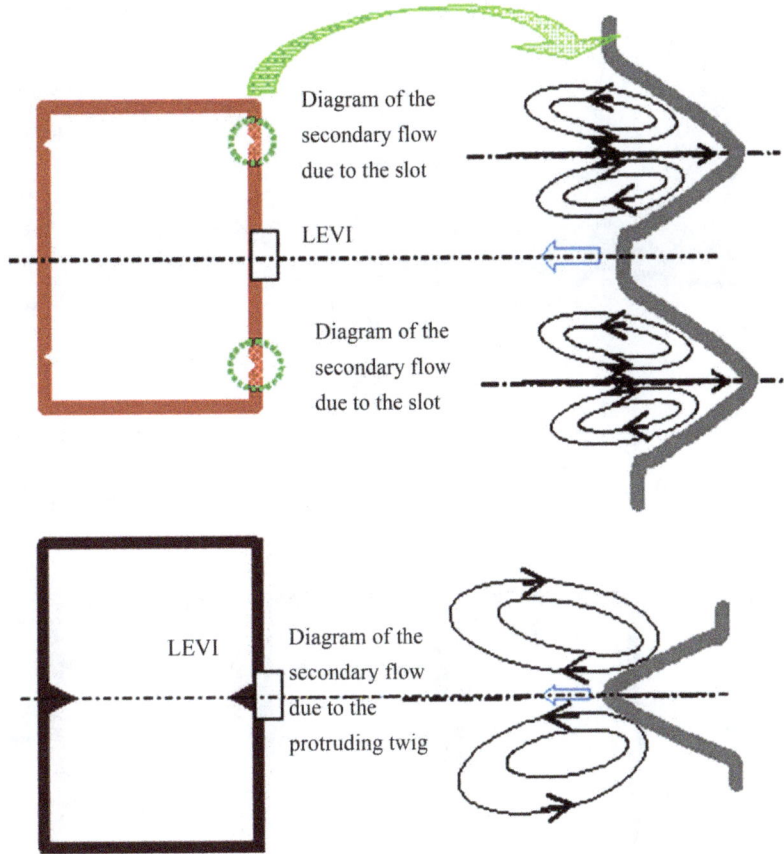

Figure 7. Schematic of the secondary flow in Pr-duct and Sv-duct.

Here $\varphi = \sigma_w t_w / \sigma_f a$ is the wall conductance ratio, σ_f and σ_w is the liquid metal conductivity and the duct conductivity, respectively. The experimentally measured dimensionless MHD pressure drop k_e is:

$$k_e = \frac{\Delta P_e}{\sigma_f V_0 B_0^2 L_0} \tag{3}$$

Here ΔP_e is measured from the sensor of pressure difference, V_0, B_0 and L_0 also is measured data, σ_f has been known.

So, after divided by a normalization factor k_{tN}/k_{tj}, the normalization experimental value is transferred to:

$$k_{e-jN} = k_{e-j} \frac{k_{t-N}}{k_{t-j}} \tag{4}$$

where subscript "j" denotes "Pr" (Pr-duct), "Sv" (Sv-duct) or "Sc" (Sc-duct), N denotes "N" (N-duct). Such as: $k_{e-\mathrm{PrN}} = k_{e-\mathrm{Pr}} k_{t-\mathrm{N}}/k_{t-\mathrm{Pr}}$.

The comparison of velocity distribution and MHD pressure drop in four case ducts are shown in **Figure 8**. It is noted that two protruding twigs at the middle of 2a side walls and four slots in verge of 2a side walls are similar in reducing MHD pressure drop.

It will be expected that MHD pressure drop will be evidently reduced in an innovation channel, a heterotypic

side wall duct, or called as "the secondary flow channel", short in "S-channel", which is combined the both effects of the Pr-duct and the Sv-duct, similarity the graph of right-1 at downer part of **Figure 9**. It may be a new hope for solving MHD pressure drop key issue of liquid metal blanket concepts.

4. Other Missions-Related

The left part of **Figure 9** shows the velocity distribution in the center plane (y = 0) of the cross section of the rectangular duct (N-duct) in different work mass of GaInSn and Nak from LMEL facility experimental data [7]. The results of Nak flow is better approach to analysis solution values. But for GaInSn flow, the velocity distribution is deviating from theory expectation. This deviation is possible that other result is shown in the right part of **Figure 9** [6]. To understand the different of velocity distribution in GaInSn flow is another important mission of investigation of liquid metal blanket MHD effect.

5. Conclusions

From above experimental results and discussions, the tentative conclusions can be deduced as below:

1) The two slots at the middle of 2a side walls of a duct

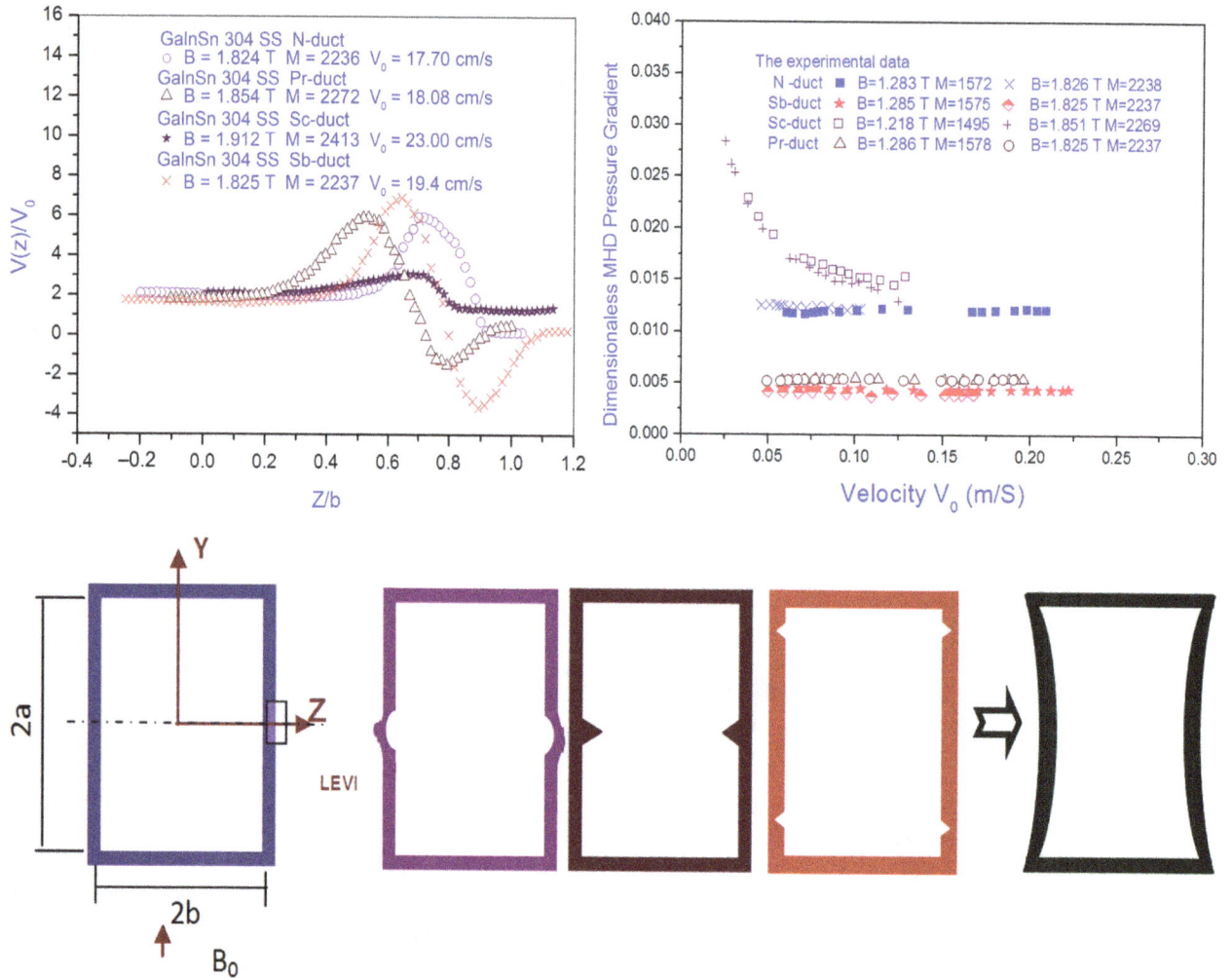

Figure 8. MHD experimental results of the velocity distribution in cross section (left) and MHD pressure drop (right) in four cases. MHD pressure drop in S-duct, Pr-duct and Sv-duct are made a normalization factor to N-duct. The right of the downer part shows a duct of combining the both characterizes of Pr-duct and Sv-duct.

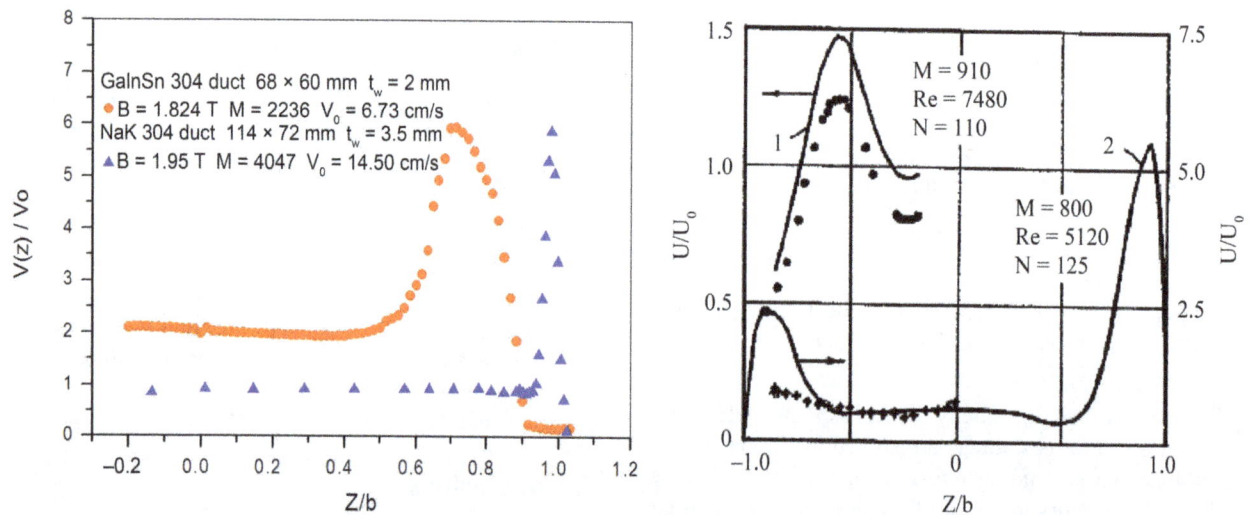

Figure 9. Comparing velocity distribution in cross section of rectangular duct for GaInSn flow with for NaK flow. The left part is the present experimental data and re-built from ref [7]) from LMEL/-U facility results, the right part is cited from ref [5], 1—for GaInSn flow, 2—for Nak flow.

makes MHD pressure drop increase;

2) The one pair of protruding twig at the middle or two pair of slots in verge of 2a side walls of a duct makes MHD pressure drop decrease;

3) The velocity distribution in cross section of normal rectangular duct for GaInSn MHD flow is different from that for NaK MHD flow;

4) An innovation S-channel concept is addressed as a reducing MHD pressure drop channel for the potential application of a liquid metal blanket system in fusion reactor. For the innovation concept, it is necessary for the more detail studied on theory and experimentally.

6. Acknowledgements

China National Nature Science Foundation grants 10775042 and Nuclear Energy Development Project in Special Program, H6603100, China, support the work and Argonne National Laboratory supplied the LEVI under the collaboration of the 1995 People's Republic of China/United States program of cooperation in magnetic fusion.

REFERENCES

[1] Z.-Y. Xu, C.-J. Pan, X.-J. Zhang, L. Zhao, *et al.*, "Geometry Sensitivity of Magnetohydrodynamic Duct Flow and Some Abnormal Phenomena," *Plasma Science and Technology*, Vol. 11, No. 3, 2009, pp. 499-503.

[2] Z.-Y. Xu, X.-J. Zhang, C.-J. Pan, X.-R. Duan and Y. Liu, "MHD Pressure Drop Geometry Sensitivity Correction Factors," *Magnetohydrodynamics*, Vol. 46, No. 3, 2010, pp. 281-288.

[3] Z.-Y. Xu, X.-J. Zhang, C.-J. Pan, X.-R. Duan and Y. Liu, "Understanding FCI Flow Magneto-Hydrodynamic Behaviors with the Secondary Flow Effect," *Advances and Applications in Fluid Mechanics*, Vol. 10, No. 2, 2011, pp. 111-222.

[4] J. C. R. Hunt, "Magnetohydrodynamics Flow in Rectangular Ducts," *The Journal of Fluid Mechanics*, Vol. 21, Part 4, 1965, pp. 577-590.

[5] I. R. Kirillov, C. B. Reed, L. Barleon and K. Miyazaki, "Present Understanding of MHD and Heat Transfer Phenomen for Liquid Metal Blankets," *Fusion Engineering and Design*, Vol. 27, No. 1, 1995, pp. 553-569.

[6] S. I. Sidorenkov and A. Shishko, "Variational Method for MHD Flow Calculation in Conducting Walls Slotted Channels," *Magnetohydrodynamics*, Vol. 27, No. 4, 1991, pp. 87-96.

[7] Z.-Y. Xu, C.-J. Pan, W.-H. Wei and X.-Q. Chen, *et al.*, "Experimental Investigation and Theoretical Analysis Two Dimensional MHD Effects in Rectangular Duct," *Fusion Technology*, Vol. 36, No. 47, 1999, pp. 47-51.

10

Nuclear Fuel Cell Calculation Using Collision Probability Method with Linear Non Flat Flux Approach

Mohamad Ali Shafii[1,2], Zaki Su'ud[1], Abdul Waris[1], Neny Kurniasih[1]
[1]Nuclear and Biophysics Research Group, Faculty of Mathematics and Natural Sciences,
Bandung Institute of Technology, Bandung, Indonesia
[2]Department of Physics, Faculty of Mathematics and Natural Sciences,
Andalas University, Padang, Indonesia

ABSTRACT

Nuclear fuel cell calculation is one of the most complicated steps of neutron transport problems in the reactor core. A few numerical methods use neutron flat flux (FF) approximation to solve this problem. In this approach, neutron flux spectrum is assumed constant in each region. The solution of neutron transport equation using collision probability (CP) method based on non flat flux (NFF) approximation by introducing linear spatial distribution function implemented to a simple cylindrical annular cell has been carried out. In this concept, neutron flux spectrum in each region is different each other because of an existing of the spatial function. Numerical calculation of the neutron flux in each region of the cell using NFF approach shows a fairly good agreement compared to those calculated using existing SRAC code and FF approach. Moreover, calculation of the neutron flux in each region of the nuclear fuel cell using NFF approach needs only 6 meshes which give equivalent result when it is calculated using 24 meshes in FF approach. This result indicates that NFF approach is more efficient to be used to calculate the neutron flux in the regions of the cell than FF approach.

Keywords: Nuclear Fuel Cell Calculation; Neutron Flux; Linear NFF Approximation

1. Introduction

Collision probability (CP) method has advantage that for relatively simple geometry, more over by its symmetrical feature such as cylindrical cell, due to the angular integration is possible to do analytically. CP method starts from the integral neutron transport equation which is usually used to calculate the flux fine structure in a heterogeneous medium or sub region of the reactor. It is assumed that the cell is a part of a repeating pattern of cells as a net of leakage into or out of the cell. The cell is divided into a number of regions within each of which the source of neutron (resulting from fission or scattering reaction) is treated as uniform and isotropic. The problem of treatment of heterogeneity exists in two parts; that is how to calculate the cell flux and how to produce equivalent homogeneous cross section [1,2]. It has two alternatives to be taken regarding the starting neutron distribution. First, use a flat source distribution within a small volume element which leads to a simple expression from collision probabilities. Second, use the non flat source distribution within a small volume element by introducing a spatial function which can improve the linear approximation of source distribution.

In the previous works [3,4], a conceptually simple numerical method to calculate cell homogenization using neutron first flight collision probabilities have been carried out. The integral transport equation is solved using CP method with the flat flux (FF) approximation. This approach explains that neutron flux spectrum is assumed constant in each region. Now, the new approach of the collision probabilities is considered. In this concept, collision probabilities can be performed using a spatial function distribution that describes a real condition inside the cell. Neutron flux spectrum in each region is different each other because of an existing of the spatial function. This assumption is called as non flat flux (NFF) approximation, so the CP matrix is performed not only by the optical path length as an exponential parameter, but also the shape of the function that describe the behavior of neutron flux spectrum in each region.

In this paper, a mathematical argument to support the new approach has been offered to solve neutron transport equation based on NFF approximation by introducing linear spatial distribution function. The computer program to calculate the neutron flux spectrum distribution in a simple cylindrical annular cell is developed by using library of SLAROM JFS-3-J33 for 70 groups.

2. Materials and Method

The infinitely long cylinder cell model is divided into several annular shells with the outer radius of the shell i is r_i and the outer radius of shell j is r_j is shown in **Figure 1**.

The probability of neutron born or travelling in region i of cell will have its next collision either in the same region i or in some other region j is defined by [5]

$$P_{ij}(E) = \frac{\Sigma_j}{4\pi V_i} \int_{V_j} dr_j \int_{V_i} dr_i' \frac{\exp(-\overline{\Sigma R})}{R^2} \quad (1)$$

If size of each cell is small compared with a neutron mean free path, exponential part in Equation (1), and the total cross section is constant then it is reasonable to assume that the neutron flux is flat therefore the neutrons are distributed uniformly in each regions of the cell [6]. In such situation the flat flux approximation is applied and the neutron transport equation for volume V_j is [5]

$$\Sigma_j(E)\int_{V_j} \phi(\overline{r},E)dr = \sum_j \int_{V_j} dr_j \int_{V_i} dr_i' \times$$

$$\left[\int_0^\infty dE' \Sigma_s(\overline{r},E'\to E)\phi(\overline{r},E') + S(\overline{r},E)\right] P(\overline{r}'\to\overline{r},E)$$
$$(2)$$

Subscript i or j denotes the space dependent of regions, $\Sigma_j(E)$ is cross sections associated to j region with energy E, $\varphi(\overline{r},E)$ is flux neutron that is assumed constant in each region, $\Sigma_s(\overline{r},E'\to E)$ is scattering cross section at point $\overline{r}$ at energy E' to E, $S(\overline{r}',E)$ is neutron source at the point $\overline{r}'$ with the energy E.

If neutron flux is not flat in each region, the neutron flux is defined as

$$\varphi(\overline{r},E) = \psi(E)\phi(\overline{r}) \quad (3)$$

Substituting Equation (3) to (2) leads to

$$\Sigma_j(E)\int_{V_j}\psi(E)\phi(\overline{r})dr$$
$$= \sum_j \int_{V_j} dr_j \int_{V_i} dr_j'$$
$$\times\left[\int_0^\infty dE'\Sigma_s(\overline{r}',E'\to E)\psi(E')\phi(\overline{r}')+S(\overline{r}',E)\right]$$
$$\cdot P(\overline{r}'\to\overline{r},E) \quad (4)$$

Integrating left side of (4) over volume V equal to sum of each region volume V_j. The neutron flux in each region is remain non-flat due to the region distribution function $\phi(\overline{r})$. For 1D case the variation of neutron flux in each region has liniear relationship, read

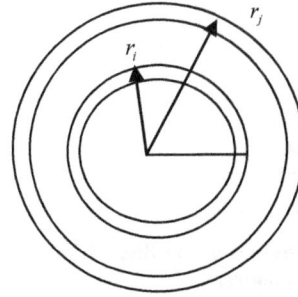

Figure 1. Modeled nuclear fuel cell.

$$\phi(x) = a + bx \quad (5)$$

Coefficients a and b are determined by using linier interpolation scheme at two adjacent nodes of the region as shown in **Figure 2** [7].

By using nodal boundary condition, the neutron flux at nodal are

$$\varphi(x) = \phi_{i-1} \text{ at } x = x_{i-1}$$
$$\varphi(x) = \phi_i \text{ at } x = x_i$$

then (5) becomes to

$$\phi_{i-1} = a + bx_{i-1} \text{ and } \phi_i = a + bx_i \quad (6)$$

where $a = \dfrac{\phi_{i-1}x_i - \phi_i x_{i-1}}{x_i - x_{i-1}}$ and $b = \dfrac{\phi_i - \phi_{i-1}}{x_i - x_{i-1}}$.

Substituting (6) into (5) yields

$$\varphi(x) = \left(\frac{\phi_{i-1}x_i - \phi_i x_{i-1}}{x_i - x_{i-1}}\right) + \left(\frac{\phi_i - \phi_{i-1}}{x_i - x_{i-1}}\right)x \quad (7)$$

In the x direction, neutron flux leads to $\varphi(E,r) = \varphi(E,x) = \varphi(x)$, that independent of energy.

Because the energy width in the group g and g' is same, therefore the neutron flux in each region in the energy grup g and g' also same accordingly. In such situation (4) can be rewritten as follows

$$\phi_{i-1g}\left\{\sum_{jg}\left(\frac{1}{2}x_j^2 x_i - \frac{1}{2}x_{j-1}^2 x_i - \frac{1}{3}x_j^3 + \frac{1}{3}x_{j-1}^3\right)\right.$$
$$\left.+\sum_{i,g}P_{ijg}\sum_{sig}\left(\frac{1}{6}x_i^3 - \frac{1}{2}x_{i-1}^2 x_i + \frac{1}{3}x_{i-1}^3\right)\right\}$$
$$+\phi_{ig}\left\{\sum_{jg}\left(\frac{1}{2}x_{j-1}^2 x_i - \frac{1}{2}x_j^2 x_{i-1} - \frac{1}{3}x_{j-1}^3 + \frac{1}{3}x_j^3\right)\right. \quad (8)$$
$$\left.+\sum_{i,g}P_{ijg}\sum_{sig}\left(\frac{1}{6}x_{i-1}^3 - \frac{1}{2}x_i^2 x_{i-1} + \frac{1}{3}x_i^3\right)\right\}$$
$$=\frac{1}{2}\sum_{i,g}P_{ijg}S_{ig}\left(x_i^3 - x_i^2 x_{i-1} - x_{i-1}^2 x_i + x_{i-1}^3\right)$$

Equation (8) can be simplified to the form of matrix equation

$$\alpha_{i,g}\phi_{ig} + \beta_{i,g}\phi_{i-1g} = Q_{ig} \quad (9)$$

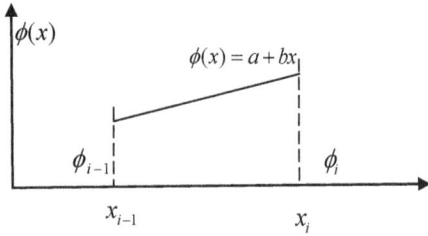

Figure 2. Element of neutron flux with the linear interpolation between two nodes.

where

$$\alpha_{ig} = \left\{ \sum_{jg} \left(\frac{1}{2} x_j^2 x_i - \frac{1}{2} x_{j-1}^2 x_i - \frac{1}{3} x_j^3 + \frac{1}{3} x_{j-1}^3 \right) \right.$$
$$\left. + \sum_{i,g} P_{ijg} \sum_{sig} \left(\frac{1}{6} x_i^3 - \frac{1}{2} x_{i-1}^2 x_i + \frac{1}{3} x_{i-1}^3 \right) \right\}$$

$$\beta_{ig} = \left\{ \sum_{jg} \left(\frac{1}{2} x_{j-1}^2 x_i - \frac{1}{2} x_j^2 x_{i-1} - \frac{1}{3} x_{j-1}^3 + \frac{1}{3} x_j^3 \right) \right.$$
$$\left. + \sum_{i,g} P_{ijg} \sum_{sig} \left(\frac{1}{6} x_{i-1}^3 - \frac{1}{2} x_i^2 x_{i-1} + \frac{1}{3} x_i^3 \right) \right\}$$

$$Q_{ig} = \frac{1}{2} \sum_{i,g} P_{ijg} S_{ig} \left(x_i^3 - x_i^2 x_{i-1} - x_{i-1}^2 x_i + x_{i-1}^3 \right)$$

Gauss elimination is used to solve (9) to explain that neutron flux in each region i on the energy group g which valid for two-diagonals matrix

$$\phi_{ig} = \frac{1}{\alpha_{i,g}} \left(Q_{ig} - \beta_{i,g} \phi_{i-1g} \right) \qquad (10)$$

Equation (10) is used to calculate the multiplication factor as an eigen value [3]

$$\Sigma_j V_j \phi_{ig} = \frac{1}{k_{eff}} \sum_i V_i P_{ij} S_i \qquad (11)$$

Collision probability is solved using Bickley-Naylor function [5]

$$K_{in}(x) = \int_0^{\pi/2} d\theta \sin^{n-1} \theta \exp\left(-\frac{x}{\sin\theta} \right) \qquad (12)$$

For the case $r_i < r_j$, collision probability becomes [5]

$$P_{ij} = \frac{2}{\Sigma_i V_i} \int_0^{r_i} d\rho \left\{ K_{i3}\left(\lambda_{ij}^1\right) - K_{i3}\left(\lambda_{ij}^1 + \lambda_i\right) - K_{i3}\left(\lambda_{ij}^1 + \lambda_j\right) \right.$$
$$+ K_{i3}\left(\lambda_{ij}^1 + \lambda_i + \lambda_j\right) + K_{i3}\left(\lambda_{ij}^2\right) - K_{i3}\left(\lambda_{ij}^2 + \lambda_i\right)$$
$$\left. - K_{i3}\left(\lambda_{ij}^2 + \lambda_j\right) + K_{i3}\left(\lambda_{ij}^2 + \lambda_i + \lambda_j\right) \right\}$$
$$(13)$$

where

$$\lambda_k = \sum_k \left(x_k - x_{k-1} \right), \ \lambda_{ij}^1 = \sum_{k=i-1}^{j-1} \lambda_k, \ \lambda_{ij}^2 = \sum_{k=1}^{i-1} \lambda_k + \sum_{k=1}^{j-1} \lambda_k$$

For the case of the shell i coincides with the shell j, the self collision probability can be written as [5]

$$P_{ii} = \frac{2}{\Sigma_i V_i} \int_0^{r_{i-1}} d\rho \left\{ 2\lambda_i - 2K_{i3}(0) + 2K_{i3}(\lambda_i) + K_{i3}(\lambda_{ii}) \right.$$
$$- 2K_{i3}(\lambda_{ii} + \lambda_i) + K_{i3}(\lambda_{ii} + 2\lambda_i) \right\} \qquad (14)$$
$$+ \frac{2}{\Sigma_i V_i} \int_{r_{i-1}}^{r_i} d\rho \left\{ 2\lambda_i - K_{i3}(0) + K_{i3}(2\lambda_i) \right\}$$

Equations (9) to (14) become a starting point for calculating the neutron spectrum and multiplication factor in the nuclear fuel cells. These are a basic formulation to obtain the solving of integral transport equation using NFF approximation.

The development of the computational program to explore the fuel cell calculation using CP method has been written in Delphi. It was written for determining the effective multiplication factor k-eff, neutron flux spectrum and CP matrix using NFF approach. The computer program in Delphi software package has been written to determine the effective multiplication factor k-eff, neutron flux spectrum and CP matrix using NFF approach. The method for calculation of the NFF collision probabilities was solved for a variety of parameter such as effective group constant and volume of cell variables. In the previous study, the calculation of neutron flux spectrum for 39 nuclides and CP matrix for 70 energy groups by using library of JFS-3-J32 have been carried out [4]. In the present work, the JFS-3-J33 library [8] provides group constants for the SLAROM code in 70 energy group structure for 383 nuclides and 8 lumped fission products is used to calculate neutron spectrum in all regions and the effective multiplication factor (k-eff) for cell homogenization. As a sample input of calculation, a nuclear fuel cell is divided in three regions; region 1 is fuel, region 2 is cladding and region 3 is coolant as shown at **Figure 3**. The fuel cell is a mixture of U-Pu nuclides. The material structure of cladding is stainless steel and Pb-Bi is as a coolant. The radii of three regions of fuel cell, cladding and coolant are 0.35, 0.46 and 0.567 cm, respectively.

In practical calculation, the cell region is composed of several sub regions, called mesh, for which the averages flux is computed. The computation was carried out for 6 meshes in a radial axis of cell. The detail of mesh composition, the radius of each region cell and number of nuclide in each region are shown in **Table 1**. In this work, 24 nuclides are used from the library with the distribution as follows; 10 nuclides in the fuel, 10 nuclides in the cladding and 4 nuclides in the coolant region.

The accuracy of the program has been examined with SRAC code for the same input sample under substantial simplified problem.

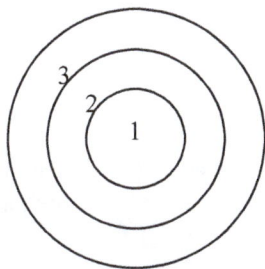

Figure 3. Three region of fuel cell, region 1 for fuel, region 2 for cladding, region 3 for coolant.

Table 1. Composition of meshes in each region of cell.

Region	Composition of 6 meshes (sub regions)	Outer radius of cell (cm)	Number of nuclide
1	3	0.350	10
2	1	0.460	10
3	2	0.567	4

3. Result and Discussion

Comparison of effective multiplication factor k-eff calculated by FF and NFF approach of the CP method and by the SRAC code is shown in **Table 2**.

As can be seen in **Table 2** that the k-eff calculated using FF and NFF approaches is not very much different when it calculated using SRAC code which difference is down to about 0.0078% only. In other words, k-eff calculated by the two approaches is in good agreement with those calculated by SRAC code.

Energy group increase to right, but the energy decreases. The fast neutron energy groups covers the cross section data for neutron energy range from 0.11109 to 10 MeV, while the thermal neutron energy group cover the neutron energy range from 0.32242 eV to 86.517 keV [9]. Comparison of the neutron spectrum for 6 meshes in the three region of cell and the energy group calculated using FF and NFF approach is shown in **Figure 4**. In the FF approach, the peak of neutron spectrum profile in a fast region (group 1 to 30) is higher than the peak in NFF approach. In the thermal region (start from group 37), however, both spectrum tend to stable. This is understood because type of reactor chosen is a fast reactor. The spectrum difference is caused by the selection of the method which assumed that neutron flux at adjacent two mesh node is no longer flat but rather follows the shape of the linear interpolation. Neutron fluxes in the three regions of cell are close each other. This suggests that the neutron distribution in each region was homogeneous in fuel, cladding and coolant.

Figure 5 shows neutron spectrum in the areas of fuel, cladding and coolant region which are reviewed using the FF approach for different mesh size. As can be seen, mesh changes have no effect in the neutron spectrum.

Different result is shown in **Figure 6** for NFF ap-

Table 2. Effective multiplication factor calculated using FF, NFF and SRAC.

Calculation with	k-eff
FF approach	1.292479
NFF approach	1.292477
SRAC code	1.292370

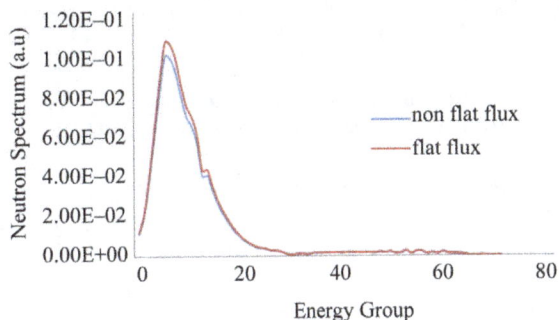

Figure 4. Neutron spectrum calculated using FF and NFF approach.

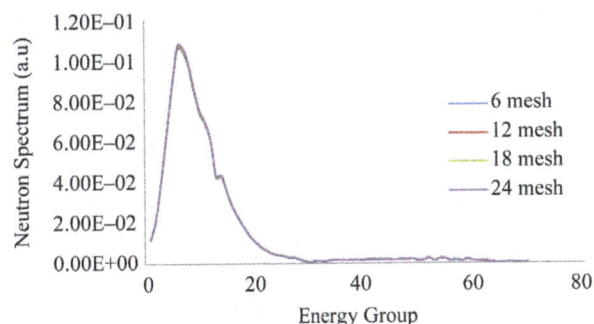

Figure 5. Neutron spectrum with variation of mesh using FF approach.

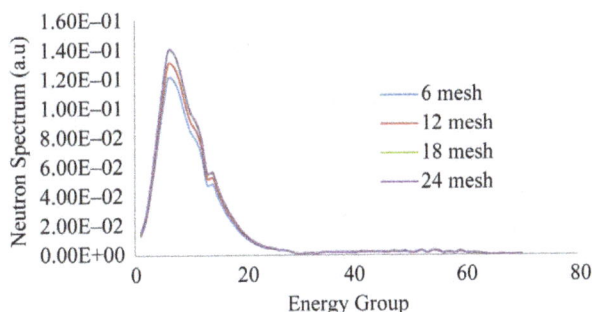

Figure 6. Neutron spectrum with variation of mesh using NFF approach.

proach. The variation of mesh affects the neutron spectrum. This event indicates that the process of nuclear fuel cell calculations with NFF approach is similar to the neutron transport processes conducted without homogenization of cells. As it is known that the heterogeneity factors are very influential in determining an anisotropic

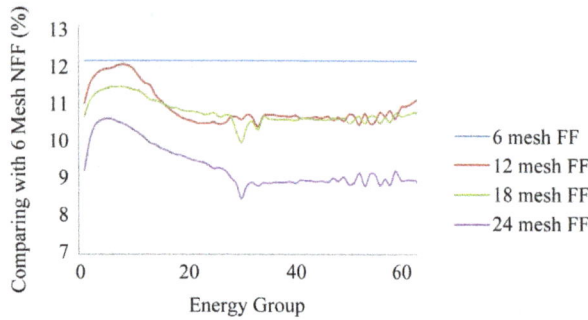

Figure 7. Difference percentage of 6 meshes using NFF approach with variation of mesh using FF approach.

scattering cross section in the nuclear fission process.

The accuracy of FF approach in term of its appropriate number of mesh used is performed by calculating the neutron flux distribution and compared it with using 6 meshes of NFF approach as the reference. To do this, the neutron flux distribution has been evaluated using 6, 12, 18 and 24 meshes of FF approach respectively which the result of the examination is shown in **Figure 7**.

Refer to **Figure 7**, it can be shown that the best accuracy is obtained when the neutron flux distribution is examined using 24 meshes of FF approach with maximum error account of 9% compared to those evaluated by 6 meshes of NFF approach. This comparison concludes that NFF approach is more effective and efficient in term of number of mesh used than FF approach for calculation of neutron flux distribution in each region within the cylindrical nuclear fuel cell. It is understood that for fast reactor the neutron flux is fluctuated in the fast energy region whereas in the thermal region the neutron flux is stable.

4. Conclusion

From this study it is concluded that solution of neutron transport equation using NFF approach is in good agreement with FF approach and SRAC code. Moreover, in term of number of mesh used, calculation of neutron flux distribution in each region of cylindrical nuclear fuel cell using 6 meshes of NFF approach gives an equivalent result when those is calculated using 24 meshes FF approach meaning that the NFF approach is more efficient in using of mesh number.

5. Acknowledgements

The first author wishes to thank Prof. T. Hazama from JAEA Japan for his deeply discussion and attention concerning this topic.

REFERENCES

[1]   Z. Su'ud, "Computer Code for Homogenization of Nuclear Fuel Cell in the Fast Reactor," *Proceeding of a Workshop in Computational Science and Nuclear Technology*, Bandung, 24-25 February 1998, pp. 110-115.

[2]   Z. Su'ud, Y. K. Rustandi and R. Kurniadi, "Parallel Computing in the Calculation of a Constant Group of Nuclear," *Proceeding of the seventh seminar of Technology and Safety of NPP and Nuclear Facilities*, Bandung, 19 February 2002, pp. 17-22.

[3]   M. A. Shafii and Z. Su'ud, "Development of Cell Homogenization Code Using General Geometry Approach," *International Conference on Advances in Nuclear Science and Engineering*, Bandung, 13-14 November 2007, pp. 403-406.

[4]   M. A. Shafii, Z. Su'ud, A. Waris and N. Kurniasih, "Development of Cell Homogenization Code with Collision Probability Method," *International Conference of Mathematics and Natural Science*, Bandung, 28-30 October 2008, pp. 169-175.

[5]   K. Okumura, T. Kugo, K. Kaneko and K. Tsuchihashi, "SRAC 2006: A Comprehensive Neutronics Calculation Code System," Japan Atomic Energy Agency, Ibaraki, 2007.

[6]   M. Nakagawa and K. Tsuchihashi, "SLAROM: A Code for Cell Calculation of Fast Reactor," Japan Atomic Energy Research Institute, Ibaraki, 1984.

[7]   S. S. Rao, "Finite Element Method in Engineering," Pergamon Press, New York, 1983.

[8]   T. Hazama, "Private Communication," 2008.

[9]   T. Hazama, G. Chiba and K. Sugino, "Development of a Fine and Ultra-Fine Group Cell Calculation Code SLAROM-UF for Fast Reactor Analyses," *Journal of Nuclear Science and Technology*, Vol. 43, No. 8, 2006, pp. 908-918.

11

Application of Bootstrap in Dose Apportionment of Nuclear Plants via Uncertainty Modeling of the Effluent Released from Plants

Debabrata Datta
Health Physics Division, Bhabha Atomic Research Centre, Mumbai, India

ABSTRACT

Nuclear power plants are always operated under the guidelines stipulated by the regulatory body. These guidelines basically contain the technical specifications of the specific power plant and provide the knowledge of the discharge limit of the radioactive effluent into the environment through atmospheric and aquatic route. However, operational constraints sometimes may violate the technical specification due to which there may be a failure to satisfy the stipulated dose apportioned to that plant. In a site having multi facilities sum total of the dose apportioned to all the facilities should be constrained to 1 mSv/year to the members of the public. Dose apportionment scheme basically stipulates the limit of the gaseous and liquid effluent released into the environment. Existing methodology of dose apportionment is subjective in nature that may result the discharge limit of the effluent in atmospheric and aquatic route in an adhoc manner. Appropriate scientific basis for dose apportionment is always preferable rather than judicial basis from the point of harmonization of establishing the dose apportionment. This paper presents an attempt of establishing the discharge limit of the gaseous and liquid effluent first on the basis of the existing value of the release of the same. Existing release data for a few years (for example 10 years) for any nuclear power station have taken into consideration. Bootstrap, a resampling technique, has been adopted on the existing release data sets to generate the corresponding population distribution of the effluent release. Cumulative distribution of the population distribution obtained is constructed and using this cumulative distribution, 95th percentile (upper bound) of the discharge limit of the radioactive effluents is computed. Dose apportioned for a facility is evaluated using this estimated upper bound of the release limit. Paper describes the detail of the bootstrap method in evaluating the release limit and also presents the comparative study of the dose apportionment using this new method and the existing adhoc method.

Keywords: Dose Apportionment; Radioactive; Effluent; Cumulative Distribution; Bootstrap; Percentile

1. Introduction

During the normal operation of any nuclear facility gaseous and liquid radioactive effluents are routinely discharged into the environment. These can result in radiation dose to the members of the public through various exposure pathways. Regulatory authority (for example, Atomic Energy Regulatory Board in India) in line with the International Commission of Radiological Protection (ICRP), has prescribed an effective dose limit of 1 mSv/y for a member of the public which requires routine radioactive release to be authorized or stipulated within a limit. The objective of dose apportionment of the nuclear facilities including power plants, fuel reprocessing plant and waste immobilization plant is to establish the discharge limit of the gaseous and the liquid radioactive effluents into the environment so that the dose received by the members of the public will be 1 mSv/year. In India, the operating Nuclear Power Plants (NPPs) are situated at various parts of the country at Tarapur, Rawatbhata, Kalpakkam, Narora, Kakrapar, and Kaiga. At some sites, there are multiple units of NPPs as well as like Tarapur and Kalpakkam, other nuclear facilities like such as fuel reprocessing plant and waste immobilisation plant are also located. Thus, at the multi-facility sites it has to be ensured that the dose to public resulting from releases (atmospheric and aquatic route) from all the facilities does not exceed 1 mSv/y. Thus, the dose received by the members of the public is basically an integral quantity in the sense that, if there exists more than one facility then the sum total of the dose apportioned to each individual should be equal to 1 mSv/year. It is also envisaged that the risk to health resulting from the discharge of radionuclides into the environment will be influenced by factors that affect the magnitude of dose the individual receives. Computation of the release limit of these radioactive effluents is an issue of ensuring safety of the facility as well as safety of the environment. The existing discharge limits for the NPPs were established when the nuclear power program was at its initial stages. However, expansion program of nuclear power in the country demands the dose apportionment to be revisited in an appropriate scientific manner, so that release limits for the existing plants can be revised and the same for new plants to be installed at the same site can be established with the similar vintage. Statistical analysis of the existing discharge (gaseous and liquid effluents) data from the facility is taken into account to establish the release limit of that facility. Statistical analysis is carried out using bootstrap technique because of the imprecise structure of the release data. This paper presents the detail of bootstrap methodology [1] of arriving at the radioactive effluent release limits of Indian Nuclear Power Plants (NPPs). Bootstrap method of estimation of the discharge limits of the gaseous and liquid effluents also provides the knowledge of the cumulative distribution of the discharge. Cumulative distribution of the gaseous and liquid effluent discharge is used to estimate the uncertainty of the same and the uncertainty is expressed in the form of 5th and 95th percentiles. Uncertainty estimate of the discharges can be used in a standard dose assessment model to compute the uncertainty of the apportioned dose to a specific facility. Recommendation of dose apportionment can be finally be made with its 95th percentile value. Computation has been carried out using a computer code developed in house using Visual Basic 6.0.

2. Regulatory Control of Radioactive Discharge

Establishment of dose apportionment of a nuclear installation is one of the important safety aspects in the sense that the task of dose apportionment will automatically provide the setting the limit of the discharge of the effluent both in the atmospheric and aquatic route. On the other hand, it will also provide the knowledge of how many more nuclear facilities can be possible to install at the same site. Dose apportionment scheme should follow the basic criteria as: the sum total of "dose apportionment of new facility", "dose apportionment of old facility" and "the reserve" should be altogether equal to 1 mSv. In order to achieve this one has to set the appropriate limit of the discharge of the radioactivity into the atmospheric and aquatic environment. Recent work on the development of standards for the radioactive discharge control includes the development of practical guidance for setting discharge limits. ICRP Publication 60 [2] recommends justifying necessary measures for protection of the public in practice situations via constrained optimization procedures [3]. Public annual effective dose limit from all controllable environmental radiation sources (except of natural background) was al-

ready established as equal to 1 mSv. Accordingly, the constraint value relevant to a single source should be a non-specified fraction of this limit. Both the dose limit and constraint apply to representatives of the highest exposed members of the public (the critical group) in the radiation conditions under consideration. According to the relevant Safety Fundamentals [4], "The objective of radioactive waste management is to deal with radioactive waste in a manner that protects human health and the environment now and in the future without imposing undue burdens on future generations."

3. Bootstrap Method

Bootstrap procedures are robust nonparametric statistical methods that can be used to construct approximate confidence limits for the population mean, to estimate the bias and variance of an estimator or calibrate hypothesis tests [5]. In these procedures, repeated samples of size n are drawn with replacement from a given set of observations. The process is repeated a large number of times (e.g., thousands), and each time an estimate of the desired unknown parameter (e.g., the sample mean) is computed. Details of bootstrap and the diversity of its recent applications can be found elsewhere [5-9]. Bootstrap procedures assume only that the sample data are representative of the underlying population. The bootstrap procedures require no assumptions regarding the statistical distribution (e.g., normal, lognormal, gamma) of the underlying population and can be applied to a variety of situations [10]. However, since they involve extensive resampling of the data and, thus, exploit more of the information in a sample, that sample must be a statistically accurate characterization of the underlying population in all respects (not just in its mean and standard deviation). Therefore, the bootstrap methods are specifically useful when: the exact population distributions of the statistics are not known; or the critical values of the test statistics are not available; Bootstrap procedures are classified as (a) standard bootstrap, (b) bootstrap-t, (c) bias corrected accelerated (BCA) bootstrap and (d) percentile bootstrap [10]. In practice, it is random sampling that satisfies the representativeness assumption. Therefore the data must represent random samples of the underlying population. If sample size is small and the distribution is not known exactly bootstrap application on that sample generates the corresponding population and the population can be used to compute the upper confidence limit (UCL) of the population distribution. Bootstrapping procedures are inappropriate for use with data that were idiosyncratically collected or focused especially on contamination hot spots. Algorithm of bootstrap used for computing upper confidence limit (UCL) [11] is written below:

Algorithm:

Let X_1, X_2, $\cdots$, X_n represent the n randomly sampled concentrations.

Application of Bootstrap in Dose Apportionment of Nuclear Plants via Uncertainty Modeling of the
Effluent Released from Plants

69

Step1: Compute the sample mean: $\bar{X} = \dfrac{1}{n}\sum_{i=1}^{n} X_i$

Step 2: Compute the sample standard deviation:

$$s = \sqrt{\frac{1}{n}\sum_{i=1}^{n}\left(X_i - \bar{X}\right)^2}$$

Step 3: Compute the sample skewness:

$$k = \frac{1}{ns^3}\sum_{i=1}^{n}\left(X_i - \bar{X}\right)^3$$

Step 4: For $b = 1$ to B (a very large number) do the following:

1) Generate a bootstrap sample data set; i.e.,
for $i = 1$ to n let j be a random integer between 1 and n and add observation X_j to the bootstrap sample data set.

2) Compute the arithmetic mean $\bar{X}_b$ of the data set constructed in step 1.

3) Compute the associated standard deviation s_b of the constructed data set.

4) Compute the skewness k_b of the constructed data using the formula in step 3.

5) Compute the studentized mean

$$W = \left(\bar{X}_b - \bar{X}\right)/s_b$$

6) Compute Hall's statistic:

$$Q = W + k_b W^2/3 + k_b^2 W^3/27 + k_b/6n$$

Step 5: Sort all the Q values computed in Step 4 and select the lower α^{th} quantile of these B values. It is the $(\alpha B)^{th}$ value in an ascending list of Q's. This value is from the left tail of the distribution.

Step 6: Compute

$$W(Q) = (3/k)\left(\left(1 + k\left(Q_\alpha - k/6n\right)\right)^{1/3} - 1\right)$$

Step 7: Compute the one-sided ($1-\alpha$) confidence limit on the mean.

$$UCL_{1-\alpha} = \bar{X} - W\left(Q_\alpha\right)s$$

4. Mathematical Background of Dose Apportionment

4.1. Atmospheric Discharges

The dose apportionment methodology is basically a scheme to compute the dose to the member of the public at exclusion distance (1.6 km from location of the plant) corresponding to the discharge limit of the radioactive gaseous and liquid effluent released due to routine operation of the facility. This estimated dose should be a fraction of the limit 1 mSv/yr as stipulated by the regulatory authority. Discharge limit is considered as the technical specification limit of the discharge. Operational experience reveals that the actual discharge is always far below the technical spe-

cification as set up during the design of the plant. With a view to this fact, it is obvious that if we properly assess the discharge limit based on the operating experience of the facility, we can have the corresponding apportioned dose in a better perspective in the sense that we can have a substantial margin to reach at 1 mSv/yr limit resulting the provision of installation of some more new nuclear power plants or other nuclear facilities. Standard environmental dose assessment model along with the site-specific meteorological data and dietary intake data are used for the computation of dose. The dose to the members of the public can be mathematically formulated as

$$\left(\sum_{p}\sum_{k=1}^{n}\text{Dose}_{k,p}\right)_{air} + \sum_{k=1}^{n}\left(\text{Dose}_k\right)_{aquatic} \le 1\,\text{mSv/yr} \quad (1)$$

where, the index "p" denotes the pathways of exposure and the index "k" denote the radionuclide. **Figure 1** gives the block diagram for the main steps in arriving at technical specification limits and dose apportionment. The dose received by the members of public is computed for all relevant pathways of exposure. The dose apportionment of fission product noble gas (FPNG) and Ar-41 is based on the dose versus release relationship and the Technical Specification limits for discharge. Mathematically the apportioned dose to FPNG and Ar-41 nuclide can be formulated as

$$\text{Appdos}\,(\text{mSv/y}) = \frac{\eta\,(\text{Bq/d}) \times DCF\,(\text{mSv/y per Bq/s})}{8.64 \times 10^4\,(\text{s/d})}$$

$$(2)$$

where, η is Tech spec limit and the other symbols have usual significance. The computation of plume doses in 22.5° sector from continuous release of noble gases (Ar-41 and fission product noble gases) can be found elsewhere in [12]. The sector receiving the highest dose is called as the

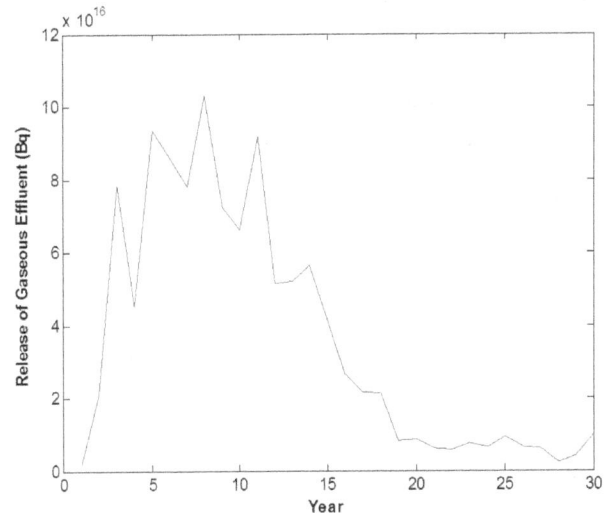

Figure 1. Distribution of yearly release of gaseous effluent from plant #1.

critical sector and subsequently sector average dose is used for computation for dose apportionment [12]. Accordingly, contributions from the adjacent two side sectors, on either side of main sector, are also consider to arrive at a relationship between release rate and dose at exclusion distance. The gamma energy used for computation is 1.28 MeV for Ar-41 and 0.65 MeV for a mixture of Xenon and Krypton isotopes. In order to compute the site specific relationship between release rate and dose the frequency (f)/wind speed (u) (m/s)) for the main sector and the two side sectors are used with which the dose rates are multiplied. Finally all these results are summed up for all weather categories [12]. The dose apportionment for internal irradiators like H-3, I-131 and particulates can be written mathematically as

$$\text{DoseApp}(\text{mSv/y})$$
$$= \frac{\eta\,(\text{Bq/d}) \times (X/Q)(\text{s/m}^3)}{\text{DAC}(\text{pub})(\text{Bq/m}^3)/\text{mSv/y} \times 8.64 \times 10^4\,(\text{s/d})} \quad (3)$$

where, η is Tech spec limit, DAC (pub) is the derived air concentration applicable to members of the public corresponding to dose limit of 1 mSv/y and X/Q is the annual average atmospheric dilution factor at the exclusion distance for the critical sector. The X/Q value applicable for the particular stack height at a distance of 1.6 km from the stack and corresponding to the critical sector is used in the calculation of dose apportionment. The derived air concentration (DAC, public) is defined as the ground level air concentration of radionuclides that results the effective dose limit of 1 mSv/y to members of the public from all-important pathways of exposure. The DAC (public) is generally calculated for both adult and infants and the more restrictive of the two are generally used for calculation of the derived limits. The computation of dose for tritium in detail is depicted elsewhere [12].

4.2. Aquatic Discharges

The dose apportionment for radionuclides discharged into the aquatic environment is computed from the technical specifications limit on concentration for gross beta and Tritium. Mathematical models used in dose apportionment for gross beta for the coastal site and for the inland site are given by

$$\text{Dose appor}(\text{mSv/y})(\text{Coastal site})$$
$$= \text{Conc}(\text{tech spec})(\text{Bq/ml}) * I_f * \Sigma(CF)_i (DF)_i P_i \quad (4)$$

where, I_f is the annual intake of fish by the population for a given NPP site, (g/y), CF_i is the concentration factor in marine fish for radionuclide i (ml/g), DF_i is the ingestion dose conversion factor for i^{th} radionuclide (Sv/Bq), P_i is the percentage composition of ith radionuclide in the gross activity.

$$\text{Dose appor}(\text{mSv/y})(\text{Inland site})$$
$$= \text{Conc}(\text{tech spec})(\text{Bq/ml}) * \Sigma\big[I_w + I_f x\,(CF)_i\big](DF)_i P_i$$
$$\quad (5)$$

where, I_w is the annual intake of drinking water (ml/y), I_f is the annual intake of fish by the population for a given NPP site (g/y), CF_i is the concentration factor in marine fish for i^{th} radionuclide, DF_i is the ingestion dose conversion factor for i^{th} radionuclide (Sv/Bq), P_i is the percentage composition of ith radionuclide in the gross activity.

5. Bootstrap Method in Dose Apportionment

Input data required for computation of dose apportionment is the arithmetic mean value of the radioactive effluents (gaseous and liquid) released into the atmospheric and aquatic environment. The mean value of the released effluent is fed into a standard dose model. Therefore, the mean value of the effluent discharged into the environment is computed from the actual release data. However, in order to compute the average value of a dataset, a-prior knowledge of the population distribution (population refers to a large set of data) of the data set is required; but the population distribution cannot be identified with small dataset. The representative sample size of the existing release data (10 years release data) from nuclear power plant is not sufficient to assume the population as normal distribution. As we do not have the sufficient years of data for concluding a specified probability distribution, we cannot adopt parametric probability distribution for the actual release data. Hence predicting average or median as the representative value of the actual release for proceeding towards the dose apportionment calculation is not at all scientific. So, obviously people will try an adhoc average value of the release based on the actual data. But basically an appropriate scientific method is searched for establishing the dose apportionment. In the new scientific method, bootstrap, the nonparametric method is adopted for identifying the probability distribution of the actual release data [14]. Actual release values for the available number of years are considered as a sample. Bootstrap when applied to the actual samples (say ~10 years release data), produces a large number of release values so called as the population. The sampling without replacement strategy has been implemented for generation of bootstrap samples [15]. Probability density function of the bootstrap generated population is constructed using non-parametric kernel smoothing density method [16-18]. Cumulative distribution function (CDF) has been constructed using the probability density function and the CDF has been used to compute the 99% confidence level (upper confidence limit) of the mean value of the release of the gaseous and liquid effluent [19]. Upper confidence limit of this release value with a defined probability of 0.99 can be accepted as the

Application of Bootstrap in Dose Apportionment of Nuclear Plants via Uncertainty Modeling of the
Effluent Released from Plants

71

appropriate limit for establishing the dose apportionment of various nuclear power stations.

6. Results and Discussion

The operating experience of a typical plant #1(considered as an installation in the coastal site) is represented in the form of actual yearly release of gaseous and liquid effluent and the same dataset is as presented in **Table 1** and **Table 2** respectively. The corresponding distribution of the gaseous and liquid effluent released from the specific nuclear power plant is as shown in **Figures 1** and **2**. It can be easily investigated from **Figures 1** and **2** that arithmetic mean or simple average of the actual release is not the representative input for the estimation of the dose apportioned for the specified nuclide for that specific plant because the shape of distribution of the release data is not the normal distribution. It is confirmed by carrying out the normality test (Kolmogorov-Smirnov Test) [20] of the represented data (**Tables 1** and **2**); the test fails confirming that the distribution of these data is not normal. Moreover,

Table 1. Yearly actual release values of gaseous effluent (plant #1).

Year	Total (Bq)	Year	Total (Bq)	Year	Total (Bq)
1	1.9E+15	11	9.19E+16	21	6.04E+15
2	2.08E+16	12	5.13E+16	22	5.94E+15
3	7.84E+16	13	5.2E+16	23	7.63E+15
4	4.52E+16	14	5.62E+16	24	6.35E+15
5	9.36E+16	15	4.14E+16	25	9.41E+15
6	8.63E+16	16	2.66E+16	26	6.56E+15
7	7.82E+16	17	2.15E+16	27	6.12E+15
8	1.03E+17	18	2.11E+16	28	2.34E+15
9	7.23E+16	19	8.11E+15	29	4.14E+15
10	6.6E+16	20	8.48E+15	30	1.01E+16

Table 2. Yearly actual release values of liquid effluent from a typical plant #1.

Year	Conc (Bq /ml)	Year	Conc (Bq/ml)	Year	Conc (Bq/ml)
1	1.11E–03	11	1.49E–02	21	1.85E–03
2	3.69E–03	12	8.51E–03	22	1.87E–03
3	2.11E–02	13	1.07E–02	23	1.69E–03
4	1.96E–02	14	8.10E–03	24	1.47E–03
5	2.07E–02	15	3.23E–03	25	1.43E–03
6	4.07E–02	16	1.99E–03	26	1.06E–03
7	2.78E–02	17	1.86E–03	27	9.18E–04
8	2.52E–02	18	2.06E–03	28	9.77E–04
9	1.51E–02	19	1.65E–03	29	5.96E–04
10	1.54E–02	20	1.56E–03	30	3.96E–04

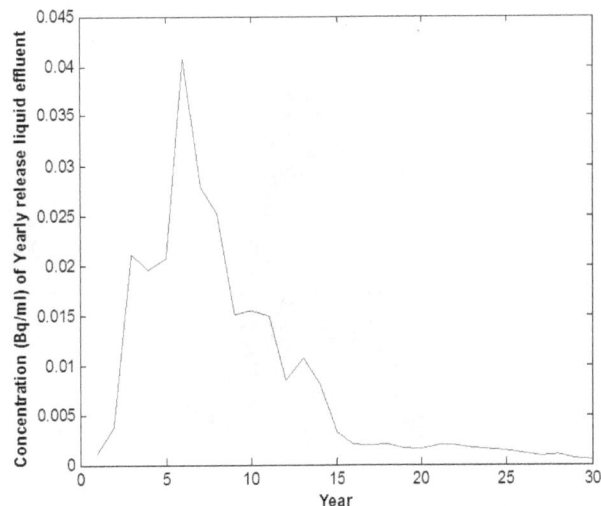

Figure 2. Distribution of yearly release of liquid effluent from plant #1.

the represented data being a time series one cannot simply take the mean without appropriate trend analysis. In order to overcome this difficulty, bootstrap method has been applied on the data set to construct the population distribution associated with the represented data. The frequency distribution of the release data after bootstrap (histogram) is shown in **Figure 3** and corresponding cumulative distribution plot is as shown in **Figure 4**. Cumulative distribution provides the knowledge of 5th and 95th percentiles of the probability density of release data set. Upper confidence limit as 95th percentiles estimated from uncertainty analysis of the effluent released data is accepted as the appropriate input for dose apportionment calculation **Figures 5** and **6** represent the frequency distribution (histogram) and the corresponding cumulative distribution of the gross beta release from plant #1. Cumulative distribution is then utilized to estimate the 5th, 50th and 95th percentiles of the release of the gaseous and liquid effluents from the plant #1.

The 95th percentiles value of the released gaseous and liquid effluent are used as the discharge limit for computing the corresponding apportioned dose for the representtative nuclide (Fission product Noble Gas (FPNG) and gross beta) for the plant #1. Results of the uncertainty estimates of both the FPNG and gross beta release from plant #1 is represented in **Table 3**. Basically dose apportionment for atmospheric release FPNG is the important and for the aquatic environment gross beta is important. Results from **Table 3** indicate that apportioned dose for atmospheric route is more than aquatic route. In order to have the probability density function of the population distribution of the actual release data for tritium in air and tritium in liquid similar approach of bootstrap is adopted. Operating experience of another nuclear power plant, Plant #2 (considered as an installation in the inland site) is taken

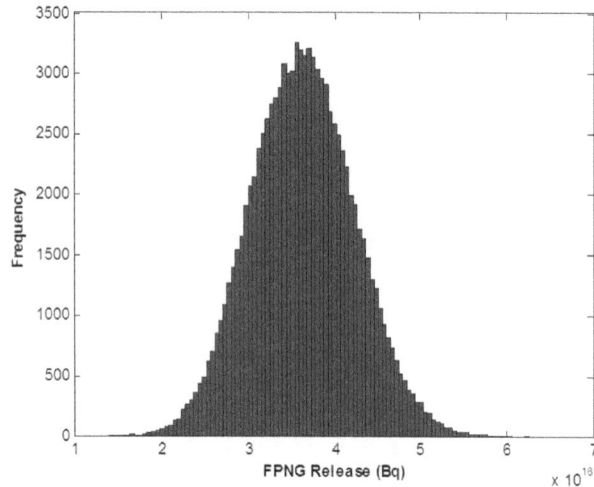

Figure 3. Population distribution of FPNG release from plant #1 after bootstrap (sample size of bootstrap: 100,000).

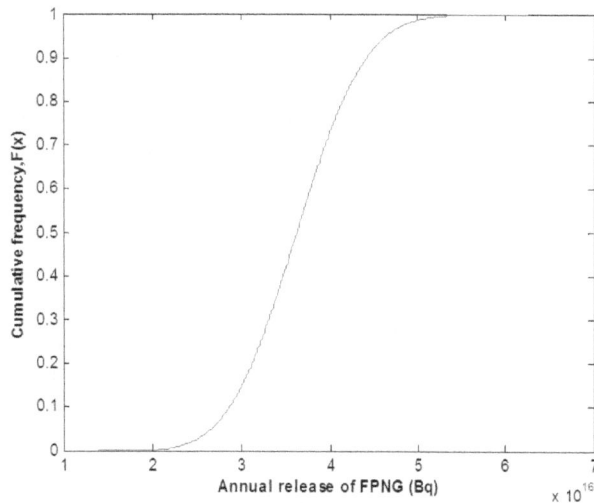

Figure 4. Cumulative distribution of FPNG release from plant #1.

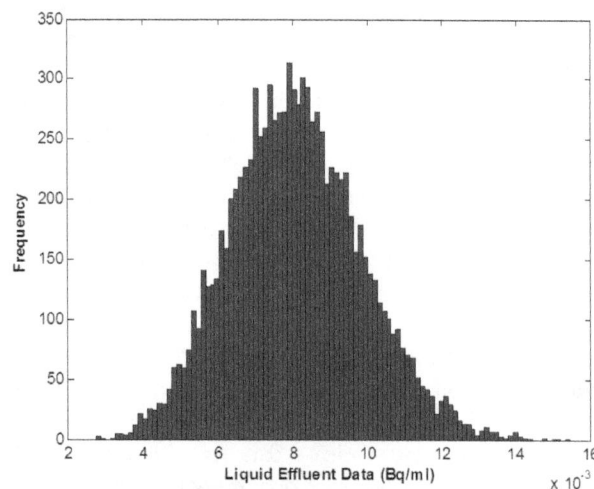

Figure 5. Population distribution of gross beta release from plant #1 after bootstrap (bootstrap sample size taken: 100,000).

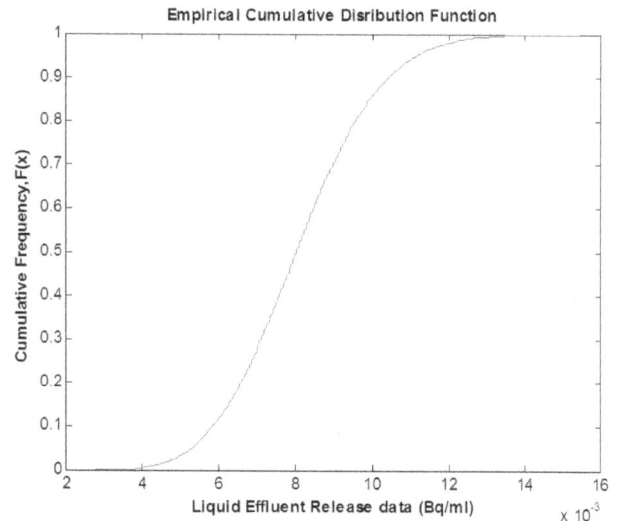

Figure 6. Cumulative distribution of gross beta release from plant #1.

Table 3. Uncertainty estimate of gaseous and liquid effluent release from plant #1.

Release per Year	5th Percentile	50th Percentile	95th Percentile
FPNG (Bq)	2.66E16	3.62E16	4.64E16
Gross Beta (Bq/ml)	0.0053	0.0080	0.0111

into consideration for this case study. Uncertainty estimates for distribution of each radionuclide using same methodology are tabulated in **Table 4**. Site specific value of the atmospheric average dilution factor (X/Q), Condenser cooling water (CCW) flow for dilution in the aquatic route, DAC (public), composition of radionuclide for gross beta, concentration factors, intake of fish and the drinking water are used as input for dose calculation. All the required input parameters are listed in **Table 5**. Apportioned dose using the input parameters listed in **Table 5** and 95th percentile value of FPNG and Gross Beta from plant #1 (coastal site) and that of FPNG, H-3 (air), H-3 (liquid) and gross beta from plant #2 are computed and the results are tabulated in **Table 6**. Comparative study of the dose apportioned for plant #1 and plant #2 from **Table 6** indicate that apportioned dose for FPNG and gross beta for plant at coastal site is more than for plant at inland site. Tritium in air and in aquatic environment for plant #1 is not representative because plant #1 is of boiling water type nuclear reactor.

7. Conclusion

Bootstrap technique has been applied to identify the probability density of the population distribution of the gaseous and liquid effluent. Uncertainty analysis of the effluent dataset has been carried to compute the 95th percenttile of the effluent released from the plant. This 95th percen-

Application of Bootstrap in Dose Apportionment of Nuclear Plants via Uncertainty Modeling of the
Effluent Released from Plants

73

Table 4. Uncertainty estimate of Gaseous and Liquid Effluent Release from Plant #2.

Release	5th Percentile	50th Percentile	95th Percentile
FPNG (Bq/d)	1.08E13	1.97E13	3.42E13
H-3 (Air) (Bq/d)	0.77E13	1.58E13	2.96E13
H-3 (Liquid) (Bq/ml)	13.77	21.05	30.7
Gross Beta (Bq/ml)	0.002	0.003	0.004

Table 5. Input Parameters used for dose computation.

Parameters	Plant #1 (Coastal)	Plant #2 (Inland)
Atmospheric average dilution factor (s/m^3)	6.92E–8	5.34E–8
CCW flow (m^3/day)	2.90E+06	44880
DAC, public (Bq/m^3)	Not representative	3294
Intake of fish (g/yr)	20000	3000
Intake of water (litre/yr)	Not representative	1.10E3
Concentration factors of important radionuclide	22 (Cs-137, Cs-134), 11.3(Sr-90), 4200 (Co-60)	13.2 (Cs-137,Cs-134), 300 (Co-60), 600 (Zn-65)

Table 6. Computed Apportioned Dose for Plant #1 and Plant #2.

Nuclide & Pathways	Dose for Plant #1 (mSv/yr)	Dose for Plant #2 (mSv/yr)
FPNG, Atmospheric	4.09E–01	1.96E–02
Gross Beta, Aquatic	3.15E–01	9.90E–02
Tritium (H-3), Atmospheric	Not Representative	5.56E–03
Tritium (H-3), Aquatic	Not Representative	6.07E–01

tile value is used as the discharge limit of the effluent and using this 95th percentile as an input to the standard dose assessment model, dose computation has been carried out. Final conclusion is that bootstrap provides a scientific basis of dose apportionment scheme. It can be further concluded that uncertainty analysis of the effluent data only is essential for estimating the apportioned dose to a nuclear facility. However, an optimization approach is essential when dose apportionment is required for many more plants located at the same site. Research is being continued to develop an optimization technique for this purpose.

REFERENCES

[1] B. Efron and R. J. Tibshirani, "An Introduction to the Bootstrap," Chapman and Hall, New York, 1993.

[2] "International Commission on Radiological Protection, 1990 Recommendations of the International Commission on Radiological Protection," ICRP Publication 60, Pergamon Press, Oxford and New York, 1991.

[3] M. Balonov, G. Linsley, D. Louvat, C. Robinson and T.

Cabianca, "The IAEA Standards for the Radioactive Discharge Control: Present Status and Future Development," *Radioprotection*, Vol. 40, Suppl. 1, S721-S726, 2005.

[4] International Atomic Energy Agency, "Regulatory Control of Radioactive Discharges to the Environment," Safety Standards Series No. WS-G-2.3, IAEA, Vienna, 2000.

[5] M. R. Chernick, "*Bootstrap Methods, A Practitioner's Guide*," Wiley, New York, 1999.

[6] P. Hall, "*The Bootstrap and Edgeworth Expansion*," Springer-Verlag, New York, 1992.

[7] G. Archer and J. M. Giovannoni, "Statistical Analysis with Bootstrap Diagnostics of Atmospheric Pollutants Predicted in the APSIS Experiment," *Water, Air, and Soil Pollution*, Vol. 106, No. 1-2, 1998, pp. 43-81.

[8] A. C. Davison and D. V. Hinkley, "Bootstrap Methods and Their Application," Cambridge University Press, Cambridge, 1997.

[9] B. Efron, "The Jackknife, the Bootstrap and Other Resampling Plans," SIAM, Philadelphia, 1982.

[10] E. J. Dudewicz, "The Generalized Bootstrap," In: K.-H. Jöckel, G. Rothe and W. Sendler, Eds., *Bootstrapping and Related Techniques*, Springer-Verlag, Berlin, 1992, pp. 31-37.

[11] T. J. DiCiccio and B. Efron, "Bootstrap Confidence Intervals (with Discussion)," *Statistical Science*, Vol. 11, No. 3, 1996, pp. 189-228.

[12] International Atomic Energy Agency, "Generic Models for Use in Assessing the Impact of Discharges of Radio- active Substances to the Environment," Safety Reports Series No. 19, IAEA, Vienna, 2001.

[13] International Atomic Energy Agency, "Basic Safety Standards Series," SS No. 115, IAEA, Vienna, 1996.

[14] A. J. Bailer and J. T. Oris, "Assessing Toxicity of Pollutants in Aquatic Systems," In: N. Lange, L. Ryan, L. Billard, D. Brillinger, L. Conquest and J. Greenhouse, Eds., *Case Studies in Biometry*, Wiley, New York, 1994, pp. 25-40.

[15] R. L. Cooley, "Confidence Intervals for Groundwater Models Using Linearization, Likelihood, and Bootstrap Methods," *Ground Water*, Vol. 35, No. 5, 1997, pp. 869-880.

[16] J. S. U. Hjorth, "Computer Intensive Statistical Methods —Validation Model Selection and Bootstrap," Chapman and Hall, New York, 1994.

[17] C. Davison and D. Y. Hinkley, "Bootstrap Methods and Their Application," Cambridge University Press, Cambridge, 1997.

[18] R. LePage and L. Billard, "Exploring the Limits of Bootstrap," John Wiley, New York, 1992.

[19] P. Barbe and P. Bertail, "The Weighted Bootstrap (Lecture Notes in Statistics)," Springer-Verlag, New York, 1995.

[20] M. H. DeGroot, "Probability and Statistics," 3rd Edition, Reading, Addison-Wesley, Boston, 1991.

12

Investigating the Effects of Neutron Source Number and Arrangement in Landmines Detection by Thermal Neutron Capture Gamma Ray Analysis

Leila Abdolahi Shiramin, **Seyed Farhad Masoudi** [*]

Physics Department, K.N. Toosi University of Technology, P.O. Box 15875-4416, Tehran, Iran

ABSTRACT

Thermal neutron capture can be used as a successful technique for detection of non-metallic landmines via the detection of their constituent like nitrogen. Recently, it has been shown that the detection of 10.829 MeV photons from the $^{14}N(n,\gamma)^{15}N$ reaction can be used for finding the landmines. In this method a high-energy neutron source like ^{241}Am-Be inside water as a moderator is used to have thermal neutron. In this paper we have investigated the effects of the number of neutron sources and their orientation on the gamma ray spectrum by using MCNP4C code. The best case for number of sources and their positions and orientations have been achieved corresponding to maximum flux of 10.829 MeV photons.

Keywords: Monte Carlo Simulation; Landmine Detection; Thermal Neutron Capture; Neutron Source; Source Arrangement

1. Introduction

During the recent years many efforts have been made to improve nuclear methods for the detection of landmines [1-8]. The most important advantage of methods based on neutron reactions is that neutrons have high penetration ability and the detection of landmines is possible even when they are hidden at the high depth. Nuclear techniques are based on that the explosives contain hydrogen, carbon, nitrogen and oxygen in variable concentration. For example, elementary compositions of some common explosives are: TNT-$C_7H_5N_3O_6$, RDX-$C_3H_6N_6O_6$, Hexogen-$C_4H_8N_8O_8$ [9]. So the problem of explosives identification reduces to the problem of identification of light elements.

One method of landmines detection is based on measurement of the 10.829 MeV gamma-ray from the $^{14}N(n,\gamma)^{15}N$ reaction [10]. This procedure is advantageous because the gamma-rays of about 11 MeV do not occur in natural background. The only exception is ^{29}Si which emits gamma-ray of 10.607 MeV via interaction with thermal neutrons, but with low percentage. The abundance of ^{29}Si in soil is very low in comparison with abundance of ^{14}N in landmine.

In present work, MCNP4C code is applied for simulation of configuration of a hidden landmine in soil and

investigation of the source number and their placement effects on gamma-ray spectrum. The maximum value for 10.829 MeV photon fluxes that depends on the numbers and orientations of the sources has been achieved by using the result of MCNP4C simulation.

2. Material and Methods

In this paper, the MCNP4C code for Monte Carlo simulation of particle transport was employed to obtain the best configuration corresponding to best values of 10.829 MeV neutron flux. An Am-Be neutron source as an isotropic source has been used with neutron spectrum shown in **Figure 1** [11]. The neutron source which is 12 cm in length and 3 cm in diameter was located horizontally in the box of water with dimensions of $60 \times 60 \times 24$ cm^3. Distance of the box of water and the surface of soil is 1 cm. A landmine (TNT) in cylindrical geometry, 5 cm in radius and 5 cm in height, was embedded at 5 cm depth from the surface of soil with dimensions of $60 \times 60 \times 11$ cm^3. NaI(Tl) 3 × 3 detector was positioned at the center of water box to take the spectrum of photons (**Figure 2**) [12]. NaI(Tl) scintillator as γ-ray detector has been suggested by considering its characteristics such as the low cost of the crystal, the good light output and a low internal background in the energy region around 11 MeV due to thermal neutron capture processes on the nuclei of the crystal [13].

---

[*]Corresponding author.

Investigating the Effects of Neutron Source Number and Arrangement in Landmines Detection by
Thermal Neutron Capture Gamma Ray Analysis

75

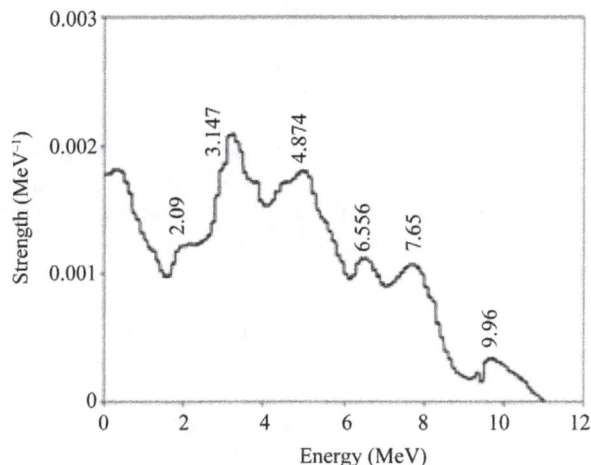

Figure 1. Neutron energy spectrum of Am-Be source.

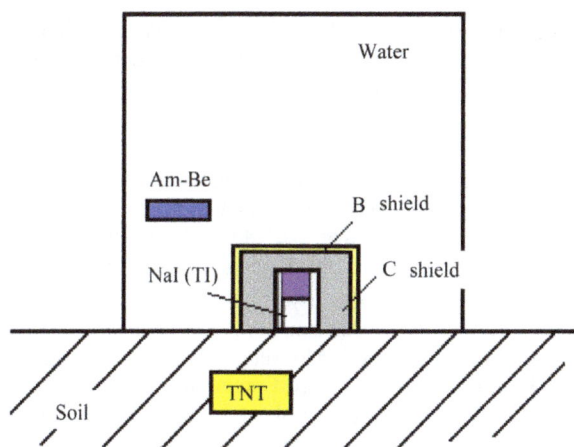

Figure 2. A schematic view of the configuration used in MCNP calculation.

In order to prevent the reaching of thermal neutrons to the detector, it was shielded by ^{12}C as a reflector of thermal neutrons [14] and was surrounded with ^{10}B to absorb thermalized neutrons [15]. The diameter and height of the carbon shield are 18 cm and 12.09 cm, respectively and that of boron shield are 19 cm and 13.09 cm, respectively.

In geometry of two sources, they were located on the both sides of the detector and in 3 sources, one of sources was positioned on one side and two other sources were located on the other side. Geometry of four sources can be explained as two of sources were placed on one side and two other sources on the other side of detector with different directions. Geometry of four sources in the presence of symmetry is the same with single source, but four sources were located in four corner of water tank.

We carried out the calculation with different configurations of sources to investigate effect of placement of sources and determine the best configuration for maximum photon flux.

3. Results and Discussions

Figure 3 shows the gamma-ray spectrum of neutron capture by soil and landmine using F5 tally in MCNP code for the best set up.

In order to investigate the effect of placement of sources in the water tank on photon flux spectrum, the sources was placed at different configurations (different rotations about z-axis and different distances from the bottom of tank). **Table1** shows some of the results of our simulations where z is the distance of source from the bottom of the water tank and α is the angle between the source axis and water tank axis. The MCNP calculations show that the number and placement of sources have effect on photon flux strongly. As shown in **Table 1**, the best set up is related to 3 sources geometry where one of the sources is on one side of the detector with a 45° rotation about the z-axis at a distance of 2 cm from the bottom of the water tank and two of them are on the other hand horizontally and vertically at a distance of 3 cm from the bottom of the water tank. Also **Table 1** shows that the symmetry in the arrangement of the sources doesn't have positive effect on increment of photon flux.

Figure 4 shows background-subtracted spectrum for the best set up of 1, 2, 3 and 4 sources. In the background spectrum a trace of 10.829 MeV peak is seen, which is smaller than 10.829 MeV peak in the spectrum of photon in the presence of TNT. This is due to neutron capture in N in air under the water tank and surrounding. As shown in **Figure 4** photon flux in presence of 4 sources is the same with flux of 3 sources because with increment of number of the source, the volume of water as a neutron moderator decreases and system can't produce more thermal neutrons and so neutron capture events and photon flux decrease.

Simulations showed background in presence of 3

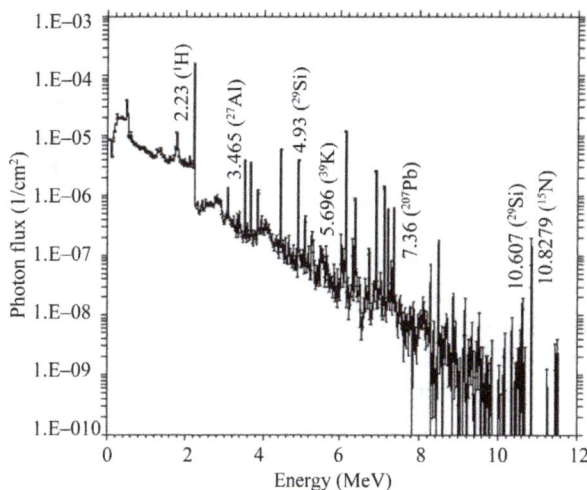

Figure 3. The gamma-ray spectrum of soil containing TNT in the presence of 3 sources.

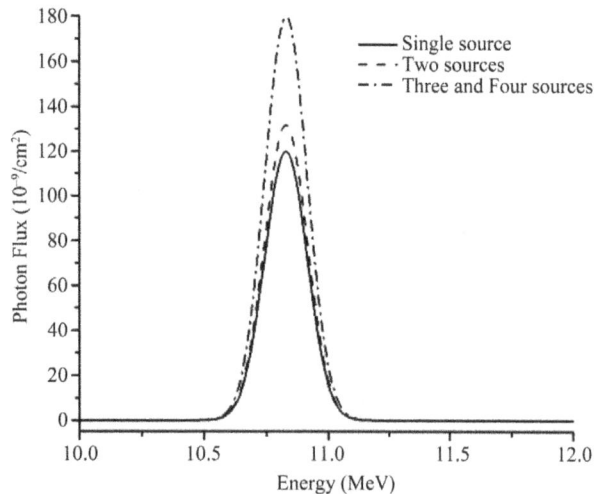

Figure 4. The photon flux for 1, 2, 3 and 4 sources corresponding to the best configurations (without background).

Table 1. The 10.829 MeV gamma-ray flux for different configurations of sources (TNT in center of soil).

Background at 10.829 MeV ($\times 10^{-9}$)	Source number	Flux of 10.829 MeV ($\times 10^{-9}$)
2.14	1	54.68
5.4	2	66.9
	3	117
8.32	3 (α:30)	91
	3 (α:60)	25.1
	3 (z = 3 cm)	49.68
	3 (z = 2.5 cm)	56.3
10.54	4	115
3.25	4 (symmetrical)	71

sources is low in comparison with 4 sources and its useful for obtaining of exact counts by NaI(Tl).

4. Conclusions

The MCNP calculations showed that the arrangement of the sources and the number of them have important influence on landmines detection by thermal neutron capture technique. In this paper the effect of the presence and arrangement of 2, 3 and 4 sources was studied on the capture events and 10.829 MeV photon flux.

Results show that the maximum photon flux is corresponding to 3 sources geometry where one of the sources is on one side of the detector with a 45° rotation about the z-axis at a distance of 2 cm from the bottom of the water tank and two of them are on the other hand horizontally and vertically at a distance of 3 cm from the bottom of the water tank Background counts in this geometry are low in comparison with 4 sources and this is advantageous for obtaining exact counts and biological effects.

REFERENCES

[1] J. C. Campbell and A. M. Jacobs, "Detection of Buried Landmines by Backscatter Imaging," *Nuclear Science and Engineering*, Vol. 110, 1992, pp. 417-424.

[2] M. Maucec and R. J. de Meijer, "Monte Carlo Simulations as a Feasibility Tool for Non-Metallic Landmine Detection by Thermal-Neutron Backscattering," *Applied Radiation and Isotopes*, Vol. 56, No. 6, 2002, pp. 837-846.

[3] K. M. Dawson-Howe and T. G. Williams, "The Detection of Buried Landmines Using Probing Robots," *Robotics and Autonomous System*, Vol. 23, No. 4, 1998, pp. 235-243.

[4] A. V. Kuznetsov, A. V. Evsenin and I. Y. Gorshkov, "Detection of Buried Explosives Using Portable Neutron Sources with Nanosecond Timing," *Applied Radiation and Isotopes*, Vol. 61, No. 1, 2004, pp. 51-57.

[5] M. Lunardon, G. Nebbia, S. Pesente, G. Viesti and V. Filippini, "Detection of Landmines by Using 14 MeV Neutron Tagged Beams," *Applied Radiation and Isotopes*, Vol. 61, No. 1, 2004, pp. 43-49.

[6] M. Maucec and C. Rigollet, "Monte Carlo Simulations to Advance Characterisation of Landmines by Pulsed Fast/ Thermal Neutron Analysis," *Applied Radiation and Isotopes*, Vol. 61, No. 1, 2004, pp. 35-42.

[7] A. Pazirandeh, M. Azizi and S. F. Masoudi, "Monte Carlo Assessment of Soil Moisture Effect on High-Energy Thermal Neutron Capture Gamma-Ray by ^{14}N," *Applied Radiation and Isotopes*, Vol. 64, No. 1, 2006, pp. 1-6.

[8] D. R. Ochbelagh, H. M. Hakimabad and R. I. Najafabadi, "The Soil Moisture and Its Effect on the Detection of Buried Hydrogenous Material by Neutron Backscattering Technique," *Radiation Physics and Chemistry*, Vol. 78, No. 5, 2009, pp. 303-306.

[9] J. Obhodas, D. Sudac, K. Nad, V. Valkovic, G. Nebbia and G. Viesti, "The Soil Moisture and Its Relevance to the Landmine Detection by Neutron Backscattering Technique," *Nuclear Instruments and Methods in Physics Research Section B*, Vol. 213, 2004, pp. 445-451.

[10] S. Pesente, M. Cinausero, D. Fabris and E. Fioretto, "Effects of Soil Moisture on the Detection of Buried Explosives by Radiative Neutron Capture," *Nuclear Instruments and Methods in Physics Research Section A*, Vol. 459, No. 3, 2001, pp. 577-580.

[11] H. M. Hakimabad, A. V. Noghreiyan and H. Panjeh, "Improving the Moderator Geometry of an Anti-Personnel Landmine Detection System," *Applied Radiation and Isotopes*, Vol. 66, No. 5, 2008, pp. 606-611.

[12] H. T. Anbaran, R. I. Najafabadi and H. M. Hakimabad, "Optimization of a Detector Collimator for Use in a Gamma-Ray Backscattering Device for Anti-Personal

Investigating the Effects of Neutron Source Number and Arrangement in Landmines Detection by
Thermal Neutron Capture Gamma Ray Analysis

77

Landmines Detection," *Applied Sciences,* Vol. 9, No. 9, 2009, pp. 2168-2173.

[13] E. Chukhaev and A. I. Melnikov, "CsI(Tl) Scintillators as *γ*-Ray Detectors for the Identification of Hidden Explosives," *Nuclear Instruments and Methods in Physics Research Section A*, Vol. 471, No. 1-2, 2001, pp. 234-238.

[14] C. P. Datema, V. R. Bom and C.W. E. van Eijk, "Experimental Results and Monte Carlo Simulations of a Land-

mine Localization Device Using the Neutron Backscattering Method," *Nuclear Instruments and Methods in Physics Research Section A*, Vol. 48, No. 1-2, 2002, pp. 441-450.

[15] H. Akkurt, J. Wagner and K. Eckerman, "Hand Held Instruments for Landmine Detection: View from Radiation Dosimetry," *Nuclear Instruments and Methods in Physics Research Section A*, Vol. 579, No. 1, 2007, pp. 391-394.

13

High-Level Nuclear Wastes and the Environment: Analyses of Challenges and Engineering Strategies

Mukhtar Ahmed Rana

Physics Division, Directorate of Science, PINSTECH, Islamabad, Pakistan

ABSTRACT

The main objective of this paper is to analyze the current status of high-level nuclear waste disposal along with presentation of practical perspectives about the environmental issues involved. Present disposal designs and concepts are analyzed on a scientific basis and modifications to existing designs are proposed from the perspective of environmental safety. A new concept of a chemical heat sink is introduced for the removal of heat emitted due to radioactive decay in the spent nuclear fuel or high-level radioactive waste, and thermal spikes produced by radiation in containment materials. Mainly, UO_2 and metallic U are used as fuels in nuclear reactors. Spent nuclear fuel contains fission products and transuranium elements which would remain radioactive for 10^4 to 10^8 years. Essential concepts and engineering strategies for spent nuclear fuel disposal are described. Conceptual designs are described and discussed considering the long-term radiation and thermal activity of spent nuclear fuel. Notions of physical and chemical barriers to contain nuclear waste are highlighted. A timeframe for nuclear waste disposal is proposed and time-line nuclear waste disposal plan or policy is described and discussed.

Keywords: High-Level Nuclear Waste; Nuclear Waste Containment and Disposal; Environment; Conceptual Model Designs; Radioactivity Damage; Chemical Heat Sink

1. Introduction

The issue of disposal of high-level radioactive nuclear waste, e.g., spent nuclear fuel (SNF), is not new and needs urgent attention due to its increasing volume worldwide. It is now one of the most important but controversial problems of nuclear technology. Only safe and successful solutions to this problem would guarantee the long-term future of nuclear power. It is extremely difficult for policy-makers worldwide to develop a consensus on final disposal of high-level nuclear waste. The disposal of high-level nuclear waste [1-3] is gaining a new momentum [4] due to the need for more electricity with minimal emission of CO_2 and other greenhouse gases to limit global warming.

Apart from disposal of safely produced SNF or high-level radioactive waste, the possibility of nuclear reactor accidents [5-8] also requires deep understanding of this issue from the perspective of failure. Forward planning [9] is the only solution of this extremely sensitive issue. The following three-pronged criterion can potentially play a significant role in achieving safety assurance on this important and near- and far-future humanity related issue. First, nuclear test [10] and accident sites can be helpful in forward planning [11,12]. Second important point which can be helpful in finding out the safe solution of this issue, is sharing of knowledge from various nuclear workplaces worldwide [13-17]. Strict critical review of policies, principles and implementation procedure for high-level radioactive waste disposal should be mandatory.

Safety of the nuclear waste containment and disposal can be assured by making the effective use of science in policy making. A policy is a set of guiding principles for making procedures of implementation of a scheme. A public policy is quite different in nature from a private policy and is complex subject. It requires the optimization of a number of technical as well as well as social parameters. Policy for the high level nuclear wastes (HLW) disposal is a multifaceted issue and it requires to resolve a number of inter-related problems. In situations like disposal of HLW, comprehensive evaluation of policy success is extremely important as implications of a failure can be smashingly serious for the present and future life at earth. Risk informed changes to the technical requirements of a HLW disposal policy is a natural solution, but stringent complications in assessment of the risks involved due to unpredictability of future geophysical events over a long time scale of more than 100,000 years are the major worries.

Main objective of this paper is to present/analyze the current status of high-level nuclear waste and/or spent nuclear fuel disposal along with practical scientific thoughts about the issue. Present disposal designs and concepts are analyzed on scientific bases and modifications to the existing designs are proposed. Next section describes an assessment of the nuclear waste disposal problem and its implementation plan. Section 4 presents analysis of current nuclear waste disposal procedures and a brief summary of a method for monitoring the radiation damage in nuclear waste containers. Section 5 is composed of status comments on different aspects of nuclear waste disposal along with a modified burial design. Paper ends with conclusions of the investigation.

2. Climate Change and Nuclear Energy

One of the biggest questions of the time is how to meet the challenges caused by escalating climate change and growing energy demand around the globe. Nuclear energy can play a central role in mitigating the global climate change by minimizing the emission of CO_2 and other greenhouse gases in commercial energy supply [18]. Public acceptance to nuclear energy is very low due to Chernobyl [9,11,19] and Three Mile Island [20] accidents. It is being realized that factual public awareness of nuclear energy and related issues, especially security and environmental safety of nuclear engineering designs needs to be raised. **Table 1** shows a foresight of energy consumption scenarios in 2050 keeping fossil-fuel carbon emissions same as at present to keep a hold on the climate change [21]. If the level of nuclear energy expected in the above mentioned scenario is considered, thoughtful and coherent research efforts around a few central themes would be needed. Nuclear reactor safety and solution to the problems associated with spent nuclear fuel

Table 1. Energy consumption scenario (Sailor *et al.* 2000).

	1997 World	1997 USA	1997 France	2050
Population (millions)	5857	268	59	9000
Total primary energy (EJ/year)	400	99	10.3	900
Fossil fuel (EJ/year)	343	85	6.2	300
Renewable (EJ/year)	30	5.2	0.7	300
Nuclear (EJ/year)	25	7.1	4.1	300
Total per capita (EJ/year)	68	371	175	100
Fossil fuel fraction (%)	86	85	61	33
Nuclear energy Generation (GW-year/year)	259	72	43	3300
Per capita (kW-year/year)	0.04	0.27	0.73	0.36
Fraction of electricity (%)	17	18	79	>50
CO_2 emission (MTC)	6232	1489	102	5500

(SNF) or HLW are major concerns.

Apart from Chernobyl and Three Mile Island, safety record of nuclear reactors has been extremely good. More than 8500 nuclear reactors built outside former Soviet Union, there has been no major radioactivity release accident. There have been considerable improvements in reactor designs after above mentioned nuclear accidents [21]. Continued commitment to the best science for improvements in safety aspects of nuclear technology, keeping the economic build low, is essentially needed. Upcoming nuclear reactors will have considerably better safety perspective compared with present. The possibility of core damage in Advanced Boiling Water Reactor (ABWR), a US design, is estimated to be 2×10^{-7} per reactor per year [22] SNF or HLW is a complicated issue. Some of nuclear countries deem SNF as a disposable waste while others as an asset which associates a kind of paradox with it. Problems associated with this paradox issue and possible solutions are discussed in the next section.

3. Problem Assessment and Implementation Plan

3.1. Composition of SNF, A Representative Radioactive Waste

Although composition of SNF is reactor, fuel and burn-up specific, a general dependence of its composition on storage time is described in this section. Spent nuclear fuel shows almost a complete spectrum of radioactivity. Some of elements in SNF will remain radioactive for hours to a few years whereas others for thousands to millions of years. Rate of change of any of radioactive nuclei in SNF can be represented by the following equation,

$$\frac{dN_i}{dt} = \frac{dN_i}{dt}\bigg|_{form} - \frac{dN_i}{dt}\bigg|_{decay} \qquad (1)$$

whereas concentration or number of a specific specie of nuclei at any time are given by the following equation,

$$N_i(t) = N_i(t_o) + \int_{t_o}^{t} \frac{dN_i}{dt}\bigg|_{form} - \int_{t_o}^{t} \frac{dN_i}{dt}\bigg|_{decay} \qquad (2)$$

where t_o is the starting time whereas t is any time afterwards. It is clear from above equations that composition of SNF will continue changing, but in a quite deterministic way assuming initial composition of SNF is known. It is an important point to be considered while selecting containment materials and disposal site.

3.2. High-Level Nuclear Waste Disposal Implementation Plan or Policy

The conceptual model of a reliable scientific investigation is shown in **Figure 1(a)**. Analysis and execution of

(a)

(b)

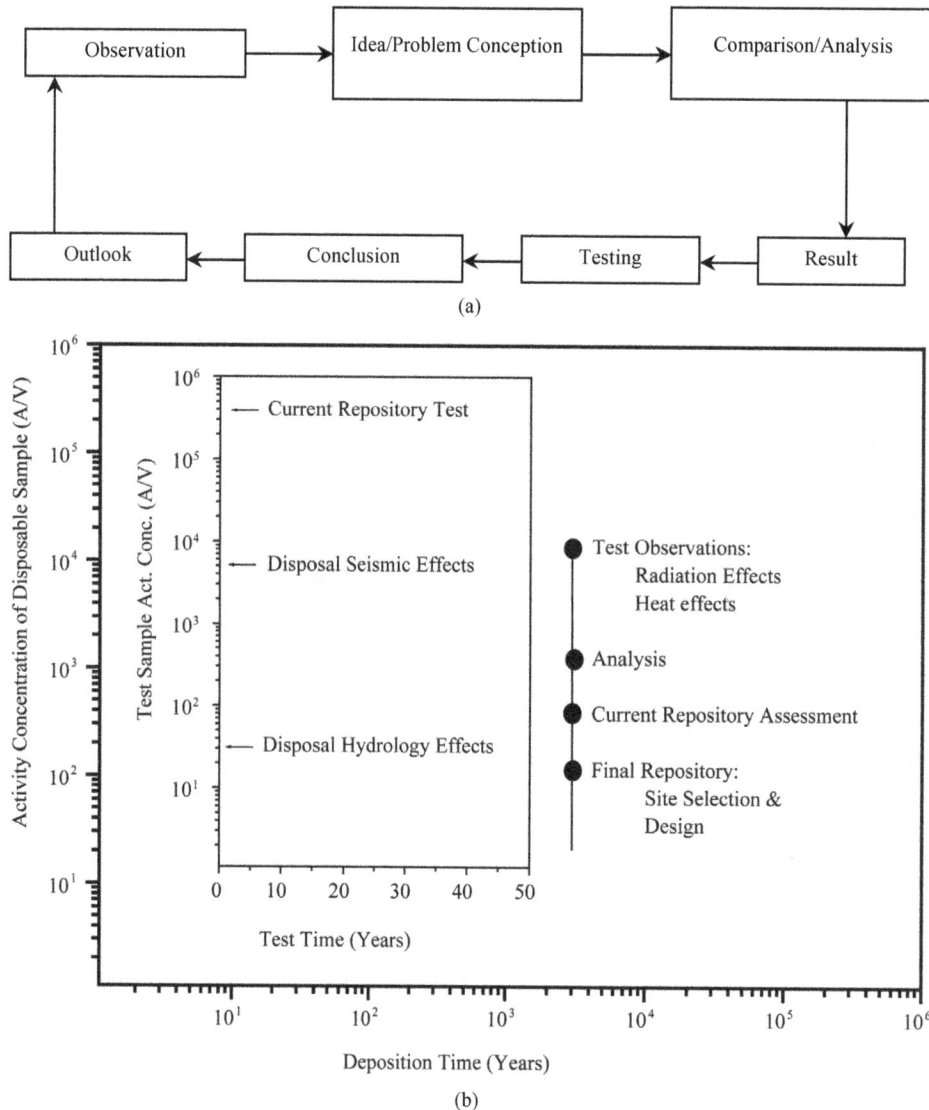

Figure 1. Conceptual model of a reliable scientific investigation (a) and high-level radioactive waste disposal plan.

close-circle coherent activities are necessary at small scale before fixing the implementation methods and techniques for a practical large scale final disposal of SNF or high-level nuclear waste. Considering millions-year long radioactive and thermal life of SNF, at least 40 - 50 years are required to start large scale disposal. Considering great difficulties and extremely high cost of retrieval of disposed nuclear waste, political and social impacts also need to be analyzed carefully [11]. **Figure 1(b)** shows implementation plan or policy proposing small scale low activity sample disposal for studies of hydrology and seismic effects on disposed nuclear waste. Suitable sites with considerable hydrology and seismic activities need to be selected for these test disposals in order to understand impacts of failures due to lack of scientific understanding about hydrology and seismic history and future evolution. **Figure 1(b)** also describes how analysis of

observations of test disposals can help in refining current repository design to achieve final practical disposal repository design and implementation plan.

Urgency for solution of final disposal of high-level nuclear waste is due to complications involved and multidisciplinary nature of the issue which will take long time 40 - 50 years to reach the stage of final disposal even after the practical selection of the final disposal site. **Figure 2(a)** shows the spent fuel cycle which is the major high-level nuclear waste. This simple schematic is based on the well-known facts and details are given by a number of authors, for example, [23-25]. **Figure 2(b)** shows general composition and forms of fission products and transuranium elements which are the most important for evaluation of a disposal activity. This figure is based on results by Buck *et al.* [26]. It is clear from **Figure 2(b)** that SNF is a very special type of waste due to high per-

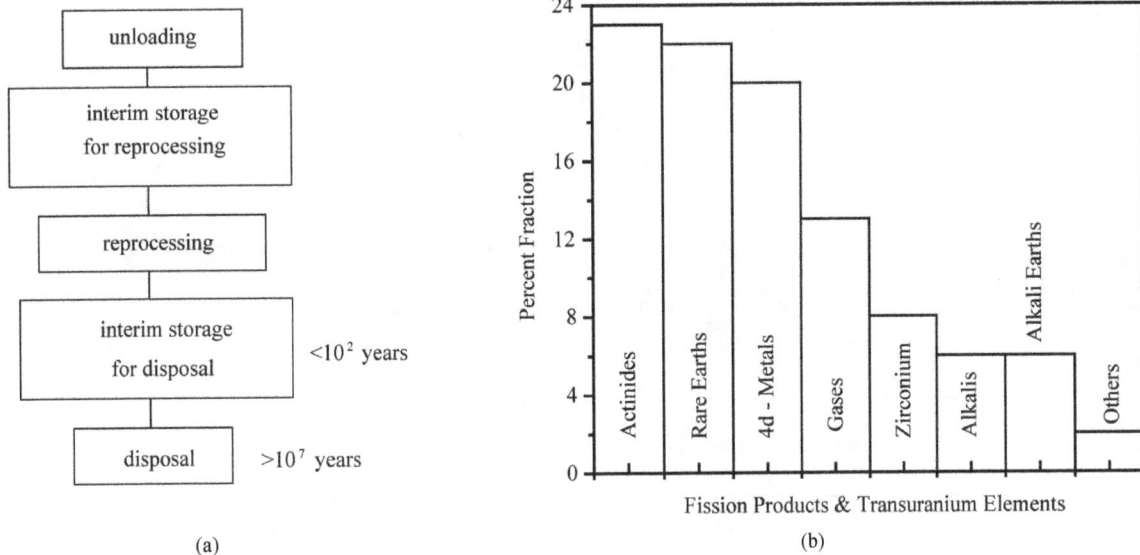

(a)

(b)

Figure 2. (a) Spent nuclear fuel cycle, and (b) General composition and forms of fission products and trans-uranium elements, which are most important in evaluation of disposal activity. Presentation is based on results by Buck *et al.* (2004).

centage of rare earth elements in it along with a quite considerable percentage of radioactive gases. These are very different characteristics from those of human safe environment. Major chemical alterations in SNF are gaseous and thermal evaporation, oxidation and dissolution of fuel pellets, and precipitation of secondary phases in changing spent fuel. These changes, based on well-known facts and results from Ref. [27] are represented by a schematic in **Figure 3**.

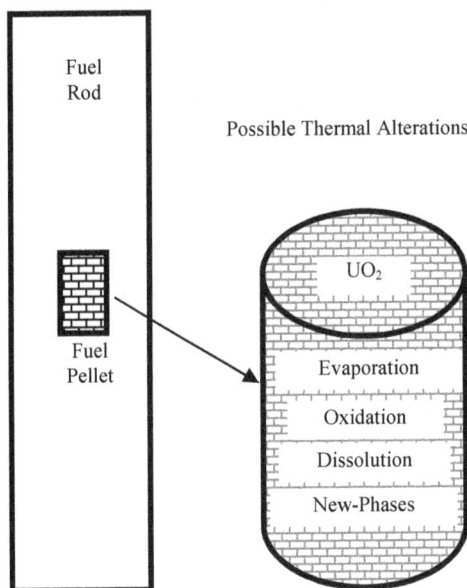

Figure 3. Thermal, structural and compositional alterations in SNF, which can cause significant consequences over long-time scale. This figure is based on generally known information in the field of nuclear engineering and that from Poinssot *et al.* (2005) and Ewing (2006).

3.3. Proof of Safety: Global Hand to Hand Policy

Can anyone on earth come up with a policy for HLW assuring a comprehensive safety of the global environment over a minimum time scale of 100,000 years? Present answer is "No". But, on the whole safe function of nuclear energy technology over a half century, despite the initial doubts about safety of nuclear technology, gives a hope. A three pronged strategy may be considered to build a trust in present and future safety of any HLW disposal policy. One is scientific basis of the disposal management policy; second the IAEA regulations for the disposal policy to assure global safety with minimum interference in any state's internal matters and third knowledge sharing among nuclear and related countries. Above mentioned strategy could provide a safety assurance with providing a chance of participation to anyone with legitimate capacity. Implementation of the above mentioned global hand to hand policy may find difficulties due to strategic nature of the issue and safety implications. This major inconvenience needs to be addressed on human grounds.

Responsibility of a failure of an HLW disposal policy and procedures, and first responders need to be defined with clarity. A comprehensive analysis is required to sort out the link between capacity and responsibility which may vary case to case and need to be carried out in the local context. But, common features of the issue of the link between capacity and responsibility should be dealt with at a global level to achieve legitimate general guidelines. Geological disposal of HLW is the best available choice. **Figure 4** shows a guidance triangle for geological disposal of HLW. The most important aspects are the failure assessment of HLW containers and hydrology

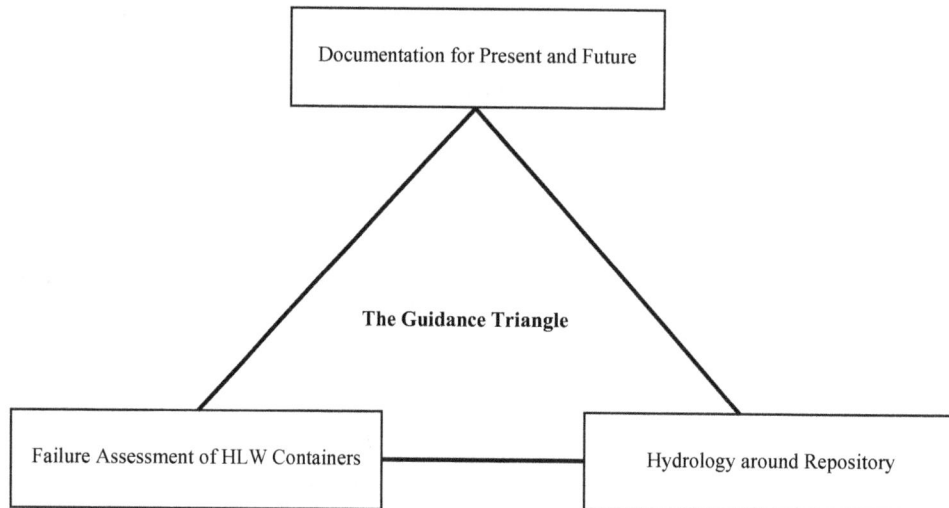

Figure 4. Guidance triangle to assure the keeping up of system for HLW geological disposal for more than 100,000 years.

around the repository site. Clear documentation of above mentioned activities, for present and possibly future, is also essentially needed. Method of lucid documentation is also an important issue.

4. Disposal of Nuclear Waste

The issue of disposal of SNF or high-level nuclear waste has been evaluated for decades now by nuclear scientists worldwide [21,25]. Ewing Considered options for SNF disposal include burial in ocean floor polar or ice hills, space disposal, keeping in interim storage facilities and more importantly, deep underground burial in special geological formations. Deep underground burial is being considered as safest in available options. Research areas involved in geological nuclear waste disposal are Materials Science & Engineering, Nuclear Geology and hydrology. Despite the investigations cited above on materials and geology, correlated research activities are required for successful geological nuclear waste disposal, especially coupled investigations on underground geological formations, seismology and hydrology. Effects of radiations on confinement materials in final disposal are very important.

It is aimed here to highlight the major problems in the disposal of high-level nuclear waste like processed or unprocessed spent nuclear fuel. Problems involved are extremely complicated and requires conceptual, materials and other technical developments. Feared by complications, it is sometimes treated as un-solvable problem, which has imposed dark shadows on the future of nuclear power. To keep nuclear technology in work in future, related scientific community is working very hard to cope with the problems. Solution of this problem will bring conceptual and material developments, which will help in overall development of science and technology.

Geological disposal of SNF can only be successful by implementing multiple barrier strategy to confine the disposed waste and its effects far from safe environment to which living being have or may need to have contact in future. **Figure 5(a)** gives an overview of possible barriers to confine the disposed high level waste. Most important of natural barriers is a solid stable crystalline rock far from earth quake related fault lines. Engineered barriers include corrosion-resistant containers possibly of copper alloys and disposal architecture. Recently, a new method has proposed by Rana [17] for monitoring the radiation damage in nuclear waste containers using ion channeling. Ion channeling measurements are possible at ion beam facilities worldwide. A 1 - 3 MeV helium ion beam can be employed to measure radiation damage in test crystalline samples placed in a section of a container wall as shown in **Figure 5(b)**. Mathematical method for determination of structure collapse rate in container wall using ion channeling measurements is given by Rana ([17] 2008a). This method can be used to monitor the radiation damage in nuclear waste containers and to predict containment failures in near and far-future. Nature of single radiation damage in bulk and surface-layer of a typical solid is recently discussed by Rana [17,18]. Total radiation damage is accumulated effect of all radiations penetrated in to a target or containment materials, with different radiations causing different magnitude and type of damage. Thermal and chemical stability [19,21] of containment materials is an important in selection of materials to be used in containment of nuclear wastes.

Status comments on major aspects of final disposal are described below in the form of a few points. 1) Initial radiation strength per unit spent nuclear fuel depends on burn-up the fuel, but extremely high for any living being without best available shielding arrangements [28]; 2) Radioactivity decay time scale for SNF of the order of

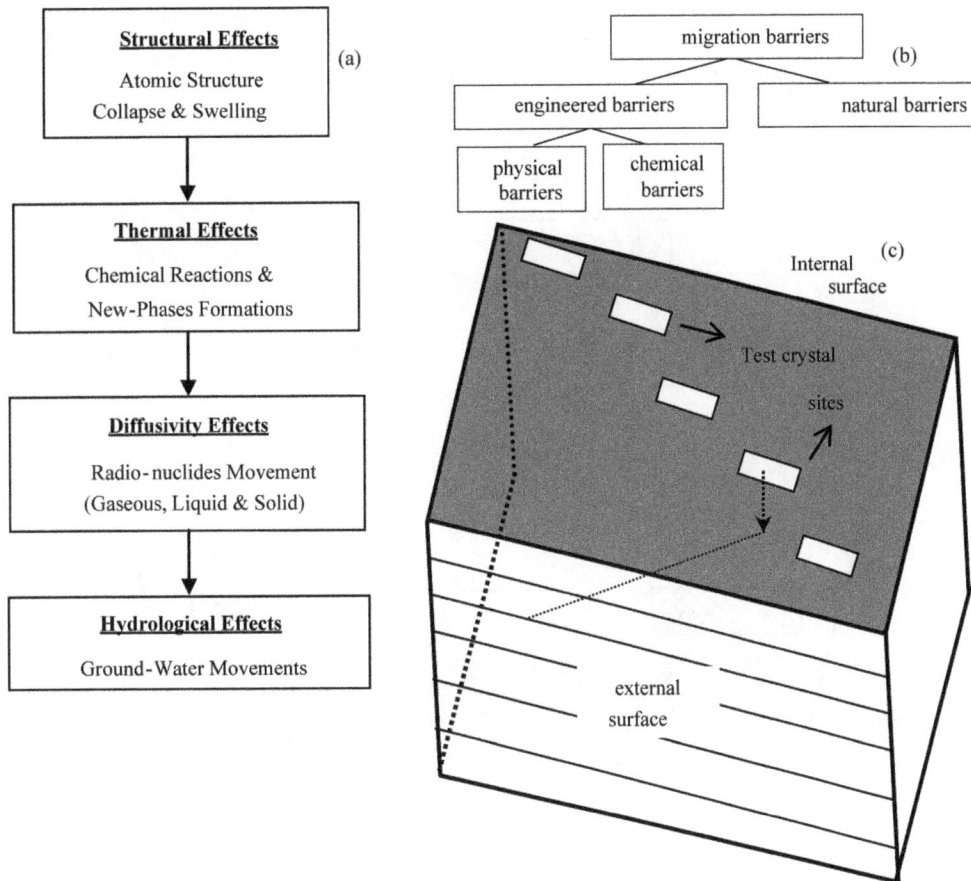

Figure 5. (a) Radiation effects on containment materials and environment; (b) Migration barriers in repository design; and (c) Section of the nuclear waste container wall for installation of radiation damage test crystal samples.

geological time scale, which is up to millions of years; 3) Forms of radiations from SNF include charged and neutral particle rays, and electromagnetic radiations; 4) Decay of radioactive elements in SNF is accompanied with the release of energy, most of which is transformed into heat. SNF is a heat source, which can harm integrity of its disposed packages; 5) Gaseous nature of radioactive products is of great concern. Thirteen percent of fission products and trans-uranium elements are gases, which has higher danger of leakage and mobility to the objectively safe environment; 6) Direct disposal of SNF will be cheaper [29], but it is like wasting potential source of energy; 7) Transmutation decreases the danger level of SNF, but does not solve the problem completely. Final disposal will still be needed [28]; 8) Ideally, retrievability after disposal is required. But, its assurance is difficult due to involvement of unexpected natural happenings like earth-quakes.

Figure 6(a) shows the outline of the rock-integration nuclear waste burial design by Maki and Ohnuma [15]. **Figure 6(b)** shows present modifications to the design shown in **Figure 6(a)** with objective to achieve improvement regarding pressure build up due to complete block-

age of underground water flow. Leaving open channels or tunnels for controlled water flow through buried waste. This water flow through open channels or tunnels will also serve as monitoring test about any leakage from waste packages. These channels will avoid water pressure build up beyond a critical limit and if a considerable leakage is observed in water through these channels, nuclear waste burial design should allow the blockage of these water channels. Another notion of chemical heat sink (**Figure 6(b)**) is introduced, which if incorporated in burial design, can keep the temperature of nuclear waste under limit. This chemical heat sink is a compound chemical material, which will decompose by absorbing heat emitted by nuclear waste. Water flow through proposed channels in the buried waste will also cool nuclear waste.

5. Radiation Effects

5.1. Radiation Damage

Degradation of spent fuel itself and containment materials due to radiation effects is a very considerable concern. Intensive radiation exposure causes dramatic degradation in structural and strength related properties of materials

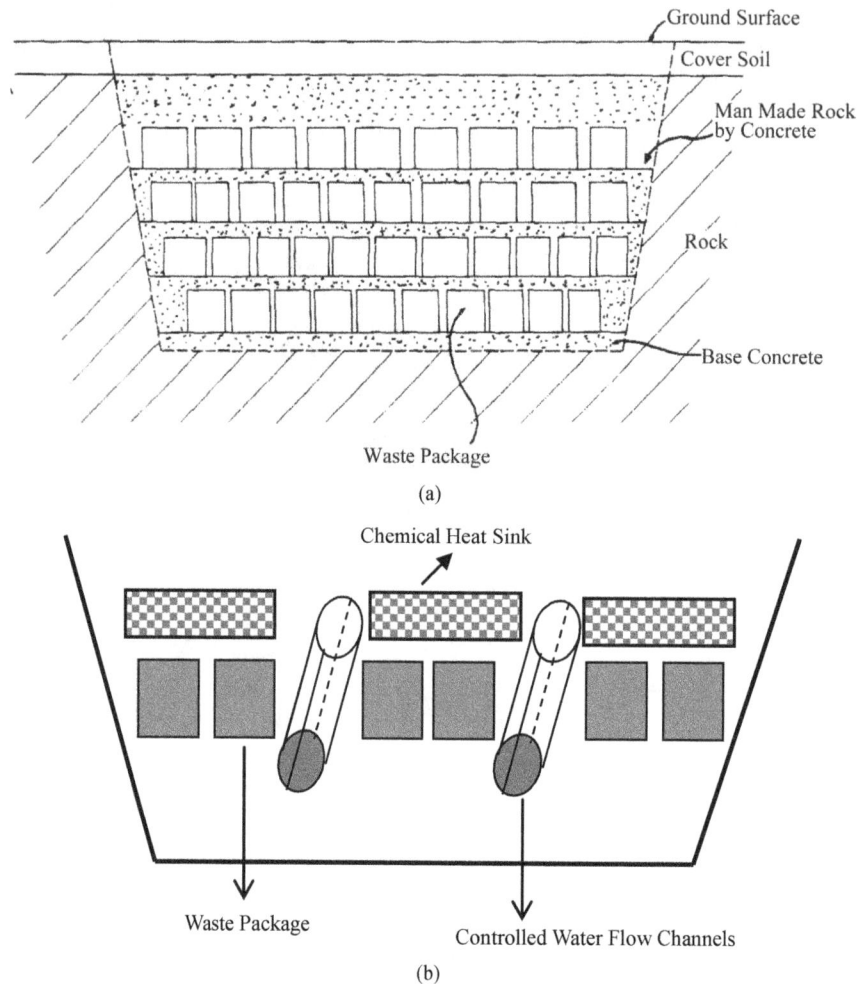

Figure 6. (a) Design of rock integration nuclear waste burial facility by Maki and Ohnuma (1992) and (b) Modifications to the above-mentioned design to assure integrity of buried waste.

leading to their failure when damage exceeds a certain limit. A number of aspects of radiation damage have been recognized and being studied over more than 60 years. Radiation damage leaves four types of effects on any material, *i.e.* electronic and optical which are not significant in nuclear waste containment, physical and chemical. Physical and chemical effects need to be considered. A variety of radiations continue penetrating waste containment and the aggregated effects over decades thus are important for determination of containment failure. A single radiation, especially energetic charged particle, causes a compound spike [30] in the target material. This compound spike arises as a consequence of a Coulomb explosion and a thermal spike, and decays very quickly within 10^{-12} s. These physical impacts result in the form of heat emitting out into the neighbouring material of the cylindrical zone through which radiation passes. The increased temperature, due to continuous radiation spikes, produce chemical changes like formation of new material phases. **Figure 7** shows the generalized view of expected

radiation effects on containment materials to be used in nuclear waste disposal.

Here, a very brief account of basic physics of radiation damage is being given which may help in implementation of the method for the radiation damage monitoring described above and interpretation of experimental observations of the method. A charged particle or radiation traveling in a solid creates a superheated cylindrical zone with a modified structure containing defects of various types and size. In the inner dotted cylindrical zone in **Figure 8**, bulk atomic flow takes place whereas in the outer shell only individual atomic flow is occurred. A fresh radiation damaged zone in a solid is highly unsteady in time and after reaching thermodynamic equilibrium it becomes an inhomogeneous structure. The energy deposited by the incident radiation in a cylindrical volume around the path is non uniform. It decreases exponentially along radial direction whereas distribution along axis of the cylinder depends on energy of the particle. For an MeV/u ion, it has a maximum at a depth into the

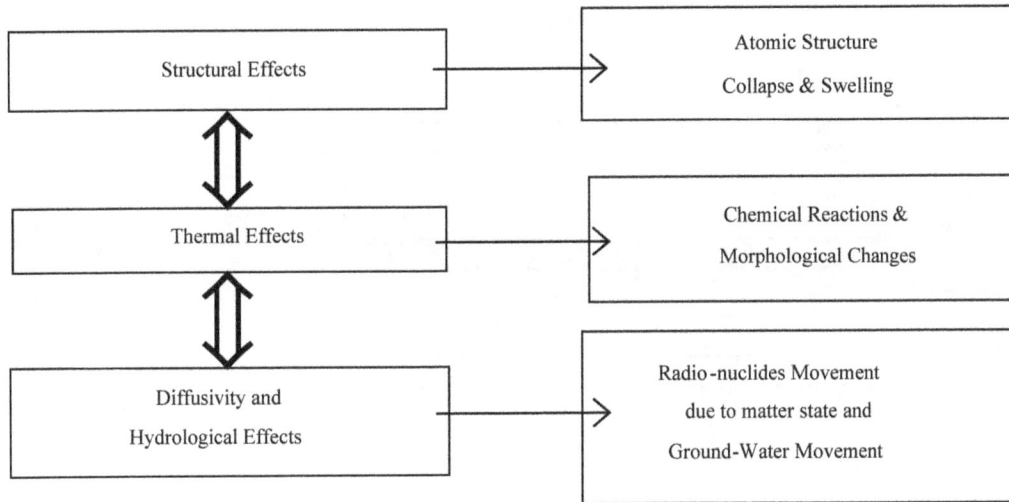

Figure 7. Radiation effects on containment materials and environment.

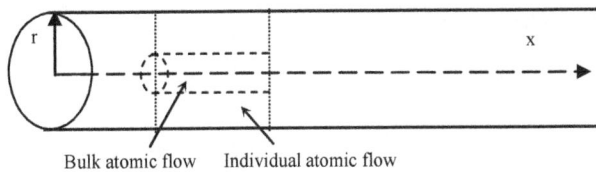

Bulk atomic flow   Individual atomic flow

Figure 8. Radiation damage produced by a charged radiation in a typical solid, showing cylindrical zones of bulk and individual atomic flows. Parameters are defined/shown in this figure for the purpose of mathematical description of the problem.

target.

Interaction of a radiation with a solid target can be treated as a compound spike including partial roles of both thermal and Coulomb explosion spikes. Fractional roles of both spikes depend on atomic and electronic structure of the target and density of deposited energy in it by the incident radiation. An incident radiation is scattered by the atoms in the target as it interacts with them and deposits energy. Weak scattering of incident radiations by light target atoms does not significantly deviate incident particles from their straight trajectories while the target atoms recoil considerably, damaging the detector. Heavier atoms scatter incident particles through wide angles, significantly deviating them from their straight paths while the target atoms recoil weakly, producing less damage. So, it is important to notice that radiation damage mechanism in a target composed of light atomic species is different from that composed of heavier atoms. Compound impacts of a number of radiations in a target, incident within a specific distance, superimpose with one another in both constructive and destructive manners. Part of the damage produced by one radiation is extended due to the damage produced by another radiation within a few hundred nanometers. Nuclear waste containers and related materials are exposed to radiations with a wide

spectrum of ionizing power including fission fragments of very high ionizing power and gamma rays of comparatively very low ionizing power.

5.2. Measurement of Radiation Damage

5.2.1. Brief Description of the Single Scattering Method

Radiation damage in a piece of a crystal (a test sample say Si, Ge, Zr or Zircon) exposed to radiations will carry information about total radiation exposure of the crystal. So, proton or helium ion channelling measurement of radiation damage in the test sample, placed in the crystalline or amorphous immobilizing containment, can in principle yield considerably complete information about total radiation exposure. A calibration between structure collapse rates of test sample and container wall material will provide the structure collapse rate of the containment material. **Figure 9** is the schematic showing components on an initially channelled proton/ion beam in a crystal. This figure is modified from the original [31]. The total random fraction of the beam $\chi_T(x)$ is the sum of the random fraction of the beam in the crystal, $\chi_R(x)$, reaching depth x and fraction of the beam randomized by the defects in the depth step $x + dx$, $\chi_D(x)$,

$$\chi_T(x) = \chi_R(x) + \chi_D(x) \tag{3}$$

The component $\chi_D(x)$ may be written as,

$$\chi_D(x) = \left[1 - \chi_R(x)\right] \frac{f n_D(x)}{n} \tag{4}$$

where $n_D(x)$ is atomic concentration of defects, n the total atomic concentration, f the defect scattering factor. The factor f accounts for the fact that all defects do not contribute equally and may have different number of scattering centres [31]. For randomly displaced atoms

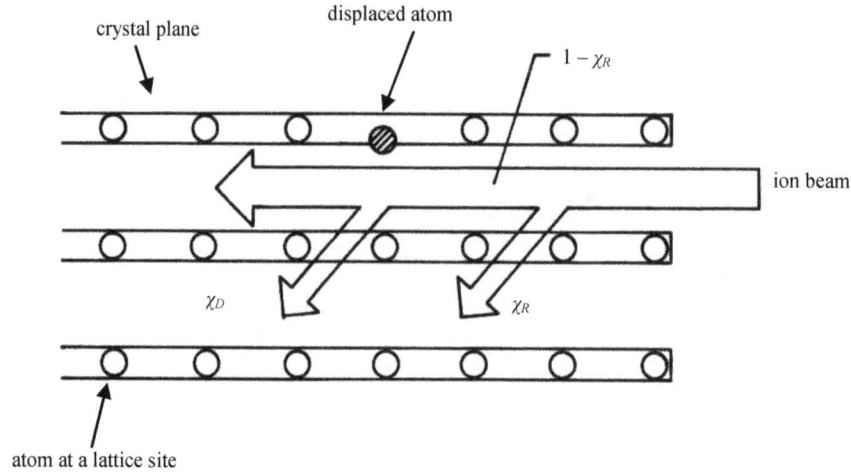

Figure 9. Dechannelling of channelled beam due to defects present at an arbitrary depth. χ_R and $1 - \chi_R$ are random and channelled fractions of the beam reaching the considered defect layer.

(called isolated interstitials), the value of f is 1. The quantity $\chi_R(x)$ is not a measurable quantity. It is the sum of random fraction in the perfect crystal $\chi_V(x)$ and the random fraction resulted from the dechanneling by all defects along the depth $0 - x$.

With single scattering approximation, $\chi_R(x)$ is given by

$$\chi_R(x) = \chi_V(x) + \left[1 - \chi_V(x)\right] \times \left[1 - \exp\left\{-\sum_i \int_0^x \sigma_D^i n_D^i(x') dx'\right\}\right] \quad (5)$$

For the case of very small defect concentration, $\chi_R(x)$ is given by,

$$\chi_R(x) = \chi_V(x) + \left[1 - \chi_V(x)\right] \times \left[\sum_i \int_0^x \sigma_D^i n_D^i(x') dx'\right] \quad (6)$$

where σD is the dechannelling factor of a certain type of defects along the beam path $0 - x$. The quantities $\chi_T(x)$ and $\chi_V(x)$ are measurable quantities in backscattering channelling experiments and yield the quantity $\chi_D(x)$ needed to determine defect density at depth x in a crystal using Equations (2) and (3). The above formulation with single scattering dechanneling approximation is only valid for small defect densities (less than ~10% of lattice sites constitute defects) [31]. For higher defect densities, the possibility of multiple scattering dechanneling needs to be incorporated.

5.2.2. Incorporation of Multiple Scattering
For multiple scattering only, the quantity $\chi_R(x)$ takes

the following shape,

$$\chi_R(x) = \chi_V(x) + \left[1 - \chi_V(x)\right] \times \left[\exp\left\{-\sum_i \int_0^x \sigma_D^i n_D^i(x') dx'\right\}\right]^{-1} \quad (7)$$

Following the recent work by [32], combining both single and multiple scattering mechanisms, $\chi_R(x)$ becomes (see Equation (8)) where $g(\eta)$ is an attenuation function and

$\eta = \int_0^x n_D(x) dx$ is the areal density of defects.

$L_n = \ln(1.29\varepsilon)$ and ε is reduced energy used in calculations of nuclear stopping power (Lindhard, 1964). The attenuation function $g(\eta)$ is given by [32],

$$g(\eta) = 1 - \exp\left[-\left(\sum_i \int_0^x 2\sigma_D^i L_n n_D^i(x') dx'\right)^{-1}\right] \quad (9)$$

Figure 10 shows selected results of 2 MeV He ions channelling along a <100> axis [33]. It is clear from this plot that single scattering dominates for low values of areal defect density η (named here Regime I) and starts losing significance after η increases beyond a certain value at the cost of increase in multiple scattering (Regime II). In Regime III, only multiple scattering takes place. Equation (6) would be valid in all three regimes of η.

5.2.3. Physical Realization of Channeling Method
Depending upon the radiation flux and temperature of the

$$\chi_R(x) = \chi_V(x) + \left[1 - \chi_V(x)\right] \times \left\{ \left[1 - \exp\left\{-\sum_i \int_0^x \sigma_D^i n_D^i(x') dx'\right\}\right] \times g(\eta) + \exp\left[-\left(\sum_i \int_0^x 2\sigma_D^i L_n n_D^i(x') dx'\right)^{-1}\right] \right\} \quad (8)$$

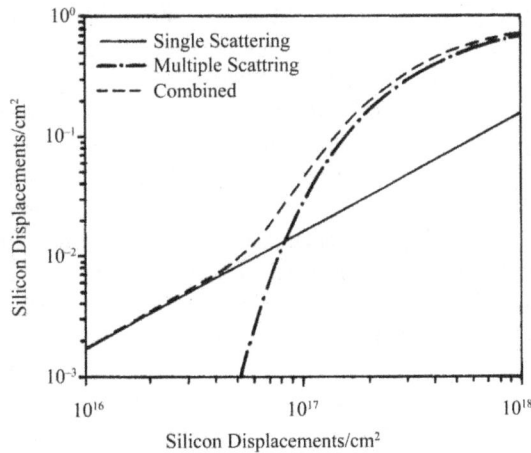

Figure 10. **Dechanneling probability due to single scattering and multiple scattering for 2 MeV He ions in silicon along <100> axis (Shao, 2008).**

containment wall, any crystal fulfilling certain conditions can be used. These crystals (Si, Ge, GaAs and GaN) are available in the market. Diamond has high melting temperature, but can not be used due to lower channeling yield. Information about these commercially available crystals is easily available (Website, University Wafers, http://www.universitywafer.com). Specific dimensions of the test crystal depend on channeling measurement facility and design of the container. Typical dimensions of a test crystal sample are shown in **Figure 11(a)**. If the container material is crystalline and a sample of the same material is used as a test sample, radiation damage in the test sample will be same as in the container material. Details about channeling measurements of defects in GaN crystals produced high temperature exposure are given by Rana *et al.* [19,21]. The same defect quantification procedure can be used for measurement of radiation damage in the test crystal sample. If container material is amorphous or a crystal on which channeling measurements are not possible, a relationship or a calibration between structure collapse rate of the test crystal and damage in amorphous containment material is required. Both test crystal and container material will undergo irradiation in the same environment, then channeling measurements will be performed on crystal, whereas some other method like X-ray photoelectron spectroscopy or XPS will be used on amorphous container material to determine the concentration of broken bonds. The relationship between concentrations of atoms displaced from lattice sites in the test crystal is determined using channeling and the broken atomic bonds in the container amorphous material will serve as a calibration. At present ion beam facilities worldwide, 1 - 3 MeV helium ion beams are available, crystal layer up to a couple of microns depth can be investigated for defect measurement using ion channeling. If it is required to determine radia-

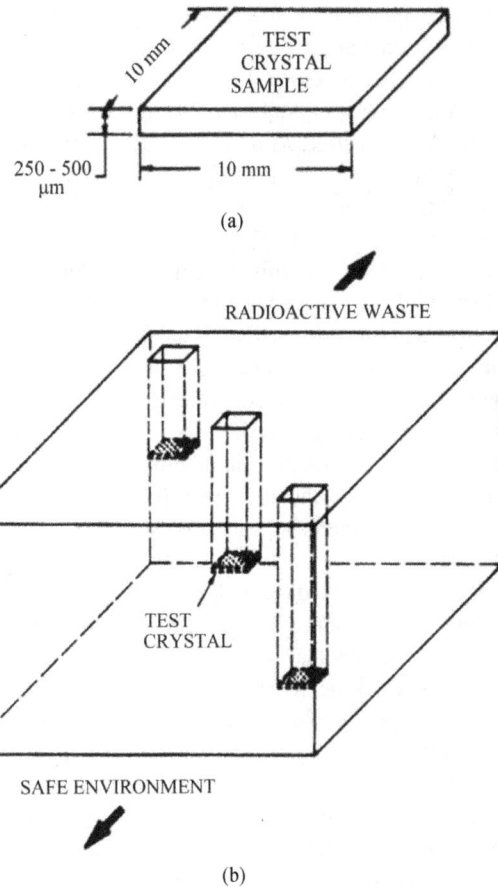

Figure 11. **(a) Dimensions of the test crystal sample; (b) Section of the nuclear waste container wall for installation of radiation damage test crystal samples.**

tion damage at 3 different sites in the container wall, five identical test crystal samples will be placed at objective sites as shown in **Figure 11(b)**. After exposure, defect concentration in surface layer of thickness 1 - 3 μm in all test samples will be measured using ion channeling and measurements will give intensity of radiation damage at crystal sample sites in the containment material. Nature of single radiation damage in bulk and surface-layer of a typical solid is recently discussed [17]. Total radiation damage is accumulated effect of all radiations penetrated in to a target, with different radiations causing different magnitude and type of damage.

5.3. Co-Use of Channeling with Other Techniques

This paper discusses which techniques can be co-used with channeling to increase the accuracy of the measurement of the radiation damage in nuclear waste containers. Nuclear magnetic resonance (NMR) is an attractive technique for radiation damage measurement as it is element specific and is sensitive to both structures in crystalline and amorphous domains in a sample [32].

Using ion channeling and NMR together will make a dual radiation damage detection and measurement system. The displacement of low Z atomic species (like hydrogen) in the test crystal (which can not be measured or can only be measured with low detection efficiency using backscattering ion channeling) can be measured using NMR. Another scheme for short and long term measurement of radiation damage in nuclear waste containers is presented here. In this scheme, ion channeling and nuclear track detection technique are used as two independent techniques for radiation damage measurement. A wide spectrum of radiations (α, β, γ and fission fragments etc.) enter the container wall from the HL nuclear waste. These defects diffuse in the material of the wall, coalesce and make extended defect structures. Production of defects and their reaction continue as radiations enter the material continuously. A typical defect structure of the container wall is three fold: A part of the material is severely damaged, another part gently damaged and the remaining undamaged. **Figure 11(a)** shows a section of the nuclear waste container wall. Channeling and nuclear track detectors can be installed in the wall as shown in **Figure 11(b)**. Both nuclear track detectors (like CR-39) and channeling detectors (test crystals like Si and Ge) will provide short term radiation damage monitor. These measurements will also provide an inter-calibration of two techniques, which would help in reliable quantification of defects in the container material at different points in time. Long term (days to years) monitoring of the radiation damage will be carried out by channeling method only.

6. Thermodynamic Equilibrium and Multi-Barrier Isolation

Thermodynamic equilibrium is a state of a system related to the minimum of the thermodynamic potential. Thermodynamic potential is the Helmholtz free energy (U - TS) for systems at constant temperature and volume whereas the Gibbs free energy (H - TS) for systems at constant pressure and temperature. U, T, H and S are, respectively, internal energy, absolute temperature, enthalpy and entropy. Minimum of thermodynamic potential is characterized by states of thermal equilibrium, mechanical equilibrium and chemical equilibrium of the system. Ideally, nuclear waste should be disposed in a way that it becomes in thermodynamic equilibrium with the environment and remains the same for almost forever without losing its original integrity.

Success probability of SNF disposal would increase by implementing multiple barrier strategy to confine the disposed waste and its effects far from safe environment to which living being have or may need to have contact in future. The definition of human vulnerability in such a

case is given in **Figure 12**. Most important of natural barriers is a solid stable crystalline rock far from seismic zones. Engineered barriers include corrosion-resistant containers possibly of copper alloys (containing mainly copper along with Al: 5% to 9%; Ni: 0.5% to 4%; Fe: 0.5% to 4%; MN: 0.1% to 3%; Ti: 0.001% to 1%, Co: 0.001% to 1%; and B: 0.001% to 0.1%) [34] and disposal architecture. Regular drilled-hole monitoring in the buffer zone and sampling the leached activity before and after earthquake can establish underground faults produced due to earthquake. Nuclear waste containment and the over-all repository environment should ideally be as close as possible to thermodynamic equilibrium, meaning unlimited stability, similar to natural metal deposits within Earth's crust [35].

7. Environment Ethics

Some of major considerations in evaluation of ethical issues related to safety of nuclear waste disposal are clarity of the policies, policy awareness of individuals involved, natural response to nuclear fear/risk factor and valid legal system to sue charges. Central specific ethical issues are summarized as a set of disposal activity start-up questions [3,4]: 1) Have the persons employed/involved been given the free informed consent to the risk involved? 2) Who bear major responsibilities in waste disposal and who is responsible for what? 3) Are the distributions of risks and benefits equitable? 4) Have individuals been informed about control over the risk? 5) Are assessment about reliability of materials and methods involved are made? 6) What are the third parties who can be held responsible for bringing in risk? 7) Evaluation of costs and benefits of intervention measures? 8) Are the plans of compensation for exposure to risk justified? 9) How will an emergency be handled? Generalizing theme build up by above questions, it may be said that issues like consent, equity, control and responsibility are essential ethical considerations for radiological protection policy [36].

It would be interesting to know how above issues or questions about nuclear waste disposal are incorporated in policy making and its implementation. Only thoughtfully critical and multiply reviewed process of policy

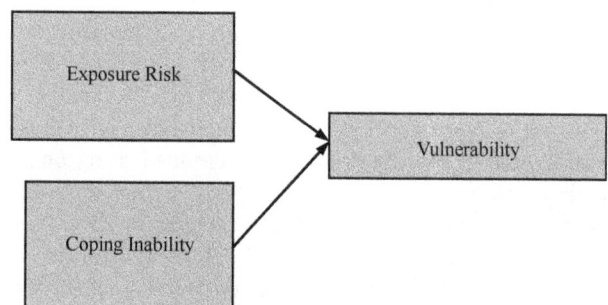

Figure 12. Definition of human vulnerability.

analysis can achieve this. Ethical issues are closely linked with scientific or technical know how about procedures involved. So, a trustworthy research is needed to finalize ethical aspects of high level nuclear waste disposal. Evaluation of risk faced by far-future generations due to present disposal of high level nuclear waste is also of great importance and equally valid ethical issue as for the case of present generation. Real problems are associated with predictions about level and nature of risks faced by future generations and their response to this problem, especially in case of disposal failures.

It would not be wise to dispense with highly radioactive material and to hope that either nature or future generations of humans will not bring it into the biosphere somehow. In principle, we should ensure that even if detail of nuclear waste disposal is lost and does not reach future generations, still they or their environment is not exposed to disposed waste at all. Nuclear waste disposal in one country can quite possibly affect biosphere in the neighbouring countries. Pakistan's two neighbouring countries are among the countries seeking sizeable future nuclear energy programs [37] whereas Russia has offered its land for a multinational nuclear waste repository [38]. These activities may pose questions of nuclear security and environmental justice which Pakistan would need to address. Nuclear waste disposal is not a solely internal matter of any country. Activity of nuclear waste disposal may have strong local, regional and even global implications. Regional and global implications would become considerable for the cases of severe failures of disposal scheme.

8. Conclusions, Final Remarks and Perspective

Present status of different aspects of spent nuclear fuel disposal is overviewed briefly, but comprehensively. Time framework and time-line plan or policy for high-level radioactive waste disposal are described and discussed. A new concept of chemical heat sink is introduced to consume the heat emitted by spent nuclear fuel without affecting the integrity of waste containment. Conceptual model description of major issues of spent nuclear fuel disposal is given along with scientific discussion and comments with focuses of materials, geology, seismic and hydrology aspects. Modifications to a proposed geological disposal of nuclear waste are proposed with scientifically supported arguments.

Paradox of HLW casts shadows on nuclear future. Best science with highest functionality can help solving this problem. Need of ultimate disposal of HLW is an inconvenient truth facing the humanity. This multi-tiered problem should be dealt with IAEA coordination among nuclear states to sort out short and long-term aspects of the HLW disposal. IAEA coordination would be aimed at sharing of the cutting edge knowledge to assure safety of the global environment. Environment is indivisible and the long-term radiological obligations require a global solution and makes it natural.

Although no repository around the globe is ready for geological disposal of nuclear wastes, some developments, mainly in conceptual and plan domains, were made in last couple of decades. **Table 2** summarizes the present plans for high-level nuclear waste repositories. Tabulated details show the sensitivity of the subject and requirement of the decades-long considerations before start up of implementation of any disposal policy. In nuclear waste disposal matters, four considerations are very important which are radiation strength, mean life, environment contamination and traditional ethical values.

Ethical values here refer to rightness or wrongness of our actions. Considering above discussion about high-level nuclear waste disposal, a long time in decades would be needed in evaluation of repository location, design and precautions before start up of disposal. Careful record

Table 2. Plans for high-level nuclear waste repositories (Andersen *et al.*, 2004).

Country	Geological medium	Estimated opening	Status
Belgium	Clay	2035 or later	Searching for site
Canada	Granite	2035 or later	Reviewing repository concept
Finland	Crystalline bedrock	2020	Site selected (Olkiluoto)
France	Granite or clay	2020 or later	Developing repository concept
Germany	Salt	Unknown	Moratorium on development
Japan	Granite or sedimentary rock	2030 or later	Searching for site
Russia	Not selected	Unknown	Searching for site
Sweden	Crystalline rock	2020	Searching for site
Switzerland	Crystalline rock or clay	2020 or later	Searching for site
United Kingdom	Not selected	After 2040	Delaying decision until 2040
United States	Welded tuff	2010	Site selected (Yucca Mountain)

keeping (including details of professionals involved) of all nuclear waste disposal evaluations should be practiced so that investigation of possible accident/emergency could be carried out with transparency.

Clear demonstration about safety aspects of nuclear waste management would help in gaining public and political confidence in any possible scheme of permanent nuclear waste disposal. A common public desire is retrievability of finally disposed wastes in case repository fails to isolate wastes from the live environment. Desire of retrievability is in direct contradiction with the principle of final disposal and adds serious complexities to the problem. Public resistance against nuclear waste repository [39] at Yucca Mountain is a typical example showing the complexities involved. **Figure 13** shows a simplified picture of the Swedish plan for geological disposal of nuclear wastes [40]. Different objects in the figure, showing steps of the disposal procedure, are explained with the inset text. The figure differentiates high radioactivity wastes from low and intermediate radioactivity wastes which require different disposal procedures.

Fukushima disaster (**Figure 14**) on March 11, 2011 and Japan's efforts (**Figure 15**) in collaboration with the whole world in dealing with it offers a knowledge and strategic framework regarding the preparedness for the next possible accident of this nature. Science, engineering, technology, social and political spheres from around

Figure 14. (a) A picture of Fukushima disaster [41] and (b) Japan's immediate response to it [42].

Figure 15. Three pictures [42] showing Japan's systematic response to the Fukushima disaster during which it collaborated with several countries from around the world.

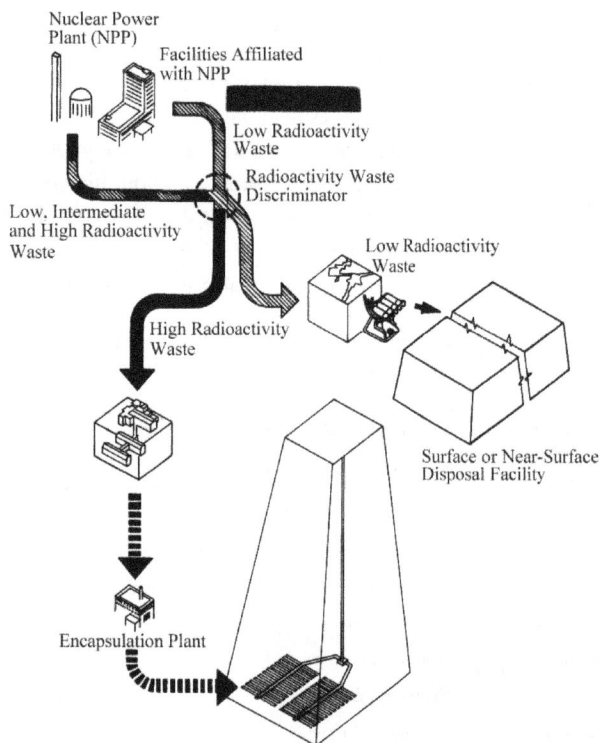

Figure 13. A simplified picture of the Swedish plan for geological disposal of nuclear wastes Thegerström [40].

the world must join together to map the response plan and its adequate execution, possibly of spontaneous nature, in facing such accident anywhere in the world. Such a grand alliance can be a solution to happenings of "type N" as it reduces the cost and the chance of failure. An incident of "type N" is a nuclear event which has a very low happening frequency but a high level of serious implications. There had been dangers in air travelling and electricity supplies, but they are technical safe and sound now. Conscious, careful and continued efforts with the joint wisdom could provide the concrete solution to the challenge of the nuclear safety by assuring the dreamlike reality line of "no failure in the nuclear technology". Putting aside the political games, this noble cause of nuclear safety is doable. Achieving this cause or goal will be a truly remarkable and historical pride for the humanity. Fukushima disaster and responding efforts have revealed strengths, weaknesses and the improvement road map for the nuclear safety.

9. Acknowledgements

Discussions with and help by colleagues Prof. E. U. Khan, Rahman Blocks, International Islamic University, Misri Water Lanes, KASHMIR Highways, H-11/4, Islamabad, Mr. Farooq Jan, Mr. Muhammad Ayub, Mr. Muhammad Ramazan, Dr. Noor Muhammad Butt, Mr. Abdul Ghani, Mr. Neik Emmal, Dr. Parveen Akhter, Mr. Muhammad Akhtar, Ms. Naila Siddique, Ms. Sana Malik, Dr. Mati, Dr. Usman Rajput, Mr. Qamar Abbas and others (Drafts-Men), Mr. Raja A. Ghaffar, the recently deceased colleague Nazir Maseeh, Mr. Yusuf, PIEAS, Islamabad, Dr. H. R. Hoorani, Shahdara GAP Valley, NCP, Islamabad, Mr. Muhammad Zulfiqar, NESCOM, Islamabad, Mr. M. Asad, KRL, Rawalpindi, Dr. Rakhshanda Bilal, SUPARCO, Islamabad, Ms. Bushra Elyas, O-Lab, Islamabad, Prof. Pervez Hoodbhoy, a retired theoretical physicist, QAU, Islamabad, LUMS, Lahore, and MIT, USA, Prof. Mark Breese, Ms. Ren Minqin, Ms. Debbie Seng, Dr. A. A. Bettiol, Dr. Markus Zmeck, National University of Singapore, Dr. Karl-Heinz Schmidt, GSI, Germany, Dr. Dietrich Hermsdorf, Dresden, Germany, Prof. Tony Peaker, University of Manchester, UK, and Miss Raaz, The National Institute of the Rehabilitation, PIMS, Islamabad, are appreciated and gratefully acknowledged. I am thankful to Mr. Paul Murray from AREVA Federal Services LLC, 7207 IBM Drive, Charlotte, NC 28262, for his comments on the manuscript and grateful to Dr. Lin Shao, from Texas A & M University, USA, for permission to adopt his plot as a figure in this manuscript. Many thanks to Prof. Faleh Abu-Jarad, Energy Research laboratory/Research Institute King Fahd University of Petroleum and Minerals Dhahran-31261, Saudi Arabia, Dr. Yajing Fu, Shanghai Institute of Applied Physics, Chinese Academy of Sciences, Shanghai 201800, China, Prof. Hideo Nakajima, The Institute of Scientific and Industrial Research, Osaka University, Ibaraki, Osaka, Japan, Prof. Harry J. Whitlow, Lund University and Institute of Technology, Lund, Sweden, and University of Jyväskylä, Finland, Dr. Walter Scandale, CERN, Switzerland, Prof. Walter Greiner, Frankfurt Institute for Advanced Studies, Johann Wolfgang Goethe-Universität, Frankfurt, Germany, Dr. Sandro Scandolo, AS-ICTP, Trieste, Italy, Dr. M. S. AlSalhi & Dr. M. R. Baig, Dept. of Phys. & Astron., King Saud University, Riyadh, Saudi Arabia, for useful exchange of views. I appreciate the inspirations from Okara & Swiss Cheese, Hamd/Naat khawan/show mediators/singers/actors, Qari Khushi Mu-hammad, Umm-e-Habibah, Qari Waheed Zafar Qasmi, Naheed Akheter, Hadiqa Kiani, Anwar Maqsood, Moeen Akhter, Bushra Ansari, Allan Fakir, Muhammad Ali Shahki, Khayal Muhammad, Musarrat Shaheen, Ghazanfer Ali, Pathaney Khan, Jamal Shah, Mah Noor Balauch, Faryal Gohar, Tahir Naeem, Sher Khan Auditorium, Shankyari Cantt. I dedicate this article to Ms. Zubaida Jalal, an educationist from Kech, Balochistan for her efforts to raise literacy in Pakistani women, Urdu poets Amjad Islam Amjad & Iftikhaar Arif due to their heartfelt poetry, Charles Dickenson, the author of the serial novel the "Great Expectations," my teachers Mr. Ahmed Tallat Fatami, Pakistan Land-Air-Water Nuclear Safety Methodologies' Initiative (PAK-LAW-NSMI), PAEC, Islamabad, Prof. Frank Watt, National University of Singapore, and Dr. S. Bashir, Dr. Fawaad & Mr. Faizan-ul-Haq (CASP, GCU, Lahore), and Singers/Actors Noor Jahan, Nabeel Bulbula, Arif Lohar, Salman Khan & Shakira. "Mr. Syed Qamar Hasnain, KANNUP/KINPOE, Karachi, Ms. Sabiha Bakhtiyar, INNUP, Islamabad, Dr. Sabiha Mansoor Sahiba, LCWL. Professional efforts by the personals from the National Nuclear Security System (NSS) which includes selections from PAEC, KRL, NESCOM, SUPARCO, PNRA, SPD, Pakistan Burri Fauj, Pakistan Fizaea, Pakistan Navy and Pakistan Services agencies are heartedly realized/appreciated. Very special thanks to the Security Guards and the Cleaning Examination Staff of NSS for help in organizing a small experiment on the safety check of a security container/box. Support from UN/IAEA in the provision of the Literature and related matters are acknowledged. Continued support from my parents (Mr. Master Saeed Ahmed & Ms. Sughra), Parents in Law (Mr. Subaidar Nazir Ahmed & Ms. Khaleida), my wife Ms. Shamila & her close friend Naurina and Dr. Maqbool Ahmed Bhatti is also heartly appreciated. Discussion with friends Mr. M. Akhtar (DCS) & S. Usman, Mr. Safdar Kayani and Mr. Tariq Azeem (IAD) are thankfully acknowledged. Inspiration from poets Muhammd Iqbal (Sialkot) & Faiz Ahmad Faiz (Narowal), John Nash, US, Pakistani Nobel Laureate A. Salam, an Egyptian born Nobel laureate Ahmed Hassan Zewail, the founder of Pakistan M.A. Jinnah, Anwar (NILOPE) & Dr. Hafiz Faisal (ICCC), Ms. Fariha Malik, Azaan N. Khan and a close friend Mr. Abdullah (Faisalabad) is recognized. Struggle/dedication by Mr. Ghulaam Ali, Nelson Mandela, boxer Muhammad Ali, Michael Jackson, Maddona, Singers Bismillah & Kaley Khan, my school teachers Munir & Afzal (Zafarwali), Pir Mehar Ali Shah, Mr. Abdul Karim, Mr. Riaz Librarian, Javed Bashir (JB), Ms. Eva (NUS). Dr. Mariyam Giorgini, Badar Mian Qawaal, actor Omar Sharif, actresses Meera, Rani & Rekha

(Ida by Mirza Ruswa), and Sir Ganga Rama is appreciated. Help of the WJNST Editorial Board & Staff Members for waving off the publication fee and the help in the improvement of figures and composing of the manuscript is very thankfully acknowledged. Possible discovery of Higgs Boson (HB) and/or HG like particle at CERN is a source of strength for long standing motivations in science." Useful interactions with several young/enthusiastic and ever-cooperative colleagues, friends Aziz, Tanwir, Tabarak, Nisar, Ayaz, Mumtaz, Allah Ditta, Islam, Aurangzeb, Fozia Imran, Najamul-Haq, Asma Latif, Zahid Munir, Atif Raza, Naveed, Dr. I. H. Bukhari, Asif Bashir, Rafia Mir, Shazia Saeed, Ms. Wasim Yawar, Miss Ishrat Rehan, Miss Rehana Mukhtar, Shahid Mukhtar, Tariq J. Sulaija, Farhat Waqar, Dr. Shahid Bilal, Syeda Jan, Mr. Shahid Riaz, Tayyab Mehmood, Tariq Mahmood, Atta Muhammad, Waqar Murtaza, Mansoor Sheikh, Ms. Shahida Waheed, Nasir Khalid, Dr. Sohaila Rahman, M. Arshad, Habib-ur-Rahman, Muhammad Siddiq, Syeda Sahar Rizvi, Athar Saeed, Yasir Faiz, Saadia Zafar Bajwa, Sumaira Naz, Hassan Waqas, Mudassir, Asif Shah, Zafar Yasin, Qamar-ul-Haq, Saira Butt, Nadeem Yaqoob, Ayesha Yameen, Jawaria Abid, Sajjad Mirza, Sh. Hussain, S. Hussain, Dr. N. Ali, Dr. M. I. Shahzad, Dr. N. U. Khattak. Eng. M. Fayyaz, Masood Anwar, Gul Sher, Dr. Samina Roohi, Rizwana Zahoor, Saima Tariq, Jamil Tariq, Dr. Khalid Saleem, Dr. Tabinda, Dr. Samina Gul, Farina Kanwal, Irum Mehboob Raja, Farid Khan, Kabul Shah, Saad Maqbool Bhatti, Dr. Shafqat Farooq, Badar Suleman, Dr. Khalid Jamil, Dr. Mansha Ch., Ansar Pervez, Mirza Brothers, Arshad Zia, Bakhtiar Majid, Arshad Zia, Munir Ahmad, Shahid Munir, Waseem Hassan, Imtiaz Rabbani, Imtiaz Abbasi, Athar Farooq, Sajjad Malik, Javaid Irfan, Abdul hameed, M. Javed, Anwar-ul-Islam, Jamshed Cheema, Abdul Hai, Mr. Zaka-ud-Din, Syed Arif Ahmad, Sher Jan, Iqbal Ali Azhar, S. H. Jaffri, Anwar Habib, M. Iqbal, Zaheer Baig, M. Ali, Shahid Mallick, Abdul Mannan, Gulam Nabi, Zia-ul-Hassan, Adnaan Kaiyani, M. M. Ashfaq, Khurshid Alam, M. Majid Azim, Khalid Mahmood, Imran Zaka, A. A. Niazi, Faiq Hanif, Jahangir Haider, Muhammad Naeem, Tariq Saleemi, Zahid Rana, Muhammad Sajid, Eng. Hashim, Qaisar Abbas, Qamar Abbas, Waqas Masood, Zulfiqar Ali, Shaukat Ali, Abbas Ali, Maj. Shabbir Sharif, Lance Naik Muhammad Mahfooz, Pilot Rashid Minhaas, Maj. Tufail Muhammad, Maj. Azeez Bhatti, Sarwar Muhammad Hussain, Capt. Muhammad Sarwar, Maj. Muhammad Akram, Hawaldar alak Jan, Capt. Karnal Sher Khan, Naik Saif Ali Janjua, Naveed Ikram Bhatti, Muhammad Kashiff, Dr. Shafkat Karim, Amjad Nisar, Library, Muhammad Farooq, Nasrullah Khan Qazi, H. A. Khan, N. Ahmad, M. Jahangir, J. I. Akhter, Eng. Nisar Ahmed, Nazar Hussain, A. H. Qureshi, Ms. Sabahat Nasir Ahmad (HP D), Dr. Nasir Ahmad, Luqman Ahmad (NCD), SGs:

Muhammad Zahoor, Basharat Hussain, Muhammad Akram; Muhammad Saeed (Zafarwal), Manzoor Hussain (Narowal), Sohail Ahmad (Baddo Malhi), Ms. Sabira Manzoor, Mukkarram Shah, Shahid Mahmood, Kaleem Haider, Tanvir Akhter, Malik Muhammad Zubair, Sartaj Ali.

Anticipation by several people around the world and in my home country contributed to shaping up my thoughts/ knowledge and even life in multi-disciplinary manners (in direct contact, indirect contact and contact less, and regular, temporary and permanent exposures of/to the world). Some of them include King Abdullah, Yasir Arafat, Sheikh Abdullah bin Zayed Al Nahyan, Queen Elizabeth, Recep Tayyip Erdoğan, Mao Zedong, Mahateer Muhammd, Lee Kuen Yew, Zakir Naik, Mother Treesa, Qari Abdul Basit, Umm-e-Habiba, Muniba Sheikh, John F. Kennedy, Michael Jackson, Jennifer Hudson/ Jennifer Lopez, Imran Khan, Jahangir Khan, Jan Sher Khan, Shahbaaz (senior & junior), Paul O'Connell, Diego Maradona, Recep Tayyip Erdoğan, Fareed Zakaria, Zubaida Khanum, Aesam-ul-Haq. I was benefited by several people at the The Punjab University (PU): Dr. Naseem Shahzad, Dr. Khadim Hussain, Ali Haider, PIEAS, Zafar Iqbal the Quaid-e-Azam University (QAU), the Govt. College University (GCU): Ms. Farzana Ashraf, M. Arshad, Ch. Arshad, Zohra Nazir, Malik A. Ghafoor, Kashiff Ahmad, Mr. Zafar Iqbal, Muhammad Humayun; Dr. Shoaib Ahmad, Church Road CASP (GCU), Dr. Zafar Iqbal, CIIT, Prof. Asghar Qadir, NUST, Col. Tanvir, Zahid, Arshad, Talib JLA, Shankyari, Tajdar Adil (LUMS); National University of Singapore: Prof. S. J. Chua (IMRE), several people at Dar-us-Salaam, Singapore, Dr. Chammika Udalagamma, Mr. Ang Kwak Te, Ms. Yvonne Seah, Prof. Thomas Osipowicz, Ms. Reshmi Rajendern, Mangi, Ms. Hasma Hamza, OSA, Gillman Heights & PGP Residences, Singapore, Prof. Ping Yuen Feng, Dr. Leszek Lewinsky, Dr. J. van Kan, Ms Zhang Fang, Dr. Huang Long; the University of Manchester (UMIST): Prof. A. R. Peaker, Dr. Huda El mubarek, Dr. Frank Podd, Dr. Leszek Majewski and several other people of several disciplines, Mathematics, Chemistry and the UM Central Library (Oxford Road); Manchester Museum, the Royal Northern College of Music, Oxford Road, Manchester. Radiation Damage Workshop (2010), the AS-ICTP, Trieste Italy; John Elis et al. & UA9 collaboration, the CERN, Switzerland; Nuclear Security Summit, Seoul 2012 participants; Prof. L. J. Van Ijzendoorn, Eindhoven University of Technology, The Netherlands, Minaal & Saim, the National University of Ireland (NUI), Prof. W. Ensinger, Institute of Material and Earth Science, Darmstadt University of Technology, Darmstadt, Germany; Alexander M. Taratin, Joint Institute for Nuclear Research, Joliot-Curie 6, 141,980, Dubna, Moscow Region, Russia, Los Almos National Laboratory (LANL): Jorgen

Randrup, Peter Moller; Stepan G. Mashnik; University of Colorodo (UC): Jerry Peterson, University of Aarhu, Denmark (UAD): Prof. Soeren Pape Moeller, Prof. J. U. Andersen; Prof, Hans Henrik Andersen, the Niels Bohr Institute of the University of Copenhagen, Denmark. The acknowledgements are partly to revisit myself in gathering thoughts/knowledge acquired at several stages of my life.

REFERENCES

[1] L. K. Hamdan, J. C. Walton and A. Woocay, "Safety Implication for an Unsaturated Zone Nuclear Waste Repository," *Energy Policy*, Vol. 38, No. 10, 2010, pp. 5733-5738.

[2] D. F. Rucker, M. T. Levitt and W. J. Greenwood, "Three-Dimensional Electrical Resistivity Model of a Nuclear Waste Disposal Site," *Journal of Applied Geophysics*, Vol. 69, No. 3-4, 2009, pp. 150-164.

[3] M. B. Schaffer, "Toward a Viable Nuclear Waste Disposal Program," *Energy Policy*, Vol. 39, No. 3, 2011, pp. 1382-1388.

[4] D. Butler, "Nuclear Power's New Dawn," *Nature*, Vol. 429, No. 6989, 2004, pp. 238-240.

[5] L. Devell, H. Tovedal, U. Bergström, A. Appelgren, J. Chyssler and L. Andersson, "Initial Observations of Fallout from the Reactor Accident at Chernobyl," *Nature*, Vol. 321, No. 6067, 1986, pp. 192-193.

[6] D. Williams and K. Baverstock, "Chernobyl and the Future: Too Soon for a Final Diagnosis," *Nature*, Vol. 440, No. 7087, 2006, pp. 993-994.

[7] M. Peplow, "Counting the Dead," *Nature*, Vol. 440, No. 7087, 2006, pp. 982-983.

[8] J. J. Bevelacqua, "Applicability of Health Physics Lessons Learned from the Three Mile Island Unit 2 Accident to the Fukushima Daiichi Accident," *Journal of Environmental Radioactivity*, Vol. 105, No. 1, 2012, pp. 6-10.

[9] G. Brumfiel, "Chernobyl and the Future: Forward Planning," *Nature*, Vol. 440, No. 7087, 2006, pp. 987-989.

[10] G. A. Cowan, "Scientific Applications of Nuclear Explosions," *Science*, Vol. 133, No. 3466, 1961, pp. 1739-1744.

[11] C. Macilwain, "Out of Sight, Out of Mind?" *Nature*, Vol. 412, No. 6850, 2001, pp. 850-852.

[12] J. Mazeika, R. Petrosius, V. Jakimaviciute-Maseliene, D. Baltrunas, K. Mazeika, V. Remeikis and T. Sullivan, "Long-Term Safety Assessment of a (Near-Surface) Short-Lived Radioactive Waste Repository in Lithuania," *Nuclear Technology*, Vol. 161, No. 2, 2008, pp. 156-168.

[13] J. E. Cantlon, "Nuclear Waste Management in the US: The Nuclear Waste Technical Review Board's Perspective," *Nuclear Engineering and Design*, Vol. 176, No. 1-2, 1997, pp. 111-120.

[14] F. Decamps and L. Dujacquier, "Overview of European Practices and Facilities for Waste Management and Disposal," *Nuclear Engineering and Design*, Vol. 176, No. 1-2, 1997, pp. 1-7.

[15] Y. Maki and H. Ohnuma, "Application of Concrete to the Treatment and Disposal of Radioactive Waste in Japan," *Nuclear Engineering and Design*, Vol. 138, No. 2, 1992, pp. 179-188.

[16] P. Poskas, R. Kilda, V. Ragaisis and T. M. Sullivan, "Impact of Spatial Heterogeneity of Source Term in Near-Surface Repository on Releases to Groundwater Pathway," *Nuclear Technology*, Vol. 161, No. 2, 2008, pp. 140-155.

[17] M. A. Rana, "A New Method for Monitoring the Radiation Damage in Nuclear Waste Containers Using Ion Channeling," *Annals of Nuclear Energy*, Vol. 35, No. 8, 2008, pp. 1580-1583.

[18] A. Verbruggen, "Renewable and Nuclear Power: A Common Future?" *Energy Policy*, Vol. 36, No. 11, 2008, pp. 4036-4047.

[19] J. Giles, "Chernobyl and the Future: When the Price is Right," *Nature*, Vol. 440, No. 7087, 2006, pp. 984-986.

[20] M. Wahlen, C. O. Kunz and J. M. Matuszek, "Radioactive Plume from the Three Mile Island Accident: Xenon-133 in Air at a Distance of 375 Kilometers," *Science*, Vol. 207, No. 4431, 1980, pp. 639-640.

[21] W. C. Sailor, D. Bodansky, C. Braun, S. Fetter and B. van der Zwaan, "A Nuclear Solution to Climate Change?" *Science*, Vol. 288, No. 5469, 2000, pp. 1177-1178.

[22] R. L. Cowan *et al.*, "The ABWR General Plant Description, GE Nuclear Energy," Nuclear Energy, San Jose, 1999.

[23] I. Farnan, H. Cho and W. J. Weber, "Quantification of Actinide α-Radiation Damage in Minerals and Ceramics," *Nature*, Vol. 445, No. 7124, 2007, pp. 190-193.

[24] J. Delay, H. Rebours, A. Vinsot and P. Robin, "Scientific Investigation in Deep Wells for Nuclear Waste Disposal Studies at the Meuse/Haute Marne Underground Research Laboratory, Northeastern France," *Physics and Chemistry of the Earth*, Vol. 32, No. 1-7, 2007, pp. 42-57.

[25] R. C. Ewing, "The Nuclear Fuel Cycle: A Role for Mineralogy and Geochemistry," *Elements*, Vol. 2, No. 6, 2006, pp. 331-334.

[26] E. C. Buck, B. D. Hanson, B. K. McNamara, R. Gieré and P. Stille, "Energy, Waste and the Environment: A Geochemical Perspective," *Geological Society of London*, Vol. 236, 2004, pp. 65-68.

[27] C. Poinssot, C. Ferry, P. Lovera, J. C. Christophe and J.-M. Gras, "Spent Fuel Radionuclide Source Term Model for Assessing Spent Fuel Performance in Geological Disposal. Part II: Matrix Alteration Model and Global Performance," *Journal of Nuclear Materials*, Vol. 346, No. 1, 2005, pp. 66-77.

[28] D. R. Wiles, "The Chemistry of Nuclear Fuel Waste Disposal," Polytechnique International Press, Montréal, 2006.

[29] M. Bun, S. Fetter, J. P. Holdren and B. van der Zwaan, "The Economics of Reprocessing Versus Direct Disposal of Spent Nuclear Fuel," Report DE-FG26-99FT4028, Harvard University, Cambridge, 2003.

[30] M. A. Rana, "A Compound Spike Model for Formation of Nuclear Tracks in Solids," *Nuclear Science Techniques*, Vol. 18, No. 6, 2007, pp. 349-353.

[31] L. C. Feldman, J. W. Mayer and S. T. Picraux, "Materials Analysis by Ion Channeling: Submicron Crystallography," Academic Press, New York, 1982.

[32] I. Farnan, H. Cho and W. J. Weber, "Identifying and Quantifying Actinide Radiation Damage in Ceramics with Radiological Magic-Angle Spinning Nuclear Magnetic Resonance," MRS Symposium Proceedings, Vol. 986, No. 1, 2007, pp. 197-206.

[33] L. Shao, "Toward High Accuracy in Channeling Rutherford Backscattering Spectrometry Analysis," Nuclear Instruments and Methods in Physics Research Section B: Beam Interactions with Materials and Atoms, Vol. 266, No. 6, 2008, pp. 961-964.

[34] S. Goto, H. Kobayashi, Y. A. Hideo, T. Kimura and H. Hayashi, "Corrosion-Resistant Copper Alloy," US Patent No. 4830825, 1989.

[35] M. O. Schwartz, "High Level Waste Disposal, Ethics and Thermodynamics," *Environmental Geology*, Vol. 54, No. 7, 2008, pp. 1485-1488.

[36] D. H. Oughton, "Ethical Values in Radiological Protection," *Radiation Protection Dosimetry*, Vol. 68, No. 3-4, 1996, pp. 203-208.

[37] F. Birol, "Nuclear Power: How Competitive down the Line?" *IAEA Bulletin*, Vol. 48, No. 2, 2007, pp. 16-20.

[38] J. I. Dawson and R. G. Darst, "Meeting the Challenge of Permanent Nuclear Waste Disposal in an Expanding Europe: Transparency, Trust and Democracy," *Environmental Politics*, Vol. 15, No. 4, 2006, pp. 610-627.

[39] A. MacFarlane, "Stuck on a Solution," *Bulletin of Atomic Scientists*, Vol. 62, No. 3, 2006, pp. 46-52.

[40] C. Thegerström, "Down to Earth and Below: Sweden's Plans for Nuclear Waste," *IAEA Bulletin*, Vol. 46, No. 1, 2004, pp. 36-38.

[41] E. Watanabe, B. Lake and M. Kuraishi, "Association for Aid and Relief," Tokyo. http://www.globalgiving.org/japan-updates

[42] D. Cyranoski, "After the Deluge: Japan Is Rebuilding Its Coastal Cities," *Nature*, Vol. 483, No. 7388, 2012, pp. 141-143.

14

Gearbox Scheme in High Temperature Reactor Helium Gas Turbine System

Sheng Liu[1], Xuanyu Sheng[2]

[1]Qinghuangdao School, North East Petroleum University, Qinghuangdao, China
[2]Institute of Nuclear and New Energy Technology, Tsinghua University, Beijing, China

ABSTRACT

Helium Turbine is used in High Temperature Reactor Helium Gas Turbine (HTR-GT) system, by which the direct helium circulation between the reactor and turbine generator system will come true. Between helium turbine and generator, there is gearbox device which reduces the turbine rotation speed to normal speed required by the generator. Three optional gearbox schemes are discussed. The first is single reduction cylindrical gearbox, which consists of one high speed gear and one low speed gear. Its advantage is simple structure, easy to manufacture, and high reliability, while its disadvantage is large volume and misalignment of input and output axle. The second is planetary gear mechanism with static planet carrier. The third is planetary gear mechanism with static internal gear. The latter two gearbox devices have similar structure. Their advantage is small volume and high reduction gear ratio, while disadvantage are complicated structure, many gears, low reliability and low mechanical efficiency.

Keywords: High Temperature Gas Cooled Reactor; Gear Box; Planetary Gear

1. Introduction

The 10 MW High Temperature Gas-cooled Test Reactor (HTR-10) has been built and come critical in 2000. HTR-10 still uses steam generator to transfer decay energy to electric power generator [1,2]. In detail, the coolant helium in primary circuit flows to evaporator, in which water vapor is produced and decay energy is transferred from helium to water vapor. The hot water vapor flows to turbine which drives the electric generator to generate power. High Temperature Reactor Helium Gas Turbine (HTR-GT) system, which will be developed based on HTR-10, will use coolant helium to drive the turbine directly, without intermediate heat transfer circuit. The project will simplify the whole energy converse system and increase the heat efficiency [3].

The thermal circuit for HTR-GT is shown in **Figure 1**. Helium with high pressure, heated in the nuclear reactor core, flows from the reactor to the turbine and drives it. The turbine drives the electric generator and compressors simultaneously. The off-gas from the turbine is still hot and will flow through recuperator, where heat is transferred to helium at high pressure side. Then the off-gas flows through precooler and the temperature is reduced lower. Helium, with low pressure and low temperature, flows through compressor group including inter cooler and is compressed to high pressure. High pressure helium

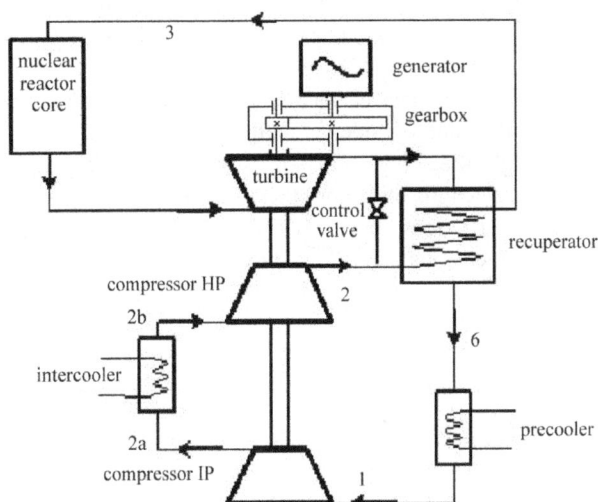

Figure 1. Thermal circuit of Helium turbine power generation system.

flows through recuperator at the high pressure side and is preheated. Finally the preheated high pressure helium flows to the reactor core and is heated again.

From **Figure 1**, it can be seen there is a gearbox between turbine and generator. In Power Conversion Unit (PCU), Turbine and generator are integrated vertical layout system. From bottom to top, there are low pressure compressor, high pressure compressor, turbine, joint coupler 1, gear box (including gear box module, flexible joint coupler 2 and thrust bearing module), and generator. The real position of gearbox in PCU is shown in **Figure 2**.

The function of gear box is transferring the rotation energy of helium turbine to rotor of generator, reducing rotation speed to normal electric network frequency, and supporting the rotor of generator. The rotation speed of helium turbine is 15,000 rpm, and 3000 rpm for generator.

The transfer power of gearbox is 2500 kW. The normal input speed is 15,000 rpm, while highest input speed is 18,000 rpm. The normal output speed is 3000 rpm, while highest output speed is 3600 rpm. The mechanical efficiency is 98%. The leakage ratio for lubricant is <5 g/day. The working atmosphere is helium. The pressure in gearbox is 0.7 MPa, while temperature is 65°C.

The differences between this gearbox and other normal industrial gear box are high rotation speed, high power transfer, and vertical layout. Most of gearboxes are installed horizontally, that is the axle of gear box is parallel to ground. The gearbox in HTR-GT is installed vertically, that is the axle of gear box is perpendicular to ground.

2. Single Reduction Cylindrical Gearbox

The main parts of single reduction cylindrical gearbox are a pair of meshing gears. The input side of high speed gear is connected to the output axle of helium turbine by flexible joint coupler. The high speed gear has small diameter, while low speed gear has large diameter. These two gears are installed side by side vertically. As the power is relatively high, the designed width of two gears is large. To ensure the length of action, there is not tooth in central part. The mate tooth becomes two parts, upper and lower. The 3 dimensional modal is shown in **Figure 3**. The 2 dimensional design drawing is shown in **Figure 4**.

Because the rotation speed is very high, the meshing parts of the two gears are lubricated by oil injection. Special oil supply line and nozzle are installed for direct oil injection on the gears. At the bottom of the gearbox, special oil return line is installed for collecting oil back to the oil station. The oil is filtered and cooled in the special oil station.

At the bottom internal side of gearbox, there is a shoe plate, by which both high speed gear and low speed gear are supported. On the contrary, the upper parts of these two gears are supported directly on gearbox case. There

Figure 3. Meshed gear part of single reduction cylindrical gearbox.

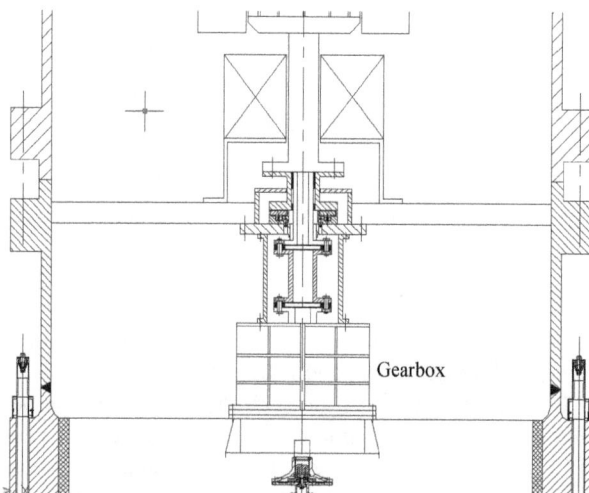

Figure 2. Gearbox device in power conversion unit.

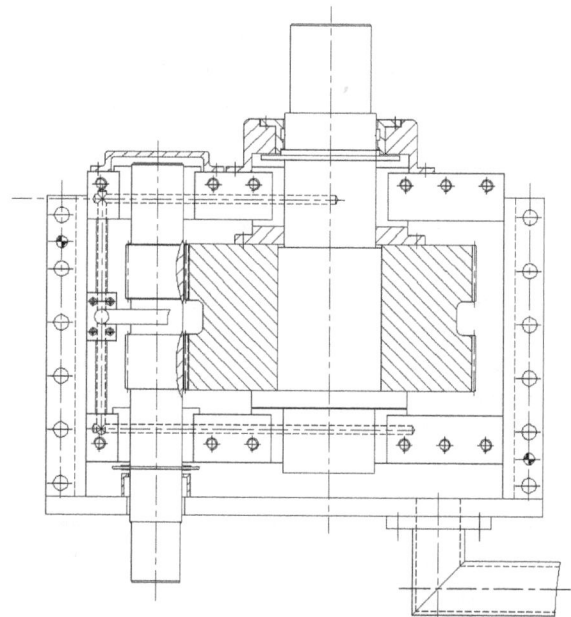

Figure 4. Assembly diagram of single reduction straight gearbox.

are high speed bearings on both ends of the high speed gear. For low speed gear, there are thrust bearings on both ends, which bear the axial force and the weight of low speed gear.

Both high speed gear and low speed gear are skew gear, with upper gear tooth skew opposite to the lower gear tooth. With this kind of skew direction, the axial force of two meshing parts is counteracted.

The whole gearbox case is divided into 2 parts, which are connected by bolts. The contacting area is sealed by special material to prevent lubricant leakage.

3. Planetary Gear Mechanism with Static Planet Carrier

For the planetary gear mechanism with static planet carrier, it is not the real planetary gear train but ordinary gear train in concept because all the gear axles are fixed. Planetary gear mechanism with static planet carrier can be divided into double-geared drive. Because there are 3 planetary gears, the gear mechanism includes 3 double-geared drives which deliver the power together.

The structure of planetary gear mechanism with static planet carrier is shown in **Figure 5**. The whole gearbox case is cylinder which is divided into 2 parts, the upper and lower. There is intermediate shoe plate between these two part cases. All 3 parts, intermediate shoe plate, upper gearbox case and lower gearbox case, are connected together by bolts. The contacted area is also sealed by special sealing material to prevent lubricant leakage. There is shoe plate under the gear box by which whole gear box is fixed in the power conversion unit (PCU).

Figure 5. 3-dimension schematic sketch of planetary gear mechanism with static planet carrier.

To prevent the lubricant leakage from the bottom of high speed gear, a special seal device is installed. There is a shoe pallet, with center part a through hole, under the intermediate shoe plate. The input high speed axle cross the central hole and connected with high speed sun wheel. A pair of high speed rolling bearings is also installed in the central hole to support the high speed axle.

Three axles of planetary gears are equally distributed and all the axles are fixed on static planet carrier. The static planet carrier is connected together with intermediate shoe plate. Three planetary gears mesh with high speed sun wheel on one side, while meshing annular wheel on the other side. The annual wheel rotates and drives output axle rotating together.

4. Planetary Gear Mechanism with Static Internal Gear

Planetary gear mechanism with static internal gear is real planetary gear. The planet carrier rotates, while the annual wheel is fixed. This kind of gear mechanism has relatively high speed reduction rate.

The structure of planetary gear mechanism is shown in **Figure 6**. The gearbox case is also cylinder, divided into 3 parts, upper, intermediate, and lower and connected by bolts. The contacting area is sealed by special material. The input oil line is installed on the upper case, while the output oil line is installed on the lower case. There is a horizontal back plate, supporting the whole planetary gear mechanism, under the lower surface of intermediate gearbox case. There is also a through hole in the horizontal back plate in which the high speed axle crosses. To support the high speed axle, a high speed bearing is installed in the through hole, too. The planet carrier, at the outer side of the high speed axle, is partially installed in the through hole in the horizontal back plate. The outer ring of high speed bearing contacts with the hole of planet carrier. Between the planet carrier and the through hole in the horizontal back plate, another rolling bearing is installed to support the carrier.

The carrier supports 3 evenly distributed planetary wheels. The inner side of 3 planetary wheels meshes with high speed gear, while outer side meshing with annular wheel. The annular wheel is fixed on the horizontal back plate. The carrier is connected with the lower speed output axle which makes them rotating simultaneously.

There is also a through hole on the center of upper gearbox case. The upper end of high speed axle is installed in this through hole, supported by a rolling bearing. To secure enough meshing length, the teeth of the high speed gear, planetary gear and annular gear are all divided into two parts, with intermediate blank.

5. Comparisons of 3 Gearbox Schemes

Single reduction cylindrical gearbox consists one high

(a)

(b)

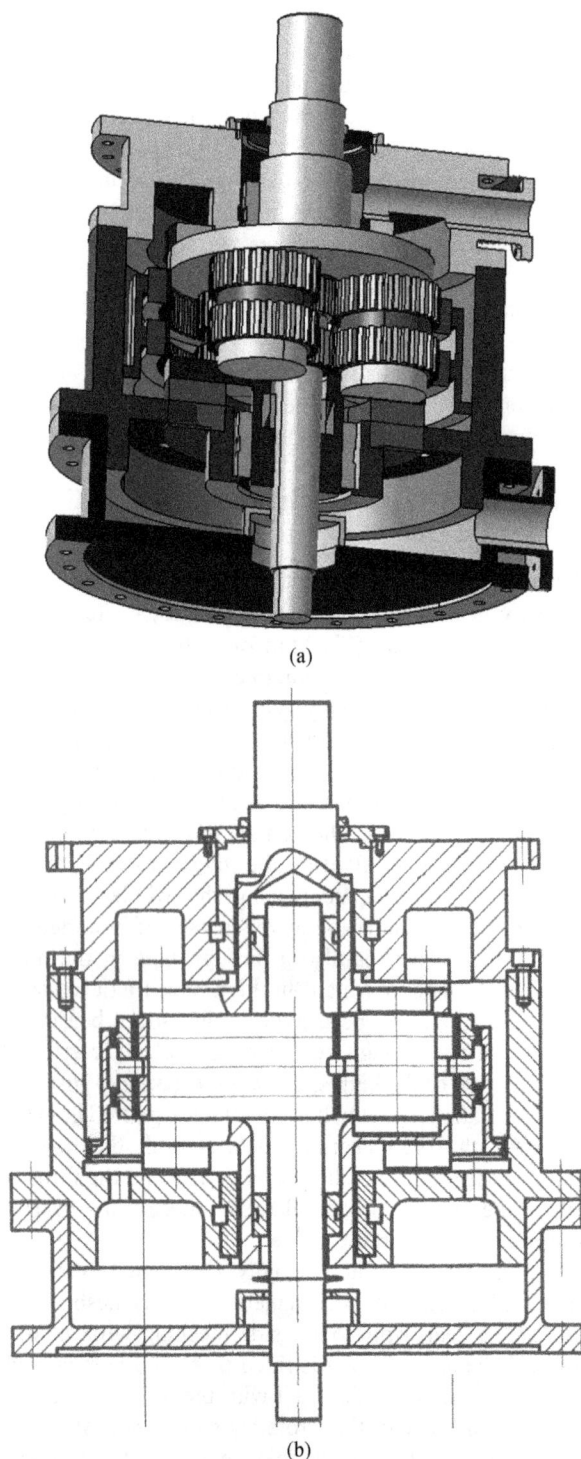

Figure 6. Planetary gear mechanism with static internal gear, (a) 3-dimension schematic sketch; (b) 2-dimension assembly diagram.

speed gear and one low speed gear. The advantage of this kind of gearbox is simple structure, easy manufactured, high reliability because of only two gears. The power lose is lower because the motion transfers between only two gears, which makes single reduction cylindrical gearbox higher efficiency. Because of relatively high reduction ratio, 5, the pith circle diameters of two gear different greats. The diameter of low speed gear is very large which leads to big volume of the gear box. As two axles lay parallel, the input and output axles are not concentricity.

There are similar structures for planetary gear mechanism with static planet carrier and planetary gear mechanism with static internal gear. As their speed reduction ratio can be very high, the volume and weight are small and the structure is compact. The supporter for gearbox in PCU can also be designed easily. The disadvantages of these two kinds of gearbox are more complicated structure, more gears and lower reliability. Any meshing teeth on 5 gears fails, the reactor should be stopped and the gearbox should be repaired or replaced. The maintenance for components in PCU is not an easy task. More gears in motion transfer lead to higher power loss and lower efficiency. Another problem of power loss is that the lost power is converted into heat. This heat increases the temperature of gearbox and lubricant and thus increases the temperature of PCU. The requirement for cooling system of PCU and lubricant would be stricter.

In summary, for 3 gearbox schemes, each has its strong point. The authors prefer to single reduction cylindrical gearbox. One main point of nuclear power plant is high reliability. Designer should try to increase the reliability of nuclear power plant in any kind of possible way. The loss is also very high for each shut-down. To increase the reliability of PCU, single reduction cylindrical gearbox should be used because of fewer gears. Although planetary gear mechanism is small and easy to install, it should be avoided because of its more gears default.

Finally, gearbox scheme in helium turbine system is successful, avoiding the high speed electric generator and high power transducer to reduce the speed. The manufacture fee of PCU is reduced. The shortage of this scheme is the introduction of lubricant subsystem into PCU.

REFERENCES

[1] Z. X. Wu, D. Lin and D. Zhong, "The Design Features of the HTR-10," *Nuclear Engineering and Design*, Vol. 218, No. 1-3, 2002, pp. 25-32.

[2] Z. Y. Zhang and S. Y. Yu, "Future HTGR Developments in China after the Criticality of the HTR-10," *Nuclear Engineering and Design*, Vol. 218, No. 1-3, 2002, pp. 249-257.

[3] Y. L. Sun, Z. K. Zhang and Y. G. Zhang, "Preliminary Study on the HTR-10 Gas Turbine Cycle Design," *High Technology Letters*, Vol. 11, No. 7, 2001, pp. 99-102.

15

Effects of Food Diet Preparation Techniques on Radionuclide Intake and Its Implications for Individual Ingestion Effective Dose in Abeokuta, Southwestern Nigeria

Nnamdi Norbert Jibiri[*], **Tolulope Hadrat Abiodun**

Radiation and Health Physics Research Laboratory, Department of Physics, University of Ibadan, Ibadan, Nigeria

ABSTRACT

The radioactivity measurements in food crops and their diet derivatives and farm soil samples from Abeokuta, one of the elevated background radiation areas in Nigeria have been carried out in order to determine the concentration levels of natural radionuclides (^{40}K, ^{226}Ra and ^{232}Th). The activity concentrations of the natural radionuclides in the samples were determined via gamma-ray spectrometry using a 76 mm × 76 mm NaI(Tl) detector. Different common food crops representing the major sources of dietary requirements to the local population were collected for the measurements. The collected food crops were prepared into their different derivable composite diets using preparation techniques locale to the population. Using available food consumption data and the activity concentrations of the radionuclides, the ingestion effective doses were evaluated for the food crops and diet types per preparation techniques. For the tuberous food crop samples, the annual ingestion effective doses in the raw and different composite diets were 0.02 - 0.04 µSv and cumulatively 0.04 - 0.05 µSv while in the non-tuberous crops the doses were 0.44 - 0.70 µSv and cumulatively greater than 1 µSv respectively. Results of the study indicate that method of diet preparation is seen to play a major role in population ingestion dose reduction especially for tuberous crops than in non-tuberous crops. The study also showed that more ingestion dose could be incurred in diets prepared by roasting techniques. The result of the study will serve as a useful radiometric data for future epidemiological studies in the area and for food safety regulations and policy implementations in the country.

Keywords: Natural Radionuclides; Radionuclide Intake; Gamma Ray Spectroscopy; Food Crops; Radiation Effective Ingestion Dose; Diet Preparation Techniques

1. Introduction

Natural radioactive elements are transferred and recycled through natural processes and between the various environmental compartments by entering into ecosystems and human food chains. The level of terrestrial radiation varies from a geographical location to another depending upon the variation of radionuclide concentration in soil which largely depend on the local geology [1,2]. Foods may be contaminated to some extent as a result of deposition of radionuclides on food crops or on pasture. However, radioactive contamination of the environment may result in the increased radiation exposure of human beings due to ingestion of radionuclides in food. In the chain of transfer processes which leads from the deposition of radionuclides onto soils and plant surfaces to their presence in the diet, food processing and the various methods of food preparation are the last processes which can affect the radionuclide content of foodstuffs. Man's ingestion of radioactive materials may also result from drinking contaminated water in the environment.

Baseline studies of terrestrial outdoor gamma dose rate levels in Nigeria had shown that the areas under investigation exhibited high concentrations of natural radionuclides [1,3]. Also recently an annual effective dose rate of 1.64 mSv due to environmental radiation has been reported for the city [4]. Metal pollution in shallow well water samples in the neighboring town to the study areas has been reported [5], radioactivity in borehole water in neighboring state [6] and in part of Ogun State [7]. It is therefore envisaged that radionuclide concentrations may be elevated in the different food crops and the diets derivable from them. For broad assessment of dose estimate, knowledge of the radiological impact on the effect of

---

[*]Corresponding author.

doses via ingestion in food and water chain as well as the effects of various preparation techniques on radionuclide content of food is therefore required, since the resulting dose from the persistence of a radionuclide in the body is a function of its radiological and biological half lives. Moreover, the carcinogenic nature and long half lives of many radionuclides make them a potential threat to human health. The objective of this work is to investigate the natural radioactivity levels (^{226}Ra, ^{40}K and ^{232}Th) in some selected major food crops and evaluate the effects of varying methods of diet preparation techniques on the radionuclide contents of the food crops and the implication on population radiation ingestion dose.

2. Materials and Methods

2.1. Food Sample Collection and Preparation

Food crops and the soil in which they were grown were collected from some selected farmlands at Abata, Fami, Obakan logun, and Ita osu villages, located respectively in Odeda and Obafemi Owode local government Area of Ogun state (**Figure 1**). Food samples of white yam (*Discorea specie*), cassava (*Manihot utilisima*), Plantain (*Plantago specie*) and maize (*Zea mays*), were collected directly from farmlands in the villages to ensure they were site specific samples. Different samples of the same

food crops were collected from three different farms across the area. Substantial quantity of food samples enough for the analysis was collected. Soil samples were also collected from the same location for analysis. All the food crop samples collected for analysis were native to these areas and represent the common food crop types and choice from where over 85% of daily food/diets needs and requirements of the population are derived. The food samples were washed and prepared fresh shortly after collection into different diets according to the major dietary preferences of the people in the locality, using varying techniques of food preparation. The source of water used for cooking and washing was from the water supply sources in the villages; such as hand dug well, bore hole etc. However, before the preparation of the samples, the non edible parts were discarded such as the peel of yam tuber, cassava as well as the infested parts in some of the food crops. To investigate the possible removal of radionuclide during diet preparation, part of the food samples were prepared using typical local practices, such as the making of yam into boiled and, roasted yam, and yam flour, cassava was made into lafun (*Cassava flour*), Cooked cassava (Amala) Flour and Fried Cassava paste-garri, plantain was prepared roasted and boiled, maize into boiled and roasted maize (**Table 1**). To ensure that the diet preparation represent true local

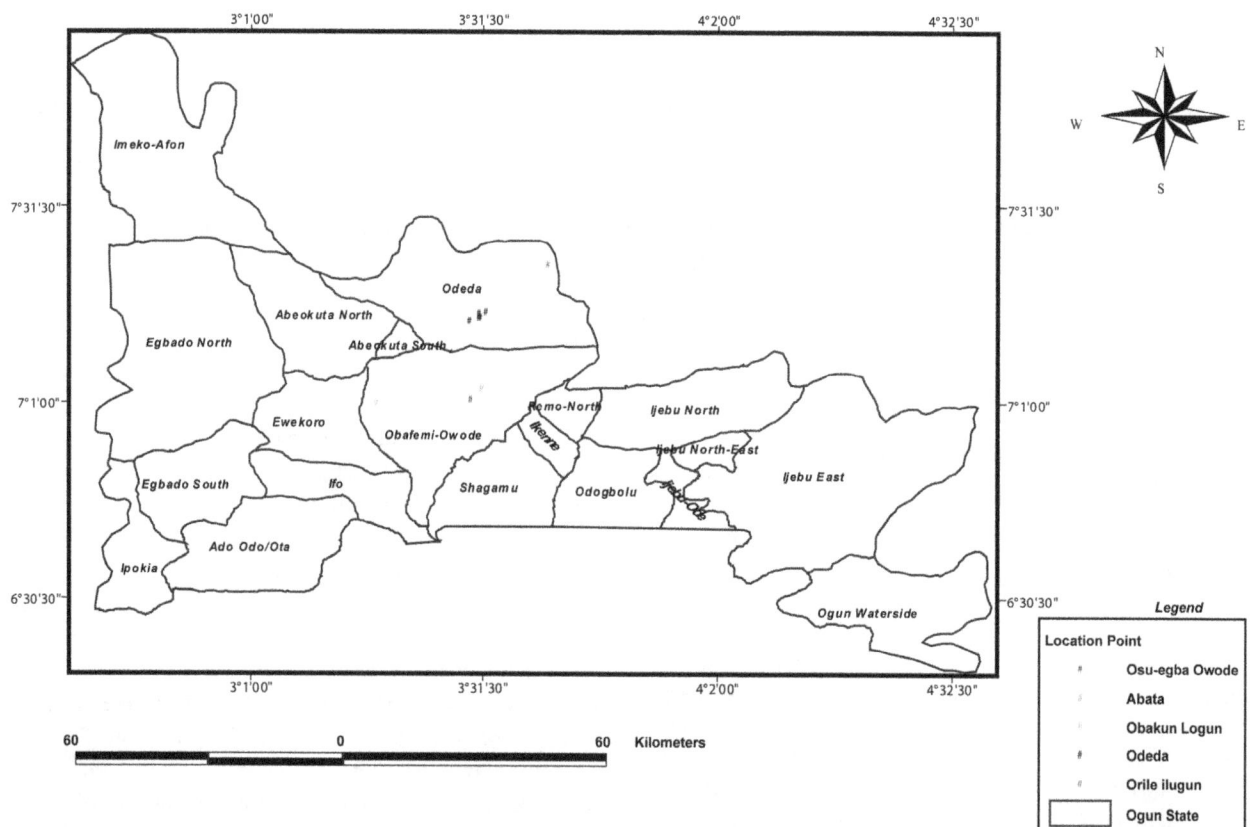

Figure 1. The map of Ogun State Nigeria showing the sample locations.

Table 1. Common food crops, their local diet derivatives and preparation techniques.

Crops	Diet Derivatives (local diets)	Preparation Techniques
Cassava	Garri	Peel the cassava, wash and grind and put in a sack for a day. Apply pressure by placing a jack on the sack for 3 days to remove moisture. Then fry in an empty pan for 30-40 mins to produce coarse grained roasted flour.
	Amala	Peel the cassava, wash and soak in warm water for 3-5 days. Remove from the water and dry in the sun. Grind to become cassava flour (lafun). Then cook with boiled water.
	Fufu	Peel the cassava, wash it clean with water. Soak it for 5 days. Remove and squeeze. Sieve off and make into molds. Then cook with water and stir.
Yam	Isu sisun	Make a charcoal fire, put raw yam on it to roast. Scrap off the burnt peel.
	Iyan	Peel the yam, cook with water and pound with mortar and pestle.
Plantain	Boli	Roasted plantain: Remove the peel or skin. Put on charcoal fire to roast, turn several times to avoid burning. Scrap off the burnt part.
	Ogede sise	Remove the peel of the plantain, cut as desired. Put into pot, cook with water for 30 - 35 minutes.
Maize	Agbado sisun	Remove the corn husk. Put on charcoal fire to roast.
	Langbe	Remove the corn husk. Put into a pot of water. Cook with water for 40 minutes.

food, the diets were prepared by the local women in the study areas. The prepared diets, the raw food crop and soil samples from the raw food stuffs were transported to the laboratory where they were oven dried at a temperature of 100°C to attain a constant weight. At the laboratory, samples after drying were crushed, homogenized and sieved to pass through 2 mm mesh size. They were transferred into uncontaminated empty cylindrical plastic containers of uniform size (60 mm height by 65 mm diameter) and were sealed for a period of 4 weeks. This was done in order to allow for Radon and its short-lived progenies to reach secular radioactive equilibrium prior to gamma spectroscopy.

2.2. Radioactivity Measurements in the Samples

2.2.1. Radioactivity Counting
Radioactivity counting in this work was carried out using a lead-shielded 76 mm × 76 mm NaI(TI) detector crystal (Model No. 802 series, Canberra Inc.) coupled to a Canberra Series 10 plus Multichannel Analyzer (MCA) (Model No. 1104) through a preamplifier. The detector has a resolution of about 8% at 0.662 MeV of ^{137}Cs. This is capable of distinguishing the gamma ray energies considered during these measurements. The Uranium –238 and Thorium –232 activities were determined indirectly through the activities of their daughter products. The choice of the reference nuclides for their activity determination was made based on the fact that NaI(Tl) detector used in this study has a poor resolution, hence, the peaks of interest to be considered would be sufficiently discriminated and intense. Based on this consideration, therefore, the ^{226}Ra content of the samples was determined from the intensity of the 1.760 MeV γ-ray peak from ^{214}Bi, the ^{232}Th content from the 2.615 MeV γ-ray

peak from ^{208}Tl and the ^{40}K content from 1.460 MeV γ-ray peak following the decay of ^{40}K. These peaks are clean, reasonably strongly with very low continuum. Caesium-137 (~30 yr) is a very common fallout radionuclide whose presence is usually an index of environmental pollution due to nuclear accidents and weapon tests. As a check for the presence of this radionuclide, an indication for any radioactive contamination in the environment following past mining activities in the area, a fourth region of interest was created around 0.662 MeV γ-ray peak. Its presence in the environment is inimical to public health as it accumulates in many types of tissues and its penetrating γ-ray reaches all body cells [8].

The samples of both soil and food were placed symmetrically on top of the detector and counting for each sample was done for a period of 10 hr (36,000 s). The net area count under the photopeaks of each of the radionuclide was computed from the memory of the MCA using the firmware algorithm which subtracts counts due to Compton scattering of higher peaks and other background effects from the total area.

2.2.2. Radioactivity Determination
The net area count after background corrections in each photopeak was used in the computation of the activity concentration of each of the radionuclides in the food and soil samples. The activity concentrations in the samples were obtained using the expression [9-11]:

$$C\left(\text{Bq}\cdot\text{kg}^{-1}\right) = \frac{C_n}{\varepsilon P_\gamma M_s} \qquad (1)$$

where C is the activity concentration of the radionuclide in the sample, C_n is the count rate under each photopeak due to each radionuclides, ε is the detector efficiency of the specific γ-ray, P_γ is the absolute transition probability of the specific γ-ray and M_s is the mass of the sample (kg). Equation (1) has been shown to be expressed as [12, 13]:

$$C\left(\mathrm{Bq\cdot kg^{-1}}\right) = \frac{C_K}{A_K}\cdot A \qquad (2)$$

C_k is the activity concentration of the radionuclide in the standard reference sample (Bq·kg^{-1}), A is the net area count after background correction in the spectrum of the radionuclide in the sample and A_K is the net area count after background correction under the spectrum of the radionuclide in the standard reference sample. The standard reference soil sample used was prepared from Rocketdyne Laboratories California; USA which is traceable to a mixed standard gamma source (Ref. No. 48722-356) by Analytic Inc. Atlanta, Georgia. For the food samples the reference standard sample was obtained from International Atomic Energy Agency traceable to source Ref No. IAEA-312.

3. Results and Discussion

3.1. Activity Concentration in the Food Samples

Using Equations (1) and (2), the activity concentration in the farm soil samples, food crops and their diet derivatives were determined. The activity concentration of the ^{40}K, ^{226}Ra and ^{232}Th in the soil samples from the farms where the food crop samples were collected are presented in **Table 2**. The activity concentrations of the radionuclides in the different food crops and their local food derivatives are presented in **Table 3** for Cassava, **Table 4** for Yam, **Tables 5** and **6** for Plantain and Maize respectively.

As could be seen from the **Tables 3** to **6** food preparation techniques were repeated thrice on each food crops for the diet derivable from each of them. Each experiment was carried out for each food crops obtained from different farms in order to determine the consistency of variations in addition or reduction of activity concentrations of the radionuclides for each preparation techniques.

Table 2. Activity concentrations of the radionuclides, absorbed dose rates and annual effective dose rates due to the farm soils from where the raw food crops were collected.

Food Samples	Farm soil samples	^{40}K (Bq·kg^{-1})	^{226}Ra (Bq·kg^{-1})	^{232}Th (Bq·kg^{-1})	Absorbed Dose (nGy/h)	Effective dose (µSv·y^{-1})
Cassava	1	527.29 ± 46.56	6.40 ± 2.03	6.54 ± 0.30	29.05	0.02
	2	550.78 ± 48.60	5.86 ± 1.51	7.01 ± 0.63	30.07	0.02
	3	326.08 ± 29.10	3.13 ± 1.04	5.01 ± 0.23	18.17	0.01
Yam	1	583.07 ± 51.41	4.88 ± 0.28	4.05 ± 0.76	29.19	0.02
	2	428.60 ± 29.29	6.25 ± 0.34	4.43 ± 0.51	23.56	0.01
	3	336.96 ± 30.03	ND	8.35 ± 0.95	19.20	0.01
Plantain	1	836.08 ± 73.49	5.55 ± 0.31	5.87 ± 0.85	31.22	0.02
	2	257.25 ± 23.16	5.35 ± 1.46	7.42 ± 1.92	17.76	0.01
	3	376.96 ± 33.42	3.75 ± 0.22	2.04 ± 0.10	18.76	0.01
Maize	1	831.92 ± 73.11	10.04 ± 3.60	11.83 ± 2.59	46.72	0.02

Table 3. Activity concentrations of the radionuclides in cassava (*Manihot esculenta*) and local diet derivatives and the annual effective dose due to ingestions.

Expt	Food Samples/Diet Derivatives	^{40}K (Bq·kg^{-1})	^{226}Ra (Bq·kg^{-1})	^{232}Th (Bq·kg^{-1})	Effective Dose (µSv·y^{-1})
1	Raw Cassava	452.23 ± 40.02	ND	ND	0.03
	Garri	506.77 ± 44.73	4.09 ± 1.17	ND	0.04
	Amala	463.64 ± 40.99	2.94 ± 0.06	8.19 ± 1.78	0.05
	Lafun	484.83 ± 42.83	4.94 ± 0.28	ND	0.04
2	Raw Cassava	534.01 ± 47.10	8.03 ± 0.42	ND	0.04
	Garri	441.57 ± 39.07	12.05 ± 2.70	3.58 ± 0.41	0.05
	Amala	379.86 ± 33.73	6.26 ± 1.84	0.21 ± 0.02	0.03
	Lafun	431.44 ± 38.19	8.16 ± 0.43	6.64 ± 0.76	0.05
	Raw Cassava	479.87 ± 42.42	2.97 ± 1.02	0.67 ± 0.08	0.04
	Garri	424.76 ± 37.61	5.12 ± 0.28	3.86 ± 0.45	0.04
	Amala	370.31 ± 32.91	10.64 ± 2.99	ND	0.04
	Lafun	429.23 ± 38.00	11.64 ± 1.73	ND	0.04

Local names: Raw cassava—Fried cassava paste—Gari; Boiled cassava flour—Amala; Dried cassava flour—lafun.

Effects of Food Diet Preparation Techniques on Radionuclide Intake and Its Implications for Individual Ingestion
Effective Dose in Abeokuta, Southwestern Nigeria

103

Table 4. Activity concentrations of the radionuclide in yam (*Dioscorea sp*) and diet derivatives and the annual effective dose due to ingestions.

Expt.	Food Samples	^{40}K (Bq·kg^{-1})	^{226}Ra (Bq·kg^{-1})	^{232}Th (Bq·kg^{-1})	Effective Dose (μSv·y^{-1})
1	Raw yam	459.04 ± 40.6	7.23 ± 2.88	2.66 ± 0.31	0.03
	Roasted yam	491.64 ± 43.44	3.07 ± 0.18	ND	0.02
	Boiled yam	473.24 ± 41.84	0.15 ± 0.01	ND	0.02
2	Raw yam	481.42 ± 42.55	1.27 ± 0.08	ND	0.02
	Roasted yam	504.07 ± 44.52	5.87 ± 0.32	ND	0.03
	Boiled yam	426.49 ± 37.55	7.04 ± 0.37	ND	0.02
3	Raw yam	490.67 ± 43.35	9.60 ± 0.49	ND	0.03
	Roasted yam	585.81 ± 51.61	4.64 ± 0.26	ND	0.03
	Boiled yam	465.01 ± 41.13	2.51 ± 0.15	ND	0.02

Local names: Raw Yam—Isu; Roasted Yam—Isu sisun; Boiled yam—Isu sise.

Table 5. Activity concentrations of the radionuclide in plantain (*Plantago sp*) and diet derivatives and the annual effective dose due to ingestions.

Expt.	Food Samples	^{40}K (Bq·kg^{-1})	^{226}Ra (Bq·kg^{-1})	^{232}Th (Bq·kg^{-1})	Effective Dose (μSv·y^{-1})
1	Raw Plantain	424.76 ± 37.64	3.13 ± 0.19	ND	0.44
	Roasted Plantain	470.98 ± 41.65	8.54 ± 1.40	ND	0.53
	Boiled Plantain	503.18 ± 44.43	7.63 ± 0.30	0.99 ± 0.12	0.54
2	Raw Plantain	410.83 ± 36.43	4.60 ± 2.03	4.81 ± 0.56	0.62
	Roasted Plantain	376.68 ± 33.47	6.57 ± 1.86	2.89 ± 0.34	0.53
	Boiled Plantain	423.83 ± 37.55	7.04 ± 0.37	ND	0.47
3	Raw Plantain	472.58 ± 41.78	2.75 ± 0.16	ND	0.48
	Roasted Plantain	417.51 ± 37.00	9.66 ± 0.49	1.94 ± 0.23	0.56
	Boiled Plantain	486.64 ± 43.01	7.93 ± 2.07	ND	0.47

Local names: Raw Plantain—Ogede; Roasted Plantain—Boli; Boiled Plantain—Ogede sise.

Table 6. Activity concentrations of the radionuclide in raw maize (*Zea mays*) and diet derivatives and the annual effective dose due to ingestions.

Expt.	Food type/Diet Derivative	^{40}K (Bq·kg^{-1})	^{226}Ra (Bq·kg^{-1})	^{232}Th (Bq·kg^{-1})	Effective Dose (μSv·y^{-1})
1	Raw maize (Agbado)	386.28 ± 34.28	1.76 ± 0.68	6.55 ± 0.74	0.70
2	Roasted maize (Agbado sisun)	276.10 ± 4.77	4.70 ± 0.77	ND	0.80
3	Boiled maize (Langbe)	403.00 ± 5.73	5.26 ± 0.29	ND	0.59

Addition or reduction of radionuclides in the derivable composite diets could be observed from the tables. It represents the radionuclide concentrations value in the diets with respect to the raw concentration value. As could be seen from **Tables 3-6**, ^{40}K showed more radionuclide addition in the diet derivatives than reduction. Potassium-40 is usually of limited interest because, as an isotope of an essential element, it is homeostatically controlled in the human cells. As a result, the body content of ^{40}K is determined largely by its physiological characteristics rather than by its intake. Of particular radiological interest is ^{226}Ra and ^{232}Th concentrations in the diets. It has been estimated that a large portion of at least one-eighth of the mean annual effective dose due to natural sources can be attributed to the intake of food. The radionuclides in the naturally occurring ^{238}U and ^{232}Th series contribute about 30% to 60% to the internal radiation dose [11,14]. From the tables it is clearly evident that there is radionuclide addition of ^{226}Ra in all the diet derivatives from the crops under consideration while reduction was observed for ^{232}Th except in cassava. The preparation techniques of exposing some of the cassava diet processing to the sun for drying may introduce soil or other debris into the food items hence influencing the radioactivity. Thorium-232 because of its low solubility, it does not biomagnify in terrestrial or aquatic food chains while ^{226}Ra solubility may therefore account for its concentrations in the raw food crops as well as in the diets. It could also be observed that the same preparation technique does not translate to the same amount of radionuclide addition or reduction in the diets.

3.2. Absorbed Dose Rates and Annual Effective Dose in Soil Samples

Absorbed Dose Rates

The absorbed dose rate, D (nGy·h^{-1}) in air at 1 m above the ground level for soils containing the concentrations of the radionuclides measured in the samples is calculated using the equation [14]:

$$D = a \cdot C_{Ra} + b \cdot C_{Th} + c \cdot C_K + d \cdot C_{Cs} \qquad (4)$$

where a is the dose rate per unit ^{226}Ra activity concentration (4.27×10^{-10} Gy·h^{-1}/Bq·kg^{-1}), C_{Ra} is the concentration of ^{226}Ra in the sample (Bq·kg^{-1}), b is the dose rate per unit ^{232}Th activity concentration (6.62×10^{-10} Gy·h^{-1}/Bq·kg^{-1}), C_{Th} is the concentration of ^{228}Th in the sample (Bq·kg^{-1}), c is the dose rate per unit ^{40}K activity concentration (0.43×10^{-10} Gy·h^{-1}/Bq·kg^{-1}), C_K is the concentration of ^{40}K in the sample (Bq·kg^{-1}), d is the dose rate per unit ^{137}Cs activity concentration (0.30×10^{-10} Gy·h^{-1}/Bq·kg^{-1}) and C_{Cs} is the concentration of ^{137}Cs in the sample (Bq·kg^{-1}). Since Cesium-137 was not detected in any of the samples, the last term in Equation (4) was taken as zero. The absorbed γ-dose rates in air are usually related to human absorbed γ-dose in order to assess radiological implications. In assessing the outdoor effective dose equivalent to members of the population, two important factors were considered. The first is a factor that converts the absorbed dose rates (Gy·h^{-1}) in air to human outdoor effective dose rates (Sv·y^{-1}) while the second factor gives the proportion of the total time for which the typical individual is exposed to outdoor or indoor radiation. The United Nations Scientific Committee on the effects of Atomic Radiation [14] has recommended 0.7 Sv·Gy^{-1} as the value of the first factor and 0.2 and 0.8 as for outdoor and indoor occupancy factors, respectively. This second factor implies that the average individual spends only 4.8 hours (about 5 hours per day) outdoors. In this work, only outdoor exposure from γ-rays sources due to the concentrations of the primordial radionuclides in the soil were considered. The effective dose rate resulting from the absorbed dose rate values was calculated using the following relation:

$$E_{air} = TfQD_{air}\varepsilon \qquad (5)$$

where E_{air} is the effective dose rate (μSv·y^{-1}), T is time being 8766 h·y^{-1}, f is the outdoor occupancy factor that corrects for the average time spent outdoors (0.2), Q is the quotient of the effective dose rate and absorbed dose rate in air (0.7 Sv·Gy^{-1}), ε is a factor converting nano (10^{-9}) into micro (10^{-6}) and D_{air} is the absorbed dose rate in air (nGy·h^{-1}). The results of the gamma absorbed dose rates and the corresponding annual outdoor effective doses due to the farm soils are also presented in **Table 2**.

These values were within the normal ranges in the world [14]. These value deviate a little from the anticipated high levels in the soil indicated in previous studies [1,4]. The reason may be attributed to the fact the villages and farmlands considered in the study are located in the remote areas which were far from industrial activities in the town of Abeokuta and its characteristic outcrop of granite rocks known for high concentrations of natural radionuclides. This is reflected on the concentrations of the radionuclide in the food crops and the soil.

3.3. Ingestion Effective Dose Evaluation

Radiation doses obtained due to the intake of food is calculated from the amount of radionuclide deposited on foodstuff, the activity concentration of particular radionuclide in food per unit deposition, the consumption rate of the food products and the dose per unit activity ingested. To summarize, the effective dose H to a certain tissue T due to intake of radionuclide r is given by [15]:

$$H_{T,r} = \sum \left(U^i \times C_r^i\right) \times g_{T,r} \qquad (6)$$

where, i denotes a food group, the coefficients U^i and C_r^i denote the consumption rate (kg·y^{-1}) and activity concentration of the radionuclide r of interest (Bq·kg^{-1}), respectively, and $g_{T,r}$ is the dose conversion coefficient for ingestion of radionuclide r (Sv·Bq^{-1}) in tissue T. For adult members of the public, the recommended dose conversion coefficient $g_{T,r}$ for ^{40}K, ^{226}Ra, ^{228}Th, and ^{137}Cs, are 6.2×10^{-9} Sv·Bq^{-1}, 2.8×10^{-7} Sv·Bq^{-1}, 7.2×10^{-8} Sv·Bq^{-1}, and 1.3×10^{-8} Sv·Bq^{-1}, respectively [16]. Presently, no site specific consumption data exist in the study area and as such we have adopted the country's mean annual consumption rate per capita values (**Table 7**) to enable us calculate the effective dose due intake of the food stuffs using Equation (6).

The results are presented in column 6 of **Tables 3-6** for each of the crops and the diet derivatives. As could be observed from these tables, for the tuberous food crop samples, the annual effective ingestion doses in the raw and different composite diets were 0.02 - 0.04 µSv and cumulatively 0.04 - 0.05 µSv while in the non-tuberous crops the doses were 0.44 - 0.70 µSv and cumulatively greater than 1 µSv respectively.

Table 7. Consumption rate for different food products.

Food stuffs	Consumption rate (kg·yr^{-1})
Cassava	116.6
Yam	76
Plantain	16.539
Maize	20.7

Source: Food Balance Sheet, Nigeria. 2002.

Effects of Food Diet Preparation Techniques on Radionuclide Intake and Its Implications for Individual Ingestion
Effective Dose in Abeokuta, Southwestern Nigeria

105

The coefficient of variation (reduction or addition) in the ingestion effective dose due to diet preparation technique could be determined using the Equation (7) below:

$$K_{i,f} = \frac{D_{T_{Ri,f}} - D_{T_{Di,f}}}{D_{T_{Ri,f}}} \qquad (7)$$

where $K_{i,f}$ is the coefficient of ingestion dose variation in the food crop i for diet type technique f, $D_{T_{Ri,f}}$ is the total ingestion dose in the raw food crop i, and $D_{T_{Di,f}}$ is the total ingestion dose in the diet type of food crop i due to preparation technique f. If $K > 0$ it implies ingestion dose reduction due to preparation technique type f for food type i, while $K < 0$ denotes addition of dose due to ingestion of diet type of food crop i due to preparation technique type f. As could be seen from the tables most of the crops and diet preparation techniques showed that $K > 0$ most especially from ^{226}Ra.

4. Conclusion

The activity concentrations in farm soils and major food crops of dietary importance to the population in Abeokuta and their diet derivatives have been determined via gamma-ray spectroscopy. Tuberous food crops and their diet derivatives resulting from different preparation techniques showed lower population ingestion dose than the non tuberous crops where ingestion doses were greater than 1 μSv·y^{-1}. The radionuclide addition or reduction due to food preparation technique was investigated and results indicate that radionuclide addition was evident in the diets than in the raw food crops. It could also be observed that the same repeated preparation technique does not translate to the same amount of radionuclide addition or reduction in the diets. Results also showed that there is more ingestion dose to the population resulting from addition of ^{226}Ra in diets due to different preparation techniques than from ^{40}K and ^{232}Th.

5. Acknowledgements

The authors wish to thank the International Foundation for Science (IFS) Stockholm, Sweden, for providing the research grant (E-3585-2) used in carrying out this study under her Food Science Programme.

REFERENCES

[1] I. P. Farai and N. N. Jibiri, "Baseline Studies of Terrestrial Outdoor Gamma Dose Rate Levels in Nigeria," *Radiation Protection Dosimetry*, 2000, Vol. 88, No. 3, pp. 247-254.

[2] N. N. Jibiri and O. S. Bankole, "Soil Radioactivity and Radiation Absorbed Dose Rates at Roadsides in High-Traffic Density Area in Ibadan Metropolis, Southwestern, Nigeria," *Radiation Protection Dosimetry*, Vol. 118, No. 4, 2006, pp. 453-458.

[3] N. N. Jibiri, "Assessment of Health Risk Associated with Terrestrial Gamma Radiation Dose Rate Levels in Nigeria," *Environment International*, Vol. 27, No. 1, 2001, pp. 21-26.

[4] I. P. Farai and U. E. Vincent, "Outdoor Radiation Level Measurement in Abeokuta, Nigeria by Thermoluminiscent Dosimetry," *Nigerian Journal of Physics*, Vol. 18, 2006, pp. 121-126.

[5] O. Fasunwon, J. Olowofela, O. Akinyemi, B. Fasunwon and O. Akintokun, "Contaminants Evaluation as Water Quality Indicators in Ago-Iwoye, Southwestern Nigeria," *African Physical Review*, Vol. 2, No. 1, 2008, pp. 10-116.

[6] A. O. Awodugba and P. Tchokossa, "Assessment of Radionuclide Concentrations in Water Supply from Bore-Holes in Ogbomoso land, Western Nigeria," *Indoor and Built Environment*, Vol. 17, No. 2, 2008, pp. 183-186.

[7] O. S. Ajayi and J. Achuka, "Radioactivity in Drilled and Dug Well Drinking Water of Ogun State, Southwestern Nigeria and Consequent Dose Estimates," *Radiation Protection Dosimetry*, Vol. 135, No. 1, 2009, pp. 54-63.

[8] R. A. Sutherland and E. de Jong, "Statistical Analysis of Gamma-Emitting Radionuclide Concentrations for Three Fields in Southern Saskatchewan, Canada," *Health Physics*, Vol. 58, No. 4, 1990, pp. 417-428.

[9] M. K. Akinloye and J. B. Olomo, "The Measurement of the Natural Radioactivity in Some Tubers Cultivated in Farmlands within the Obafemi Awolowo University Ile-Ife, Nigeria," *Nigerian Journal of Physics*, Vol. 12, 2000, pp. 60-63.

[10] I. P. Farai and J. A. Ademola, "Population Dose Due to Building Materials in Ibadan, Nigeria," *Radiation Protection Dosimetry*, Vol. 95, No. 1, 2001, pp. 69-73.

[11] N. N. Jibiri, I. P. Farai and S. K. Alausa, "Activity Concentration of ^{226}Ra, ^{228}Th and ^{40}K in Different Food Crops from a High Background Radiation Area in Bitsichi, Jos Plateau Nigeria," *Radiation and Environmental Biophysics*, Vol. 46, No. 1, 2007, pp. 53-59.

[12] N. N. Jibiri and A. O. Ajao, "Natural Activities of ^{40}K, ^{238}U and ^{232}Th in Elephant Grass (*Pennisetum purpureum*) in Ibadan Metropolis, Nigeria," *Journal of Environmental Radioactivity*, Vol. 78, No. 1, 2005, pp. 105-111.

[13] N. N. Jibiri, I. P. Farai and S. K. Alausa, "Estimation of Annual Effective Dose Due to Natural Radioactive Elements in Ingestion of Foodstuff in Tin Mining Area of Jos-Plateau, Nigeria," *Journal of Environmntal Radioactivity*, Vol. 94, No. 1, 2007, pp. 31-40.

[14] UNSCEAR, "Sources, Effects and Risks of Ionizing Radiation," *Report to the General Assembly*, United Nations Committee on the Effects of Atomic Radiation, New York, 2000.

[15] International Commission on Radiological Protection (ICRP), "Age-Dependent Doses to Members of the Public

from Intake of Radionuclides: Part 5 Compilations of Ingestion and Inhalation Dose Coefficients (ICRP Publication 72)," Pergamon Press, Oxford, 1996.

[16] International Commission on Radiological Protection (ICRRP), "Protection of the Public in Situations of Pro-
longed Radiation Exposure (ICRP Publication 82)," Per-
gamon Press, Oxford, 2000.

Determination of Full Energy Peak Efficiency of NaI(Tl) Detector Depending on Efficiency Transfer Principle for Conversion Form Experimental Values

Mohamed Abd-Elzaher[1]*, Mohamed Salem Badawi[2], Ahmed El-Khatib[2], Abouzeid Ahmed Thabet[1]

[1]Department of Basic and Applied Science, Faculty of Engineering, Arab Academy for Science,
Technology and Maritime Transport, Alexandria, Egypt
[2]Physics Department, Faculty of Science, Alexandria University, Alexandria, Egypt

ABSTRACT

In this work we calibrated the NaI(Tl) scintillation detectors (5.08×5.08 cm^2 and 7.62×7.62 cm^2) and the Full Energy Peak Efficiency (FEPE) for these detectors have been calculated for point sources placed at different positions on the detector axis using the analytical approach of the effective solid angle ratio. This approach is based on the direct mathematical method reported by Selim and Abbas [1,2] and has been used successfully before to calibrate the cylindrical, parallelepiped, and 4π NaI(Tl) detectors by using point, plane and volumetric sources. In addition, the present method is free of some major inconveniences of the conventional methods.

Keywords: Full Energy Peak Efficiency; NaI(Tl) Scintillation; Efficiency Transfer

1. Introduction

Determination of detector efficiency is very important in various scientific and industrial fields. Because the experimental work is tedious and even difficult for extended sources, many researches have been focused on the development of computational techniques to determine these efficiencies. There are three famous methods used in this field, the semi-empirical, the Monte Carlo, and the direct mathematical methods. One of these computational techniques is the efficiency transfer method in which the computation of the detector efficiency for various geometrical conditions is derived from the known efficiency for a reference source-detector geometry. The main advantage of the Efficiency Transfer approach with a point calibration source located at a sufficient distance from the detector is that one may neglect coincidence summing effects and obtain a coincidence free efficiency curve [3]. The efficiency transfer method is particularly useful due to its insensitivity to the inaccuracy of the input data, e.g. to the uncertainty of the detector characterization [4,5].

Change in efficiency under conditions of measurement different from those of calibration can be determined on the basis of variation of the geometrical parameters of the source-detector arrangement. By calculation, it is possi-

ble to determine the efficiency corresponding to non-point samples and/or different distances. The basic case corresponds to calibration with known efficiency for a point source located at position, P_o, at energy, E, the efficiency can be expressed as:

$$\varepsilon\left(E, P_o\right) = \varepsilon_i\left(E\right) \cdot \Omega_{eff}\left(P_o\right) \qquad (1)$$

where, $\varepsilon_i\left(E\right)$, represents the intrinsic efficiency of the detector for energy, E, and, $\Omega_{eff}(P_o)$, is the solid angle subtended by point, P_o, and the active surface of the detector, this geometrical factor must include absorbing factors, taking into account the attenuation effects in the materials between the source and the active part of the crystal [6].

For a point source located at a different distance, P, the efficiency can be written, in a similar manner, as:

$$\varepsilon\left(E, P\right) = \varepsilon_i\left(E\right) \cdot \Omega_{eff}\left(P\right) \qquad (2)$$

So we can establish the basic relationship which makes it possible to express the efficiency as a function of the reference efficiency, known at the same energy, E:

$$\varepsilon\left(E, P\right) = \varepsilon\left(E, P_o\right) \frac{\Omega_{eff}\left(P\right)}{\Omega_{eff}\left(P_o\right)} \qquad (3)$$

In general, by knowing the source-detector geometry, we can compute the detector efficiency for different positions using the principle of efficiency transfer by com-

---

*Corresponding author.

puting the relevant solid angle and absorbing factors [7].

2. Mathematical Treatment

Selim and co-workers using the spherical coordinate system derived direct analytical elliptic integrals to calculate the detector efficiencies (total and full-energy peak) for any source-detector configuration [8].

The solid angle, (Ω), subtended by the detector at the source point has been given by Abbas [9], and it is defined as

$$\Omega = \iint_{\theta\ \varphi} \sin\theta \, d\varphi \, d\theta \qquad (4)$$

The effective solid angle is defined as:

$$\Omega_{eff} = \iint_{\theta\ \varphi} f_{att} \cdot \sin\theta \, d\varphi \, d\theta \qquad (5)$$

where, f_{att}, factor determins the photon attenuation by all absorbers between source and detector and it is expressed as:

$$f_{att} = e^{-\sum_i \mu_i \delta_i} \qquad (6)$$

where, μ_i, is the attenuation coefficient of the ith, absorber for a gamma-ray photon with energy, E_γ, and, δ_i, is the average gamma photon path length through the ith absorber.

The location of an arbitrarily positioned axial point source is specified by, (h, θ, φ) where, h, is the source-detector distance, see **Figure 1**, and the polar, θ, and the azimuthal, φ, angles at the point of entrance of the considered surface define the direction of the incidence of a gamma-ray photon. Where, the polar angles can be expressed as, Abbas [9]

$$\theta_1 = \tan^{-1}\left(\frac{R}{h+L}\right) \quad \& \quad \theta_2 = \tan^{-1}\left(\frac{R}{h}\right) \qquad (7)$$

And the azimuthal angles (φ) will be from 0 to 2π, therefore the effective solid angle can be expressed as:

$$\Omega_{eff} = 2\sum_{i=1}^{n=2} Y_i \qquad (8)$$

where:

$$Y_1 = \int_0^{\theta_1}\int_0^{\pi} f_{att} \sin\theta \, d\varphi \, d\theta, \quad Y_2 = \int_{\theta_1}^{\theta_2}\int_0^{\pi} f_{att} \sin\theta \, d\varphi \, d\theta \qquad (9)$$

The previous integrations calculated numerically by using the trapezoidal rule in a BASIC program.

3. Experimental Setup

The Full Energy Peak Efficiency values will determined for NaI(Tl) Scintillation Detector Model number 802-made by Canberra USA in this work two NaI (Tl) scin-

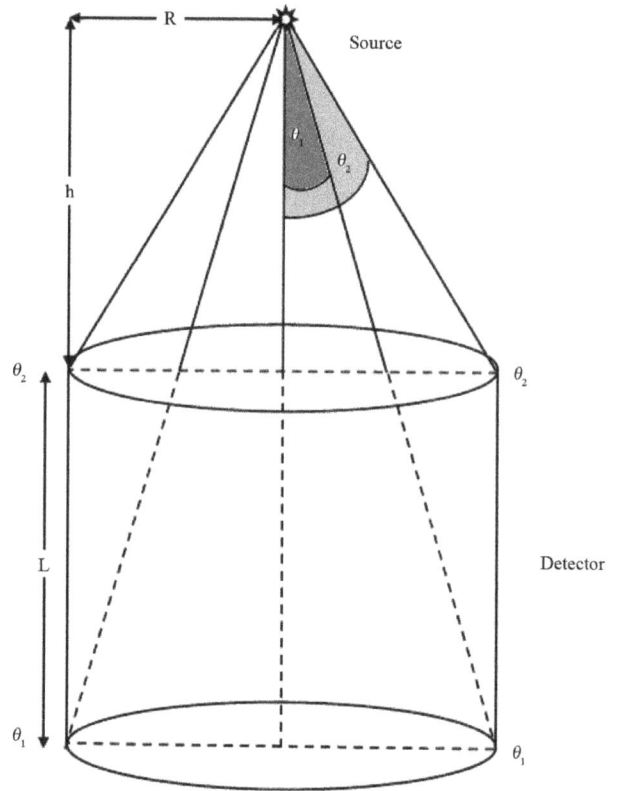

Figure 1. An axial point source with cylindrical detector.

tillation detectors (5.08 × 5.08 cm^2) detector (D1) with resolution 8.5% which specified at the 661 keV, and (7.62 × 7.62 cm^2) detector (D2) with resolution 7.5% which also specified at the 661 keV were used. The details of these detectors setup parameters with acquisition electronics specifications are listed in **Table 1** supported by the (serial & model) number.

In these measurements, the standard point sources ^{241}Am, ^{133}Ba, ^{152}Eu, ^{137}Cs and ^{60}Co where used (these point sources were purchased from The Physikalisch-Technische Bundesanstalt (PTB) in Braunschweig and Berlin). The certificates give the sources activities and their uncertainties for (PTB) sources are listed in **Table 2**. The data sheet states values of half lives, photon energies and photon emission probabilities per decay for the all radionuclides used in the calibration process are listed in **Table 3**, which available from the National Nuclear Data Center Web Page or on the IAEA website.

The calibration process was done by using (PTB) point sources, measured these sources at seven different axial distances starts from 20 cm up to 50 cm by step 5 cm each time from the detectors surface, using the home-made Plexiglas holder which placed directly on the detector entrance window as an absorber to avoid the effect of β- and x-rays and to protect the detector heads, so there is no correction was made for x-gamma coincidences, since in most cases the accompanying x-ray were

Determination of Full Energy Peak Efficiency of NaI(Tl) Detector Depending on Efficiency Transfer Principle
for Conversion Form Experimental Values

109

Table 1. Detectors setup parameters with acquisition electronics specifications for Detector (D1) and Detector (D2).

Items	Detector (D1)	Detector (D2)
Manufacturer	Canberra	Canberra
Serial Number	09L 654	09L 652
Detector Model	802	802
Type	cylindrical	cylindrical
Mounting	vertical	vertical
Resolution (FWHM) at 662 keV	7.5%	8.5%
Cathode to Anode voltage	+1100 V dc	+1100 V dc
Dynode to Dynode	+80 V dc	+80 V dc
Cathode to Dynode	+150 V dc	+150 V dc
Tube Base	Model 2007	Model 2007
Shaping Mode	Gaussian	Gaussian
Detector Type	NaI(Tl)	NaI(Tl)
Crystal Diameter (mm)	50.8	76.2
Crystal Length (mm)	50.8	76.2
Top Cover Thickness (mm)	Al (0.5)	Al (0.5)
Side Cover Thickness (mm)	Al (0.5)	Al (0.5)
Reflector-Oxide (mm)	2.5	2.5
Weight (Kg)	0.77	1.8
Outer Diameter (mm)	57.2	80.9
Outer Length (mm)	53.9	79.4
Crystal Volume (cm^3)	103	347.64

Table 2. PTB point sources activities and their uncertainties.

PTB-Nuclide	Activity (KBq)	Reference Date	Uncertainty (KBq)
^{241}Am	259.0		±2.6
^{133}Ba	275.3		±2.8
^{152}Eu	290.0	00:00 Hr 1. June 2009	±4.0
^{137}Cs	385.0		±4.0
^{60}Co	212.1		±1.5

soft enough to be absorbed completely before entering the detector [10]. The source-detector separations start from 20 cm to neglect the coincidence summing correction.

The spectrum was recorded as example P4D1, where, P, refers to the source type (point) measured on detector (D1) at the distance number (4), which means (h = 20 cm).

The spectrum acquired with winTMCA32 software made by ICx Technologies, were analyzed with (Genie 2000 data acquisition and analysis software) made by Canberra using its automatic peak search and peak area calculations, along with changes in the peak fit using the interactive peak fit interface when necessary to reduce

Table 3. Half lives, photon energies and photon emission probabilities per decay for the all radionuclide's used in this work.

PTB-Nuclide	Energy (keV)	Emission Probability %	Half Life (Days)
^{241}Am	59.52	35.9	157861.05
^{133}Ba	80.99	34.1	3847.91
	121.78	28.4	
	244.69	7.49	
	344.28	26.6	
^{152}Eu	778.9	12.96	4943.29
	964.13	14.0	
	1408.01	20.87	
^{137}Cs	661.66	85.21	11004.98
^{60}Co	1173.23	99.9	1925.31
	1332.5	99.982	

the residuals and error in the peak area values. The live time, the run time and the start time for each spectrum were entered in the spread sheets. Those sheets were used to perform the calculations necessary to generate the experimental full energy peak efficiency (FEPE) curves with their associated uncertainties as a function of the photon energy for all cylindrical NaI(Tl) detectors listed in **Table 1** and with different point sources positions.

The ETNA program used to convert the Full Energy Peak Efficiency (FEPE) curve from point sources at position (P4) to the FEPE at positions (P5, P6, P7, P8, P9 and P10). These calculations extended for two cylindrical NaI(Tl) detectors (D1 & D2).

4. Results and Discussion

This part shows the comparisons between the efficiency transfer theoretical method (ETTM) with the experimental work which is done at Younis S. Selim Laboratory for Radiation Physics, Faculty of Science, Alexandria University. This laboratory uses several coaxial NaI(Tl) scintillation detectors (5.08 × 5.08 cm^2 and 7.62 × 7.62 cm^2) which used in the present work. The detectors were calibrated by measuring low activity point sources, which previously described. The theoretical Full Energy Peak Efficiency (FEPE) can obtain as described in Equation (3).

Another method of calibration is by using ETNA program (an acronym standing for Efficiency Transfer for Nuclide Activity measurements) developed in the Laboratoire National Henri Becquerel (BNM/LNHB) CEA/Saclay, France by Marie Christine [11]. The percentage error between the measured and the calculated efficiencies is given by:

$$\Delta\% = \frac{\varepsilon_{Cal} - \varepsilon_{meas}}{\varepsilon_{meas}} \times 100 \qquad (10)$$

where, ε_{Cal} and ε_{meas}, are the calculated and experimentally measured efficiencies, respectively.

The measured efficiency values as a function of the photon energy, $\varepsilon(E)$, for all NaI(Tl) scintillation detectors were calculated by:

$$\varepsilon(E) = \frac{N(E)}{T \cdot A_S \cdot P(E)} \prod C_i \qquad (11)$$

where, $N(E)$, is the number of counts in the full-energy peak which can be obtained using Genie 2000 software, T, is the measuring time (in second), $P(E)$, is the photon emission probability at energy, E, A_S, is the radionuclide activity and, C_i, are the correction factors due to dead time, radionuclide decay.

In these measurements of low activity sources, the dead time always less than 3%, so the corresponding factor was obtained simply using ADC live time. The statistical uncertainties of the net peak areas were smaller than 1.0% since the acquisition time was long enough to get number of counts at least 10,000 counts. The background subtraction was done. The decay correction, C_d, for the calibration source from the reference time to the run time was given by:

$$C_d = e^{\lambda \cdot \Delta T} \qquad (12)$$

where, λ, is the decay constant and, ΔT, is the time interval over which the source decays corresponding to the run time. The main source of uncertainty in the effi-

ciency calculations was the uncertainties of the activities of the standard source solutions. Coincidence summing effects were negligible in the reference measurement geometries.

The uncertainty in the full-energy peak efficiency, σ_ε, was given by:

$$\sigma_\varepsilon = \varepsilon \cdot \sqrt{\left(\frac{\partial \varepsilon}{\partial A}\right)^2 \cdot \sigma_A^2 + \left(\frac{\partial \varepsilon}{\partial P}\right)^2 \cdot \sigma_P^2 + \left(\frac{\partial \varepsilon}{\partial N}\right)^2 \cdot \sigma_N^2} \quad (13)$$

where, σ_A, σ_P, and, σ_N, are the uncertainties associated with the quantities, A_S, $P(E)$, and, $N(E)$, respectively, assuming that the only correction made is due to the source activity decay.

In order to study the effect of the detector volume, and the source-to-detector distance on the full-energy peak efficiency of NaI(Tl) detectors (D1 and D2), comparing the measured efficiency for different source detector arrangement were done.

For D1 the maximum measured FEPE value of detector measured with point sources placed at P4 and the minimum one which measured at P10. Also we found that D2 obey the same behavior, by comparison between D1 and D2 results we found that D2 FEPE is greater than it for D1, P4D2 has the maximum FEPE and P10D1 has the minimum one as shown in **Figure 2**. This phenomenon related to that, the gamma-ray intensity emanating from a source falls off with the distance according to the inverse square law. In addition to the larger detector in dimensions is the more efficient one.

The full-energy peak efficiency calculated by the pre-

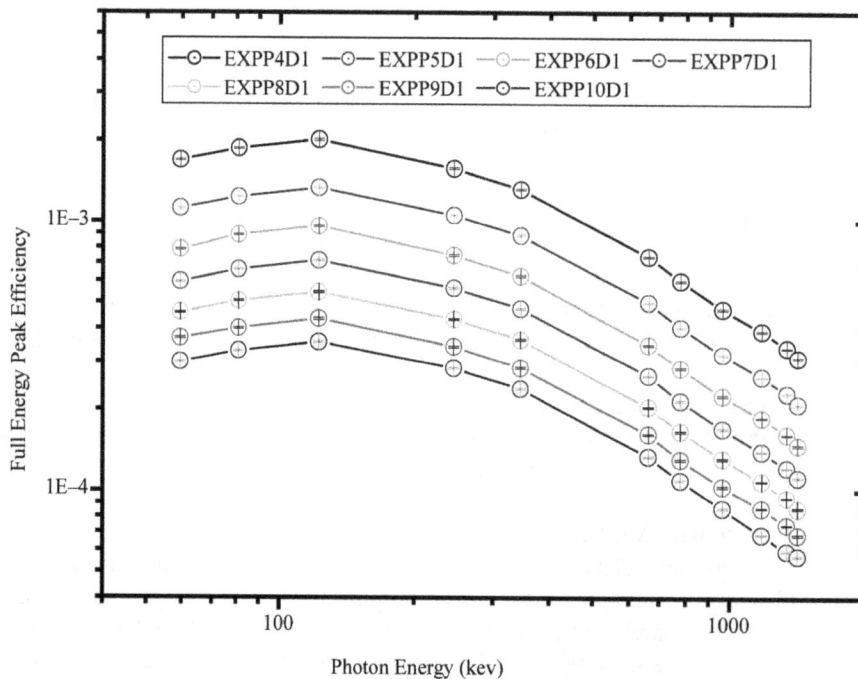

(a)

Determination of Full Energy Peak Efficiency of NaI(Tl) Detector Depending on Efficiency Transfer Principle
for Conversion Form Experimental Values

111

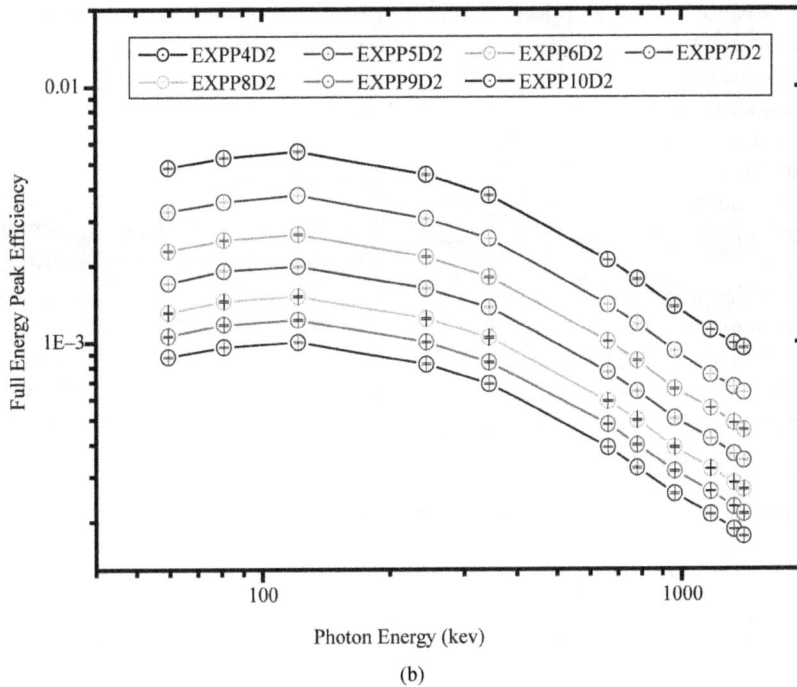

(b)

Figure 2. Comparison between various experimental (FEPE) efficiency results for measured point sources at different positions (P4 up to P10) by using detectors (D1 and D2). (a) Detector (D1) Experimental results; (b) Detector (D2) Experimental results.

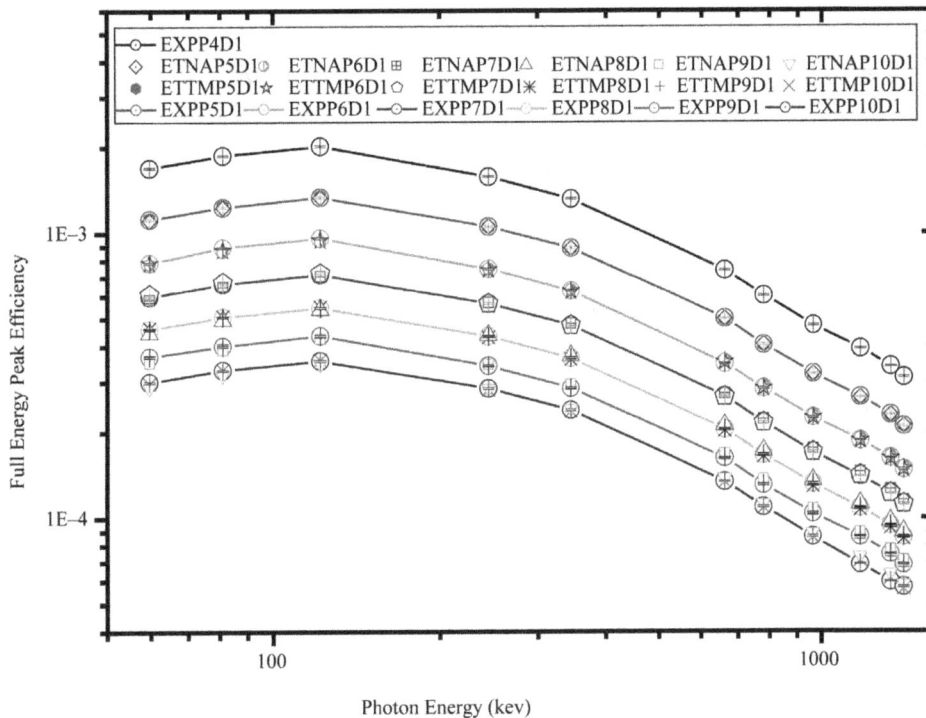

Figure 3. Comparison between ETNA, ETTM, and experimental (FEPE) efficiencies for conversion from point sources at (P4 up to P10) using (D1) detector.

sent ETTM and ETNA program over a wide energy range and have been tested against various data sets obtained by the experimental method using point sources measured from position P4 up to P10 by (D1 and D2), respectively, and good agreement was obtained from the comparisons, see **Figures 3** and **4**.

The efficiency of the detectors is higher at low source energies (absorption coefficient is very high) and decreases as the energy increases (fall off in the absorption coefficient) because the photoelectric is dominant below 100 keV, which mean in other words that it is higher for the bigger detector than the smaller one and it is higher for lower source energy than higher source energy because of the dominance of the photoelectric at lower source energies.

The present work provides a great understanding to several aspects of gamma-ray spectroscopy and will provide us with useful tools ETTM for efficiency calculation for detectors (D1 and D2). This method constitute good approach for the efficiency computation for laboratory routine measurements and can save time in avoiding experimental calibration for different position geometries, where the values of the efficiency calculations using ETTM was compared with the measured ones **Tables 4** and **5**.

5. Conclusion

This work show the way to a simple method (ETTM) to compute the full-energy peak efficiency over a wide energy range, which deal with different detector types using isotropic axial point sources. The present work can be extensive to calculate the FEPE for more complicated geometries. The discrepancies in wide-ranging for all the measurements were found to be less (8%) in case of ETNA program and our ETTM expressions with the

Table 4. Point sources theoretical full energy peak efficiency for D1 (ETTM), and the Discrepancy percentage ($\Delta\%$) with experimental values.

Nuclide	Energy	P5 (Exp)	P5 (ETTM)	$\Delta\%$	P6 (Exp)	P6 (ETTM)	$\Delta\%$	P7 (Exp)	P7 (ETTM)	$\Delta\%$
Am-241	59.53	3.249E–03	3.252E–03	0.099%	2.285E–03	2.295E–03	0.437%	1.707E–03	1.739E–03	1.842%
Ba-133	80.99	3.552E–03	3.555E–03	0.072%	2.521E–03	2.514E–03	–0.309%	1.907E–03	1.903E–03	–0.214%
Eu-152	121.78	3.760E–03	3.753E–03	–0.183%	2.651E–03	2.658E–03	0.283%	1.982E–03	2.010E–03	1.423%
Eu-152	244.69	3.046E–03	3.046E–03	–0.012%	2.158E–03	2.161E–03	0.155%	1.625E–03	1.632E–03	0.411%
Eu-152	344.28	2.549E–03	2.528E–03	–0.819%	1.798E–03	1.795E–03	–0.169%	1.368E–03	1.355E–03	–1.017%
Cs-137	661.66	1.401E–03	1.411E–03	0.704%	1.006E–03	1.003E–03	–0.283%	7.646E–04	7.565E–04	–1.063%
Eu-152	778.9	1.177E–03	1.184E–03	0.596%	8.460E–04	8.422E–04	–0.451%	6.425E–04	6.347E–04	–1.215%
Eu-152	964.13	9.249E–04	9.257E–04	0.088%	6.576E–04	6.588E–04	0.185%	5.030E–04	4.963E–04	–1.327%
Co-60	1173.23	7.452E–04	7.468E–04	0.221%	5.514E–04	5.317E–04	–3.587%	4.195E–04	4.004E–04	–4.542%
Co-60	1332.5	6.679E–04	6.639E–04	–0.598%	4.828E–04	4.728E–04	–2.085%	3.649E–04	3.560E–04	–2.445%
Eu-152	1408.01	6.369E–04	6.368E–04	–0.009%	4.539E–04	4.535E–04	–0.077%	3.448E–04	3.415E–04	–0.963%
Nuclide	Energy	P8 (Exp)	P8 (ETTM)	$\Delta\%$	P9 (Exp)	P9 (ETTM)	$\Delta\%$	P10 (Exp)	P10 (ETTM)	$\Delta\%$
Am-241	59.53	1.311E–03	1.314E–03	0.227%	1.059E–03	1.086E–03	2.560%	8.795E–04	8.980E–04	2.096%
Ba-133	80.99	1.452E–03	1.440E–03	–0.779%	1.175E–03	1.180E–03	0.387%	9.589E–04	9.732E–04	1.495%
Eu-152	121.78	1.516E–03	1.523E–03	0.489%	1.227E–03	1.240E–03	1.110%	1.002E–03	1.021E–03	1.871%
Eu-152	244.69	1.239E–03	1.239E–03	–0.018%	1.000E–03	1.001E–03	0.133%	8.226E–04	8.217E–04	–0.101%
Eu-152	344.28	1.043E–03	1.029E–03	–1.367%	8.345E–04	8.292E–04	–0.633%	6.883E–04	6.794E–04	–1.284%
Cs-137	661.66	5.880E–04	5.752E–04	–2.166%	4.773E–04	4.610E–04	–3.413%	3.887E–04	3.767E–04	–3.087%
Eu-152	778.9	4.948E–04	4.828E–04	–2.436%	3.971E–04	3.865E–04	–2.672%	3.232E–04	3.156E–04	–2.365%
Eu-152	964.13	3.875E–04	3.777E–04	–2.544%	3.136E–04	3.018E–04	–3.754%	2.563E–04	2.463E–04	–3.937%
Co-60	1173.23	3.204E–04	3.048E–04	–4.872%	2.612E–04	2.432E–04	–6.871%	2.128E–04	1.983E–04	–6.797%
Co-60	1332.5	2.821E–04	2.710E–04	–3.928%	2.278E–04	2.161E–04	–5.112%	1.859E–04	1.761E–04	–5.243%
Eu-152	1408.01	2.656E–04	2.600E–04	–2.114%	2.140E–04	2.072E–04	–3.185%	1.748E–04	1.689E–04	–3.395%

Determination of Full Energy Peak Efficiency of NaI(Tl) Detector Depending on Efficiency Transfer Principle for Conversion Form Experimental Values

113

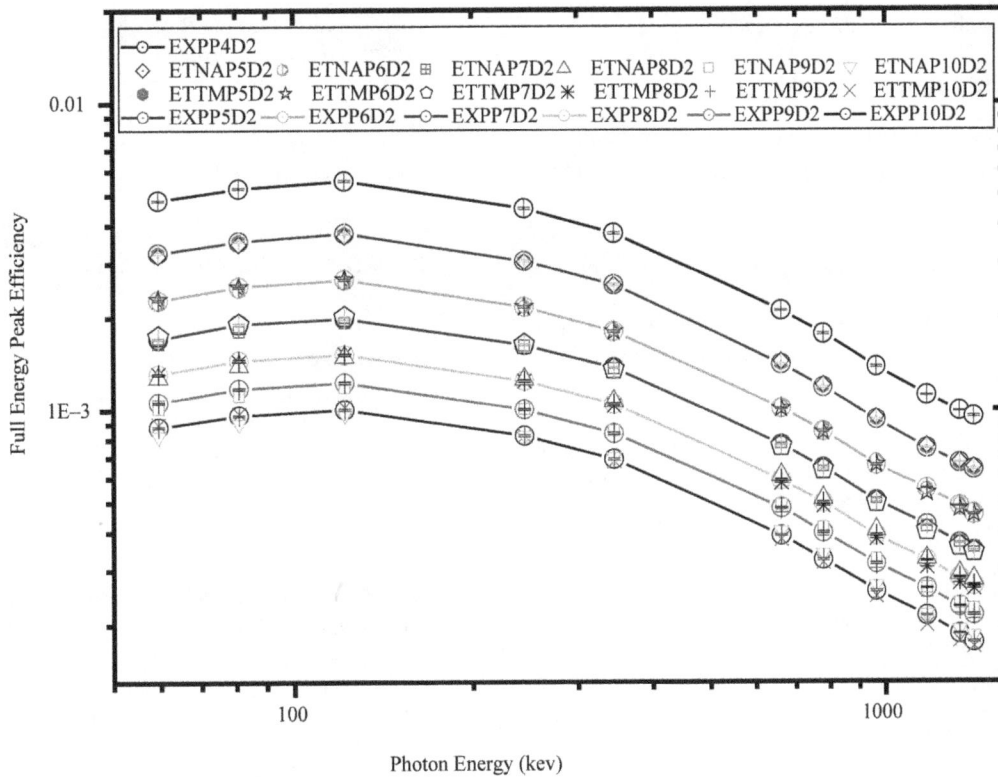

Figure 4. Comparison between ETNA, ETTM, and experimental (FEPE) efficiencies for conversion from point sources at (P4 up to P10) using (D2) detector.

Table 5. Point sources theoretical full energy peak efficiency for D2 (ETTM), and the Discrepancy percentage (Δ%) with experimental values.

Nuclide	Energy	P5 (Exp)	P5 (ETTM)	Δ%	P6 (Exp)	P6 (ETTM)	Δ%	P7 (Exp)	P7 (ETTM)	Δ%
Am-241	59.53	1.122E–03	1.133E–03	0.951%	7.868E–04	7.849E–04	−0.238%	5.963E–04	6.094E–04	2.188%
Ba-133	80.99	1.234E–03	1.250E–03	1.313%	8.911E–04	8.727E–04	−2.068%	6.625E–04	6.715E–04	1.358%
Eu-152	121.78	1.335E–03	1.349E–03	1.058%	9.611E–04	9.453E–04	−1.647%	7.146E–04	7.236E–04	1.270%
Eu-152	244.69	1.054E–03	1.054E–03	0.024%	7.481E–04	7.421E–04	−0.809%	5.644E–04	5.654E–04	0.183%
Eu-152	344.28	8.864E–04	8.825E–04	−0.435%	6.263E–04	6.222E–04	−0.659%	4.738E–04	4.731E–04	−0.144%
Cs-137	661.66	4.999E–04	4.959E–04	−0.800%	3.476E–04	3.507E–04	0.875%	2.668E–04	2.658E–04	−0.398%
Eu-152	778.9	4.038E–04	4.024E–04	−0.351%	2.847E–04	2.848E–04	0.010%	2.153E–04	2.156E–04	0.165%
Eu-152	964.13	3.195E–04	3.166E–04	−0.908%	2.249E–04	2.242E–04	−0.290%	1.695E–04	1.696E–04	0.061%
Co-60	1173.23	2.647E–04	2.620E–04	−1.012%	1.863E–04	1.857E–04	−0.318%	1.396E–04	1.404E–04	0.517%
Co-60	1332.5	2.295E–04	2.268E–04	−1.162%	1.616E–04	1.609E–04	−0.432%	1.217E–04	1.215E–04	−0.116%
Eu-152	1408.01	2.093E–04	2.080E–04	−0.590%	1.473E–04	1.475E–04	0.181%	1.117E–04	1.114E–04	−0.212%

Nuclide	Energy	P8 (Exp)	P8 (ETTM)	Δ%	P9 (Exp)	P9 (ETTM)	Δ%	P10 (Exp)	P10 (ETTM)	Δ%
Am-241	59.53	4.590E–04	4.638E–04	1.047%	3.683E–04	3.733E–04	1.369%	3.004E–04	2.970E–04	−1.122%
Ba-133	80.99	5.067E–04	5.108E–04	0.812%	4.004E–04	4.090E–04	2.152%	3.298E–04	3.307E–04	0.275%
Eu-152	121.78	5.458E–04	5.503E–04	0.821%	4.348E–04	4.390E–04	0.967%	3.557E–04	3.586E–04	0.810%
Eu-152	244.69	4.325E–04	4.297E–04	−0.652%	3.422E–04	3.413E–04	−0.264%	2.847E–04	2.818E–04	−0.997%
Eu-152	344.28	3.634E–04	3.594E–04	−1.108%	2.856E–04	2.850E–04	−0.233%	2.388E–04	2.364E–04	−1.004%
Cs-137	661.66	2.042E–04	2.018E–04	−1.205%	1.619E–04	1.594E–04	−1.548%	1.340E–04	1.334E–04	−0.464%
Eu-152	778.9	1.661E–04	1.637E–04	−1.485%	1.305E–04	1.292E–04	−0.951%	1.090E–04	1.083E–04	−0.661%
Eu-152	964.13	1.313E–04	1.287E–04	−1.950%	1.035E–04	1.015E–04	−1.909%	8.603E–05	8.531E–05	−0.839%
Co-60	1173.23	1.084E–04	1.065E–04	−1.738%	8.598E–05	8.395E–05	−2.363%	6.864E–05	7.067E–05	2.952%
Co-60	1332.5	9.408E–05	9.220E–05	−1.996%	7.456E–05	7.263E–05	−2.593%	5.963E–05	6.122E–05	2.662%
Eu-152	1408.01	8.582E–05	8.455E–05	−1.471%	6.842E–05	6.659E–05	−2.678%	5.705E–05	5.616E–05	−1.562%

experimental values at all energy region.

6. Acknowledgements

The authors would like to introduce a special thanks to Prof. Mahmoud I. Abbas, Faculty of Science, Alexandria University, for fruitful help.

REFERENCES

[1] Y. S. Selim and M. I. Abbas, "Source-Detector Geometrical Efficiency," *Radiation Physics and Chemistry*, Vol. 44, No. 1-2, 1994, pp. 1-4.

[2] Y. S. Selim and M. I. Abbas, "Direct Calculation of the Total Efficiency Cylindrical Scintillation Detectors for Extended Circular Sources," *Radiation Physics and Chemistry*, Vol. 48, No. 1, 1996, pp. 23-27.

[3] V. Tim, B. Vodenik and M. Necemer, "Efficiency Transfer between Extended Sources," *Applied Radiation and Isotopes*, Vol. 68, No. 12, 2010, pp. 2352-2354.

[4] M. C. Lepy, *et al.*, "Intercomparison of Efficiency Transfer Software for Gamma-Ray Spectrometry," *Applied Radiation and Isotopes*, Vol. 55, No. 4, 2001, pp. 493-503.

[5] T. Vidmar, *et al.*, "An Intercomparison of Monte Carlo Codes Used in Gamma-Ray Spectrometry," *Applied Radiation and Isotopes*, Vol. 66, No. 6-7, 2008, pp. 764-768.

[6] F. Piton, *et al.*, "Efficiency Transfer and Coincidence Summing Corrections for Gamma-Ray Spectrometry," *Applied Radiation and Isotopes*, Vol. 52, No. 3, 2000, pp. 791-795.

[7] S. Jovanovic, *et al.*, "ANGLE: A PC-Code for Semiconductor Detector Efficiency Calculations," *Radiation Physics and Chemistry*, Vol. 218, No. 1, 1997, pp. 13-20.

[8] M. S. Badawi, "Faculty of Science," Ph.D. Thesis, Alexandria University, Alexandria, 2009.

[9] M. I. Abbas, "HPGe Detector Absolute Full-Energy Peak Efficiency Calibration including Coincidence Correction for Circular Disc Sources," *Journal of Physics D: Applied Physics*, Vol. 39, No. 18, 2006, pp. 3952-3958.

[10] K. Debertin and U. Schotzig, "Coincidence Summing Corrections in Ge(Li)-Spectrometry at Low Source-to-Detector Distances," *Nuclear Instruments and Methods*, Vol. 158, 1979, pp. 471-477.

[11] M.-C. Lépy, M.-M. Bé and F. Piton, "ETNA (Efficiency Transfer for Nuclide Activity measurements) Software for Efficiency Transfer and Coincidence Summing Corrections in Gamma-Ray Spectrometry," CEA-SACLAY, DA-MRI-LNHB, Bât.602, 91191 GIF-SUR-YVETTE CEDEX, France, 2004.

17

Characterization of ^{137}Cs in Riyadh Saudi Arabia Soil Samples

Abdulaziz S. Alaamer

Department of Physics, Al-Imam Muhammad Ibn Saud Islamic University, KSA

ABSTRACT

The current study was conducted primarily to investigate and estimate ^{137}Cs activity concentrations and the external dose rate due to fallout radionuclide ^{137}Cs. Soil samples were collected from different 25 locations at Riyadh Province and analyzed using low level γ-spectrometry equipped with HPGe-detector. ^{137}Cs activity concentrations and calculated dose rate were found in the range of 0.8 - 3.1 Bq·kg^{-1} and 0.05 to 0.8 nSv·h^{-1} with an average value of 1.70 ± 0.7 Bq·kg^{-1} and 0.11 ± 0.05 nSv·h^{-1} respectively. The measured ^{137}Cs activity concentration range was compared with the reported ranges in the literature from some of the other locations in the world. Results obtained in this study show that ^{137}Cs concentration is of a lower level in the investigated area. However, the range of ^{137}Cs concentrations observed in this study is significantly high relative to similar data reported from Libya. The average value of estimated external effective dose rate is found far below the dose rate limit of of 1.0 mSv·y^{-1} for members of the general public recommended by ICRP as well as the external gamma radiation dose of 0.48 mSv·y^{-1} received per head from the natural sources of radiation assessed by (UNSCEAR, 2000). It is concluded that ^{137}Cs soil contamination does not pose radiation hazards to the population in the investigated areas.

Keywords: ^{137}Cs; Riyadh; Soil; Activity Concentration; Dose Rate; Annual Effective Dose

1. Introduction

Radiation is present in every environment of the Earth's surface, beneath the Earth and in the atmosphere. According to [1] about 87% of the radiation dose received by mankind is due to natural radiation sources and the remaining is due to anthropogenic radiation. The presence of artificial radionuclides in the environment is an important source of radiation exposure for human beings [15]. Artificial radioisotopes may be released into the environment during the testing of nuclear weapons, nuclear explosions and discharge of effluents from nuclear facilities. Artificial radioisotopes released from these sources are retained by environmental materials, including soil [10]. Worldwide contamination from artificial radioisotopes was partially caused by nuclear tests conducted by different countries from time to time and nuclear accidents such as Chernobyl nuclear power plant disaster which took place in 1986.In Chernobyl accident about 3.8×10^{16} Bq of ^{137}Cs was reportedly released into the environment. Activity concentration of ^{137}Cs in soil and emission of gamma rays from ^{137}Cs therefore contributes to the external gamma radiation exposure levels [11]. Measurements of the artificial radionuclides in soil samples provide a better understanding of the causes of

fluctuation in dose rates of environmental radiation in the investigated areas and help tailor public radiation protection programmers on sound scientific bases. This study is part of nationwide programme to survey environmental radioactivity with the aim of building up abroad database on natural and man-made radionuclides for producing a radiation map of the country to be used as a reference in the event of any radiological accident of global dimesion.

2. Materials and Methods

2.1. The Study Area

Riyadh is the capital city of the Kingdom of Saudi Arabia with an area of about 1554 km². It is located centrally in the Najd region with a population that is expected to reach over 5.2 million in 2007, (**Figure 1**). Riyadh is the major part of Riyadh Province. It is situated in the centre of the Arabian Peninsula on latitude 34° - 38° north and longitude 46° - 43° east 600 m above sea level.

2.2. Samples Collection and Preparation

A total of 25 Soil samples were collected from different sites of. Riyadh Province All sampling sites were preferred to be undisturbed and none eroded without any

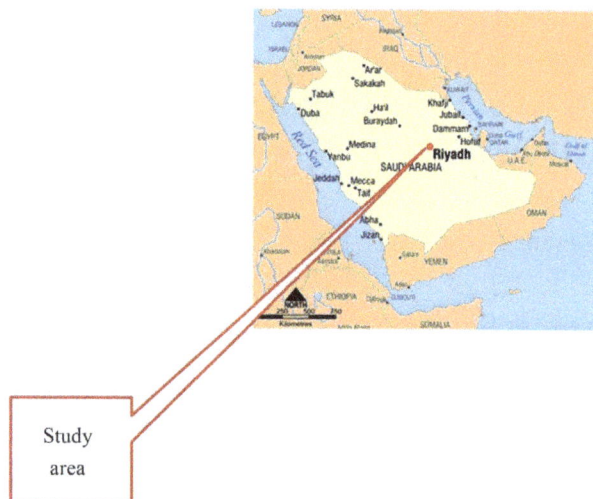

Figure 1. Map Kingdom of Saudi Arabia showing the study area.

influence of man-made structures to ensure that samples were representatives of the sites from where they were taken. From each site, 3 - 5 soil samples were collected from an area of 0.5 m × 0.5 m up the depth of 25 cm. In this way, a total of 105 soil samples were collected from all 25 sites using a clean trowel, placed in plastic bags and labeled. Soil samples were passed through a 2-mm mesh sieve to remove stones and other materials. Samples were then air-dried and sieved through a 1-mm mesh sieve. Each sample containing soil grain weighing about 200 g was stored in the standardized polyethylene containers.

2.3. Gamma-Spectrometric Measurement

^{137}Cs activity concentration was measured using high-purity germanium-detector based gamma ray spectrometer with an efficiency of 25% was employed. HPGe detector was coupled with a Canberra multi-channel analyzer (MCA). The resolution (FWHM) of the spectrometry system was 1.8 keV at 1332 keV gamma-ray line of ^{60}Co. Spectrum of every sample was collected for 54,000 second (15 h). Spectrum analysis was done with help of computer software and activity concentration ^{137}Cs was determined. To reduce the background effect, the detector was shielded in 10 cm wall lead covering lined with 2 mm copper and 2 mm cadmium foils. Standard reference materials obtained from IAEA were used for calibration of the spectrometer. The system was calibrated for energy and efficiency on regular basis. ^{137}Cs activity concentrations were measured directly using their respective photo peaks at 1460 and 662 keV. The environmental gamma background at the laboratory site was determined with an empty plastic container washed with dilute HCl and distilled water. The background was measured under the same conditions of sample measurement. It was later

subtracted from the measured gamma-ray spectra of each sample. The net integral counts under selected photo-peaks were determined by subtracting from the total counts under these photo-peaks the integral count above the baseline over the same region obtained from background runs. The activity concentration of ^{137}Cs was finally measured in Bq·kg^{-1}. The total uncertainty of the measured activity concentration, absorbed dose rate and other indices were within 3% - 8%.

3. Results and Discussion

The activity concentrations of fallout radionuclide ^{137}Cs in soil samples are shown in **Table 1**. The activity concentrations of ^{137}Cs ranged from 0.8 - 3.1 Bq·kg^{-1} with an average value of 1.70 ± 0.69 Bq·kg^{-1}. Spatial distribution of the measured ^{137}Cs concentrations **Figure 2** the results obtained indicate that the location 10, 11, 13, 14, 15 and 25, the concentration of ^{137}Cs is extremely on the lower side having values even less than the mean value of 1.70 ± 0.69 Bq·kg^{-1}. However, as appears from **Figure 2**, activity concentrations of ^{137}Cs vary from location to location and are not uniform. It is well documented in the literature that ^{137}Cs concentration in soil varies due to topographic differences, geomorphology ,and meteorological conditions of the region .Upon comparing the results with global data it was found that the obtained values are far below the reported range from Spain 10 - 60 Bq·kg^{-1} [5], Egypt 1.6 - 19.1 Bq·kg^{-1} [6], Yugoslavia 1.5 - 28.4 Bq·kg^{-1} [7], USA 5 - 58 Bq·kg^{-1} [8], 1.1 - 5.3 Bq·kg^{-1} [4] and Sudan 2 - 26 Bq·kg^{-1} [9]. However, the range of ^{137}Cs concentrations observed in this study is significantly high relative to similar data reported from Libya 0.9 - 1.7 Bq·kg^{-1} study conducted by Shenber [10].

Table 1. Statistical summary of ^{137}Cs activity concentration Bq·kg^{-1} and absorbed dose rate nGy^{-1}.

Distractive	Sample size	Mean ± Std	Minimum	Maximum
^{137}Cs	25	1.70 ± 0.7	0.8	3.1
Dose	25	0.14 ± 0.15	0.05	0.8

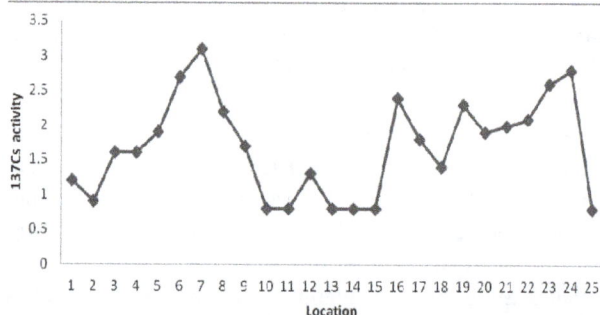

Figure 2. Location-wise distribution of fallout radionuclide ^{137}Cs concentration.

To compute the gamma-ray dose rate owing to external exposure on account of ^{137}Cs activity concentration in the ground soil, consider a smooth and uniform semi-infinite plane made of soil containing ^{137}Cs contamination and a volume element of soft tissue is under the influence of a constant photon flux (φ) originating from ^{137}Cs source concentration in the soil plane. Considering the photon flux due to single prominent energy peak of 661.6 keV with 85.12% abundance, the estimated external effective dose rate in a volume element of soft tissue at 1 m height from the ground is given by using the formula (1) and (2) [8,11,12].

$$D(E) = 0.576 \times E \times \phi(E) \times \left(\frac{\mu a(E)}{\rho}\right)^{tissue} \quad (1)$$

where $D(E)$, in Equation (1), represents external effective dose rate in nSv·h^{-1}, E is gamma photon energy in (MeV), $\phi(E)$ is the mean gamma-ray flux at energy E and $(\mu_{a(E)}/\rho)^{tissue}$ is the energy dependent mass absorption coefficient in cm^2·g^{-1} for a small volume element of soft tissue. For the conservative estimate of the maximum dose on the ground surface owing to source concentration in the soil, the photon flux $\phi(E)$ is defined as under [8,13].

$$\phi(E) = \frac{\gamma FA \times A}{2\left(\left\{\frac{\mu a(E)}{\rho}\right\}^{soil}\right)} \quad (2)$$

In Equation (2), γ_{FR} is fractional abundance of gamma rays, A is experimentally determined activity concentration in Bq·kg^{-1} of a radionuclide and $(\mu_{s(E)}/\rho)^{soil}$ represents the energy-dependent mass attenuation (scattering) coefficient in cm^2·g^{-1} of soil containing gamma-ray emitting radionuclide of energy 661.6 keV. Soils vary considerably in chemical composition, but their relative shielding effectiveness depends mainly on density and water content and, therefore, mass attenuation coefficient varies with different water contents (%) of soil and gamma-photon energy. Mass attenuation coefficient for a typical soil of density 1.625 g·cm^3 containing 30% water content by weight and 20% air by volume against gamma-ray energy of 661.6 keV for ^{137}Cs has been calculated to be 0.0780 cm^2·g^{-1} from the given data available in the literature [14]. The x of constant photon flux considered in the present study, the mass absorption coefficient is determined to be 0.0316 cm^2·g^{-1} at 661.6 keV energy using the reported data [12]. Obviously dose build up factor is not considered here as the same is very small for air thickness of one meter and small volume element of soft tissue. Calculated values of external dose rate falls within the range of 0.05 nSv·h^{-1} to 0.21 nSv·h^{-1} with an average value of 0.11 ± 0.05 nSv·h^{-1} (**Figure 3**).

To avoid the radiation hazards, the dose rate limit recommended by the International Commission on Radio-

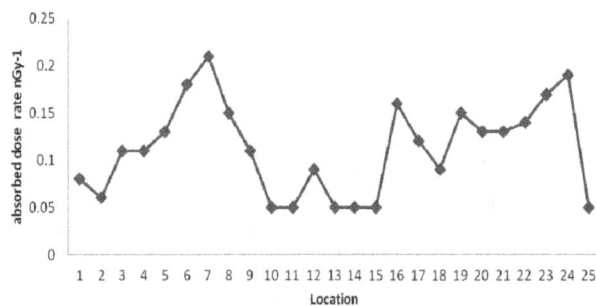

Figure 3. Calculated external dose.

logical Protection (ICRP)for members of the general public is 1 mSv·y^{-1} [15]. The average external gamma-ray dose rate computed in this study is found to be very small as compared with the dose rate limit of 1.0 mSv·y^{-1} as well as the average external gamma dose of 0.48 mSv·y^{-1} received per caput from natural radiation sources assessed by UNSCEAR [16].

4. Conclusion

There is no abnormal elevation seen in the level of ^{137}Cs activity concentrations in the soil of the Riyadh. The values are typical of a normal background level. ^{137}Cs activity concentrations in Riyadh area is lower than the worldwide data .The ranges of ^{137}Cs concentrations in soils are fairly normal compared with those reported for most of the regions of the world. However, the range of ^{137}Cs concentrations observed in this study is significantly high relative to similar data reported from Libya. Further study is necessary in order to draw a detailed radiation map of Saudi Arabia. There is a need to initiate a comprehensive nation-wide program of environmental radioactivity monitoring to bring under regulatory control all the associated sources.

5. Acknowledgements

The authors are extremely grateful to the staff Al-Imam Muhammad Ibn Saud Islamic University, Saudi Arabia for their kind assistance and support.

REFERENCES

[1] UNSCEAR (United Nations Scientific Committee on the Effects of Atomic Radiation), "Sources and Effects of Ionizing Radiation," Report to the General Assembly with Scientific Annexes, United Nations, New York, 1993.

[2] UNSCEAR (United Nations Scientific Committee on the Effects of Atomic Radiation), "Effects and Risks of Ionizing Radiation," Report to the General Assembly, with Annexes, English, Publishing and Library Section, United Nations Office, 1988.

[3] A. Noureddin, B. Baggoura, J. J. Larosa and N. Vajda,

"Gamma and Alpha Emitting Radionuclides in Some Algerian Soil Samples," *Applied Radiation and Isotopes*, Vol. 48, No. 8, 1997, pp. 1145-1148.

[4]   S. N. A. Tahir, K. Jamil, J. H. Zaidi, M. Arif and N. Ahmed, "Activity Concentration of 137Cs in Soil Samples from Punjab Province (Pakistan) and Estimation of Gamma Ray Dose Rate for External Exposure," *Radiation Protection Dosimetry*, Vol. 118, No. 3, 2006, pp. 345-351.

[5]   E. Gomez, F. Garcias, M. Casas and V. Cerda, "Determination of 137 Cs and 90Sr in Calcareous Soils: Geographical Distribution on the Island of Majorca," *Applied Radiation and Isotopes*, Vol. 48, No. 5, 1997, pp. 699-704.

[6]   R. H. Higgy and M. Pimpl, "Natural and Man-Maderadioactivity in Soils and Plants around the Researchreactor of Inshass," *Applied Radiation and Isotopes*, Vol. 49, No. 12, 1998, pp. 1709-1712.

[7]   P. Vukotic, G. I. Borisov, V. V. Kuzmic, N. Antovic, S. Dapcevic, V. V. Uvarov and V. M. Kulakov, "Radioactivity on the Montenegrin Coast, Yugoslavia," *Journal of Radioanalytical and Nuclear Chemistry*, Vol. 235, No. 1-2, 1998, pp. 151-157.

[8]   B. Karakelle, N. Ozturk, A. Kose, A. Varinlioglu, A. Y. Erkol and F. Yilmaz, "Natural Radioactivity Insoil Samples of Kocaeli Basin, Turkey," *Journal of Radioanalytical and Nuclear Chemistry*, Vol. 254, No. 3, 2002, pp. 649-651.

[9]   E. H. Bashier, I. Salih and A. K. Sam, "Gis Predictive Mapping of Terrestrial Gammaradiation in the Northern State, Sudan," Radiation Protection Dosimetry Advance Access, 15 March 2012.

[10]   M. A. Shenber, "Fallout 137 Cs in Soils from North Western Libya," *Journal of Radioanalytical and Nuclear Chemistry*, Vol. 250, No. 1, 2001, pp. 193-194.

[11]   K. Jamil, S. Ali, M. Iqbal, A. A. Qureshi and H. A. Khan, "Measurements of Radionuclides in Coal Samples from Two Provinces of Pakistan and Computation of External Gamma-Ray Dose Rate in Coal Mines," *Journal of Environmental Radioactivity*, Vol. 41, No. 2, 1998, pp. 207-216.

[12]   J. R. Lamarsh, "Introduction to Nuclear Engineering," Addison-Wisley, New York, 1983.

[13]   S. Ali, M. Tufail, K. Jamil, A. Ahmad and H. A. Khan, "Gamma-Ray Activity and Dose Rate of Brick Samples Fromsome Areas of North West Frontier Province (NWFP), Pakistan," *Science of the Total Environment*, Vol. 187, No. 3, 1996, pp. 247-252.

[14]   P. Jacob and H. G. Paretzke, "Gamma-Ray Exposure from Contaminated Soil," *Nuclear Science and Engineering*, Vol. 93, No. 3, 1986, pp. 248-261.

[15]   International Commission on Radiological Protection, "1990 Recommendations of the International Commissionon Radiological Protection," *Annals of the ICRP, ICRP Publication* 60, Vol. 21, No. 1-3, Pergamon Press, Oxford, 1991.

[16]   United Nations Scientific Committee on the Effects of Atomic Radiation, "Sources and Effects of Ionizing Radiation," Report to the General Assembly, New York, 2000.

18

An Approximation for the Doppler Broadening Function and Interference Term Using Fourier Series

Alessandro da C. Goncalves[1], Daniel A. P. Palma[2], Aquilino S. Martinez[1]

[1]Department Nuclear Engineering, Federal University of Rio de Janeiro, Rio de Janeiro, Brazil

[2]Brazilian Nuclear Energy Commission, Rio de Janeiro, Brazil

ABSTRACT

The calculation of the Doppler broadening function $\psi(x,\xi)$ and of the interference term $\chi(x,\xi)$ are important in the generation of nuclear data. In a recent paper, Goncalves and Martinez proposed an analytical approximation for the calculation of both functions based in sine and cosine Fourier transforms. This paper presents new approximations for these functions, $\psi(x,\xi)$ and $\chi(x,\xi)$, using expansions in Fourier series, generating expressions that are simple, fast and precise. Numerical tests applied to the calculation of scattering average cross section provided satisfactory accuracy.

Keywords: Doppler Broadening Function; Fourier Series; New Formulation for the Interference

1. Introduction

The phenomenon of thermal motion of the nuclei inside a nuclear reactor is well represented by the microscopic cross section of the neutron-nucleus interaction through the effect Doppler broadening. The precise determination of the Doppler broadening function and interference term are important for the calculation of the resonance integrals [1,2], self-shielding factors and for corrections of the measurements of the microscopic cross sections with the use of the activation technique [3].

The evaluation of the Doppler broadening function $\psi(x,\xi)$ and of the interference term $\chi(x,\xi)$ have a great importance in the generation of nuclear data and there are several methods for the calculation of both functions. This paper presents a new approximation for interference term applied to the calculation scattering average cross section [4] using expansions in Fourier series. The results have shown satisfactory accuracy and do not depend on the type of resonance considered. In thermally balanced medium at temperature T the velocity of the target nucleus is distributed by the Maxwell-Boltzmann distribution [5] and the expression for the average scattering cross sections is written, using to the one level formalism of Briet-Wigner, as:

$$\bar{\sigma}_s(E,T) = \sigma_0 \frac{\Gamma_n}{\Gamma}\psi(x,\xi) + \sigma_0 \frac{R}{\lambda}\chi(x,\xi) + \sigma_{pot} \quad (1)$$

where, the interference term and the Doppler broadening function are written, according with approximations pro-

posed by Gonçalves *et al.* [6], by:

$$\chi(x,\xi) = 2\int_0^\infty e^{-\frac{w^2}{\xi^2}-w}\sin(wx)\,dw \quad (2)$$

$$\psi(x,\xi) = \int_0^\infty e^{-\frac{w^2}{\xi^2}-w}\cos(wx)\,dw \quad (3)$$

The Equations (2) and (3) can be interpreted as sine and cosine Fourier transforms.

2. Mathematical Formulation

The integrals expressed by Equations (2) and (3) it is possible to find new representations for functions $\psi(x,\xi)$ and $\chi(x,\xi)$ using the Fourier series technique. In order to turn its use easily, Equations (2) and (3) can be re-written as:

$$\begin{Bmatrix}\psi\\\chi\end{Bmatrix}(x,\xi) = \int_0^\infty G(w)e^{-w}\begin{Bmatrix}\cos(wx)\\2\sin(wx)\end{Bmatrix}dw \quad (4)$$

where, $G(w) = e^{-\frac{w^2}{\xi^2}}$.

Analyzing the function $G(w)$ can be observed that it is a continuous and even function, which ensures it has a Fourier series representation. Thus, its Fourier series representation is given by:

$$G(w) = \frac{a_0}{2} + \sum_{n=1}^\infty a_n \cos\left(\frac{n\pi w}{L}\right) \quad (5)$$

where

$$a_0 = \frac{\xi\sqrt{\pi}}{L} erf\left(\frac{L}{\xi}\right) \tag{6}$$

$$a_n = \frac{\xi\sqrt{\pi}}{2L} e^{-\left(\frac{n\pi\xi}{2L}\right)^2} erf\left(\frac{2L + n\pi\xi^2 i}{2\xi L}\right) + \\ + \frac{\xi\sqrt{\pi}}{2L} e^{-\left(\frac{n\pi\xi}{2L}\right)^2} erf\left(\frac{2L - n\pi\xi^2 i}{2\xi L}\right) \tag{7}$$

Replacing the Equation (5) in the Equation (4) and applying the properties of the error functions with an imaginary argument [7], one can write the following expression for the functions $\psi(x,\xi)$ and $\chi(x,\xi)$:

$$\psi_{Fourier}(x,\xi) = \frac{\xi\sqrt{\pi}}{2L(1+x^2)} erf\left(\frac{L}{\xi}\right) + \\ + \frac{\xi\sqrt{\pi}}{L} \sum_{n=1}^{\infty} F_n(x,\xi,L) \, Re\left[Z(\xi,L)\right] \tag{8}$$

$$\chi_{Fourier}(x,\xi) = \frac{\xi\sqrt{\pi}}{L(1+x^2)} erf\left(\frac{L}{\xi}\right) + \\ + \frac{2\xi\sqrt{\pi}}{L} \sum_{n=1}^{\infty} f_n(x,\xi,L) \, Re\left[Z(\xi,L)\right] \tag{9}$$

where,

$$F_n(x,\xi,L) = \frac{\left[(n\pi)^2 + L^2(1+x^2)\right]e^{-\left(\frac{n\pi\xi}{2L}\right)^2}}{L^2(1+x^2)^2 + (n\pi)^2\left(2 - 2x^2 + (n\pi/L)^2\right)} \tag{10}$$

$$f_n(x,\xi,L) = \frac{x\left[L^2(1+x^2) - (n\pi)^2\right]e^{-\left(\frac{n\pi\xi}{2L}\right)^2}}{L^2(1+x^2)^2 + (n\pi)^2\left(2 - 2x^2 + (n\pi/L)^2\right)} \tag{11}$$

$$Z(n,\xi,L) = erf\left(\frac{n\pi\xi^2 i + 2L^2}{2\xi L}\right) \tag{12}$$

Replacing Equations (8) and (9) in Equation (1) one obtains the following expression for the average scattering cross section:

$$\bar{\sigma}_s(E,T) = \frac{\sigma_0\xi\sqrt{\pi}}{2L(1+x^2)} erf\left(\frac{L}{\xi}\right)\left(\frac{\Gamma_n}{\Gamma} + \frac{4Rx}{\lambda}\right) + \\ + \frac{\sigma_0\xi\sqrt{\pi}}{L}\left\{\frac{\Gamma_n}{\Gamma}\sum_{n=1}^{N_{max}} F_n(x,\xi,L)Re\left[Z(n,\xi,L)\right] + \sigma_{pot} \right. \\ \left. + \frac{4R}{\lambda}\sum_{n=1}^{N_{max}} f_n(x,\xi,L)Re\left[Z(n,\xi,L)\right]\right\} + \sigma_{pot} \tag{13}$$

3. Numerical Test

This section contains the results obtained with Equations

(8) and (9) for the calculation interference term and of the Doppler broadening function with $L = \pi$ and $N_{max} = 50$. In order to validate the obtained expression in this paper for $\chi(x,\xi)$ and $\psi(x,\xi)$, a systematic compareson was carried out with the comparison between the method presented in this paper and the 4-pole Padé approximation method, whose functional form is:

$$\psi_{Padé}(\xi,x) = \frac{a_0 + a_2(hx)^2 + a_4(hx)^4 + a_6(hx)^6}{b_0 + b_2(hx)^2 + b_4(hx)^4 + b_6(hx)^6 + b_8(hx)^8} \tag{14}$$

$$\chi_{Padé}(\xi,x) = \frac{2h\left(a_1(hx) + a_3(hx)^3 + a_5(hx)^5 + a_7(hx)^7\right)}{b_0 + b_2(hx)^2 + b_4(hx)^4 + b_6(hx)^6 + b_8(hx)^8} \tag{15}$$

The coefficients in Equations (14) and (15) are given by [8,9].

Figures 1 to **8** show the relative errors for the calculation of $\psi(x,\xi)$ and $\chi(x,\xi)$, using the proposed method paper, Equations (8) and (9), and the 4-pole Padé method, Equations (14) and (15), considering the benchmark results from Gauss-Legendre quadrature method that is well described in the literature [10].

From the **Figures 1** and **2** is possible to see that when the variable ξ increases, keeping the variable x constant, the relative deviations of the Padé approximation increases and are systematically higher than those of the proposed method, Equation (8), in the calculation of the function $\psi(x,\xi)$.

From the **Figures 3** and **4** it is possible to see that when the variable x increases, keeping the variable ξ constant, the relative deviations of the Padé approximation increases and are systematically higher than those of the proposed method, Equation (8), in the calculation of the function $\psi(x,\xi)$.

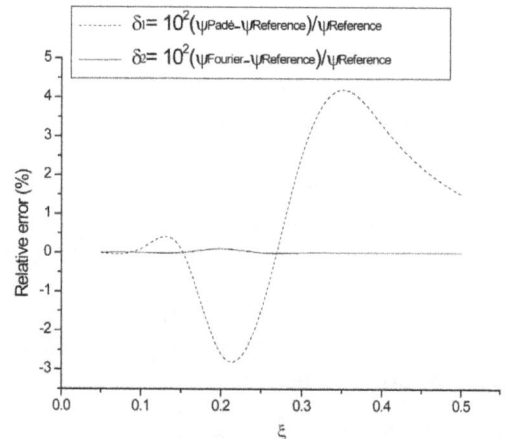

Figure 1. Relative error for the 4-pole Padé approximation, Equation (14), and for the proposed method, Equation (8), for $x = 20$.

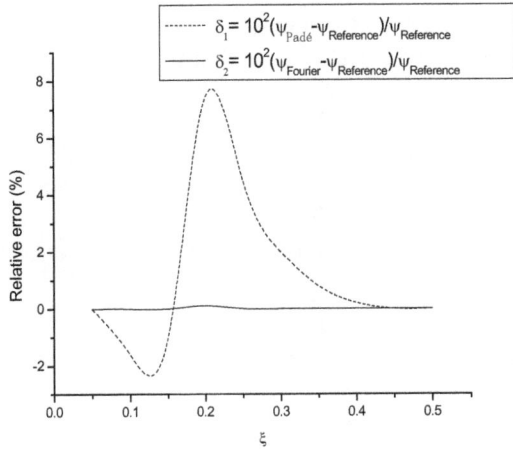

Figure 2. Relative error for the 4-pole Padé approximation, Equation (14), and for the proposed method, Equation (8), for $x = 35$.

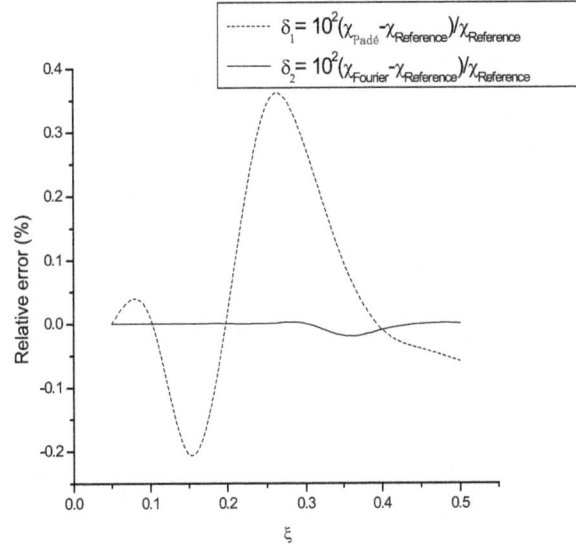

Figure 3. Relative error for the 4-pole Padé approximation, Equation (14), and for the proposed method, Equation (8), for $\xi = 0.10$.

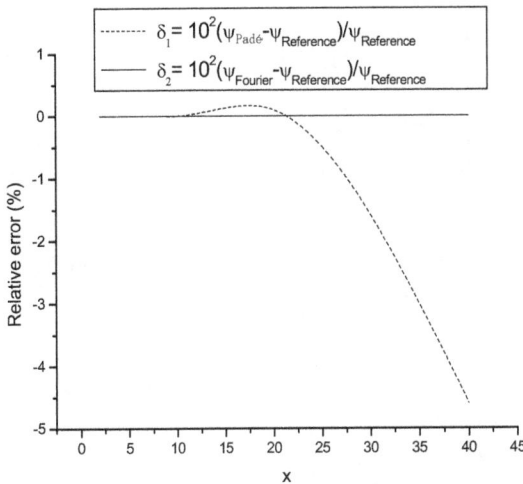

Figure 4. Relative error for the 4-pole Padé approximation, Equation (14), and for the proposed method, Equation (8), for $\xi = 0.20$.

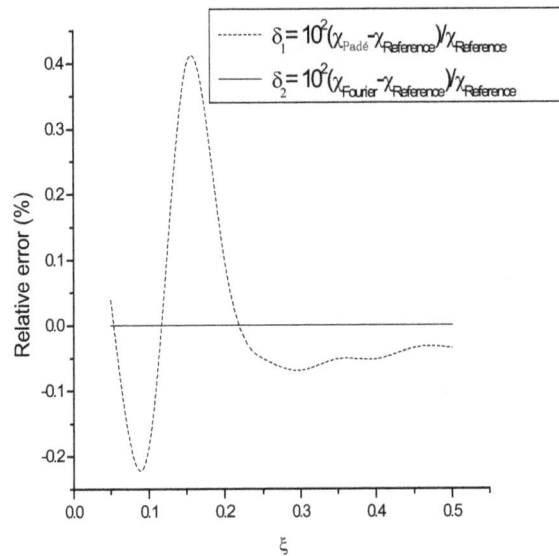

Figure 5. Relative error for the 4-pole Padé approximation, Equation (15), and for the proposed method, Equation (9), for $x = 20$.

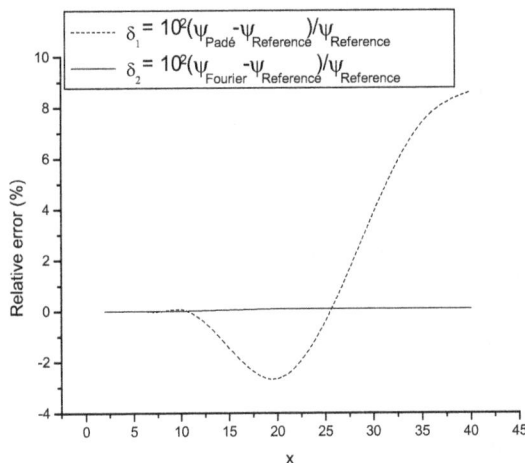

Figure 6. Relative error for the 4-pole Padé approximation, Equation (15), and for the proposed method, Equation (9), for $x = 35$.

From the **Figures 5** and **6** it is possible to see that when the variable ξ increases, keeping the variable x constant, the relative deviations of the Padé approximation increases and are systematically higher than those of the proposed method, Equation (8), in the calculation of the function $\chi(x,\xi)$.

From the **Figures 7** and **8** it is possible to see that when the variable x increases, keeping the variable ξ constant, the relative deviations of the Padé approximation increases and are systematically higher than those of the proposed method, Equation (8), in the calculation of the function $\chi(x,\xi)$.

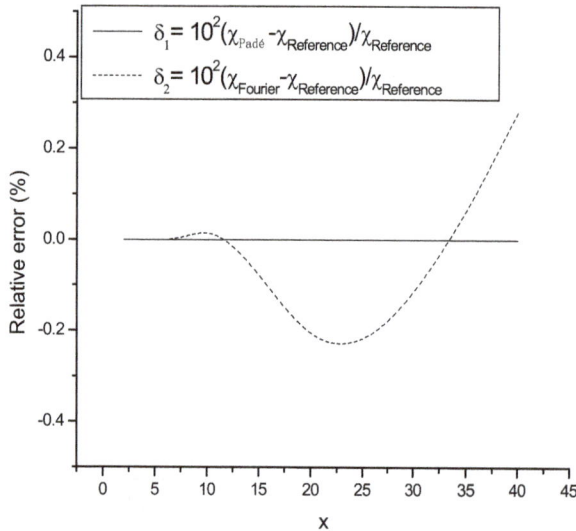

Figure 7. Relative error for the 4-pole Padé approximation, Equation (15), and for the proposed method, Equation (9), for $\xi = 0.15$.

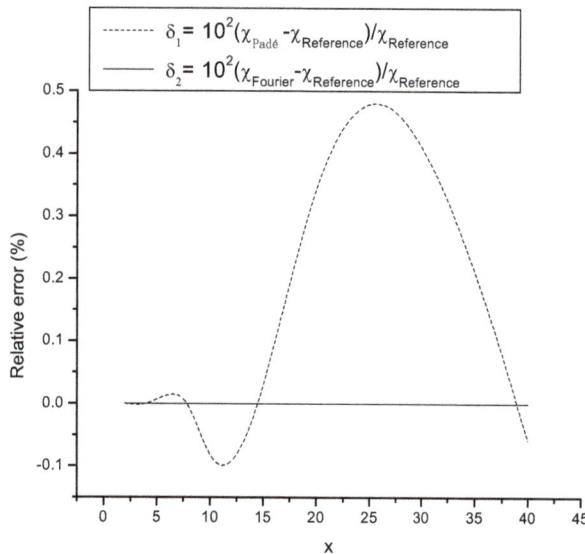

Figure 8. Relative error for the 4-pole Padé approximation, Equation (15), and for the proposed method, Equation (9), for $\xi = 0.25$.

The analysis of the results showed in **Figures 1-8** lead to the conclusion that the proposed method proved to be very precise and stable, having a 0.1% maximum relative error margin, when compared to reference values. From these results is possible to apply the approximate formalism presented in this paper in the calculation of the Doppler broadening function $\psi(x,\xi)$ and the Interference Term $\chi(x,\xi)$ in the determination of the microscopic average scattering cross sections.

4. Results

The average scattering cross section obtained from Equa-

tion (13) are found in **Figures 9-11** and **Figures 12-14** they shows their relative errors for the calculation of the average scattering cross section. The nuclear parameters used can be found in **Table 1** [2].

From the **Figures 9-11** it is possible to see that the results obtained with the method presented, Equation (13), overlapped those obtained from the numerical reference method, being compatible with the results obtained with the method proposed by Padé.

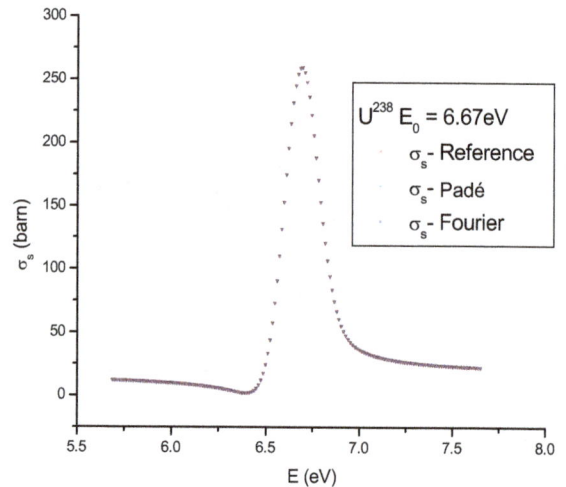

Figure 9. Average scattering cross sections of the $E_0 = 6.67$ eV resonance for the ^{238}U isotope and T = 1500 K.

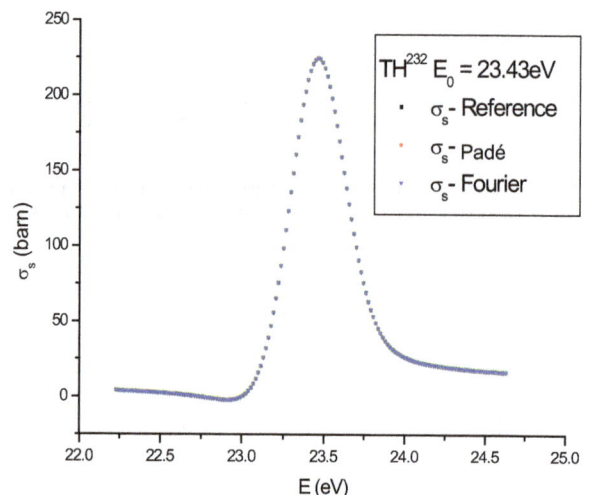

Figure 10. Average scattering cross sections of the $E_0 = 23.43$ eV resonance for the ^{232}Th isotope and T = 1500 K.

Table 1. Parameter used in the calculation of for average scattering cross sections for the ^{238}U, ^{232}Th and ^{240}Pu isotope, $\sigma = 10$ barn and T = 1500 K.

Isotope	E_0(eV)	Γ_n(eV)	Γ_γ(eV)	ξ	$\lambda_0(m)$	$\sigma_0(b)$
^{238}U	6.67	0.0015	0.0230	0.20	177.14	2.4×10^4
^{232}Th	23.43	0.0039	0.0261	0.13	94.51	1.5×10^4
^{240}Pu	20.45	0.0027	0.0322	0.17	101.16	1.0×10^4

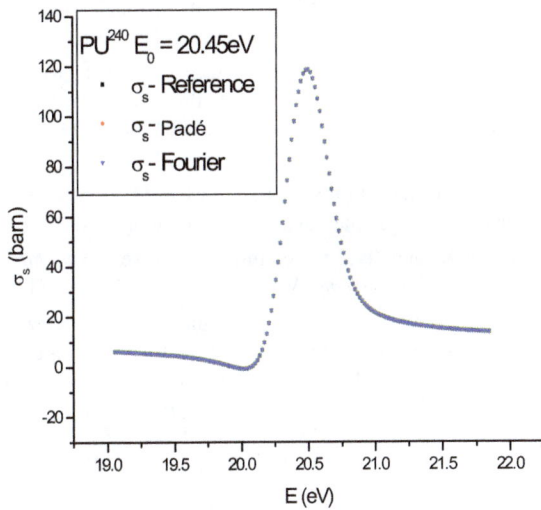

Figure 11. Average scattering cross sections of the E_0 = 20.45 eV resonance for the ^{240}Pu isotope and T = 1500 K.

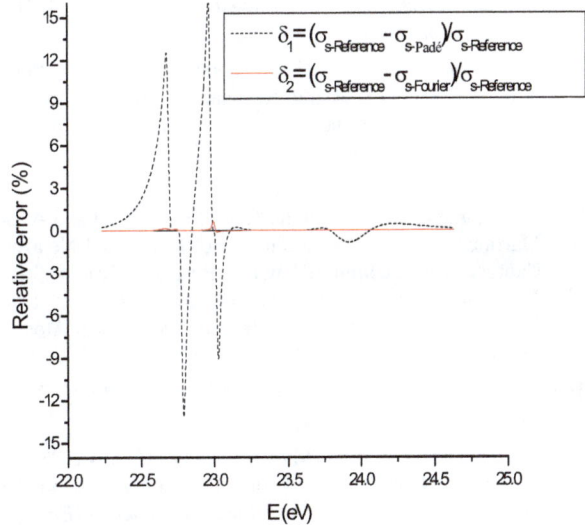

Figure 12. Relative error for average scattering sections of the E_0 = 6.67 eV resonance for the ^{238}U isotope and T = 1500 K.

From the **Figures 12** to **14** it is possible to conclude that the expression proposed in this paper presents results that overlap the numerical reference method.

5. Conclusion

This paper presents a simple and precise formulation for the Doppler broadening function $\psi(x,\xi)$ and of the interference term $\chi(x,\xi)$ based in sine and cosine Fourier transforms proposed by Goncalves *et al.* Expanding the function $G(w)$ in Fourier Series was possible to obtain an accurate analytical expression for the average scattering cross section, Equation (13), that can be an alternative to other methods existing in the literature.

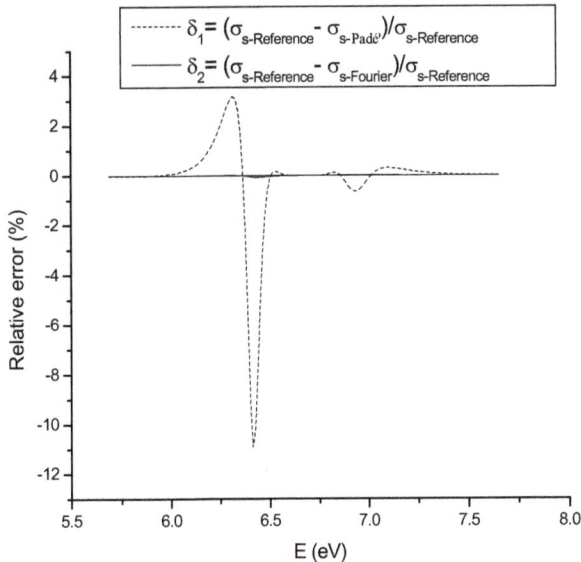

Figure 13. Relative error for average scattering cross sections of the E_0 = 23.43 eV resonance for the ^{232}Th isotope and T = 1500 K.

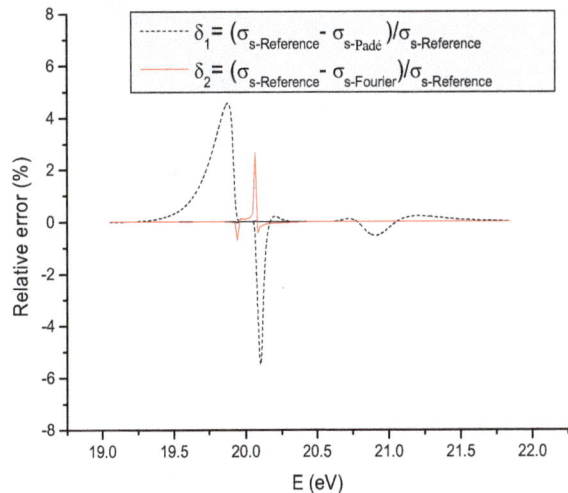

Figure 14. Relative error for average scattering cross sections of the E_0 = 20.45 eV resonance for the ^{240}Pu isotope and T = 1500 K.

6. Acknowledgements

The authors acknowledge the support provided by Brazilian Council for Scientific and Technological Development (CNPq) in the developing of this research.

REFERENCES

[1] D. A. P. Palma and A. S Martinez, "A Faster Procedure for the Calculation of the $J(\xi,\beta)$," *Annals of Nuclear Energy*, Vol. 36, No. 10, 2009, pp. 1516-1520.

[2] A. Talamo, "Analytical Calculation of the Fuel Temperature Reactivity Coefficient for Pebble Bed and Prismatic High Temperature Reactors for Plutonium and Uranium-

Thorium Fuels," *Annals of Nuclear Energy*, Vol. 34, No. 1-2, 2007, pp. 68-82.

[3]   S. G. Hong and K. S. Kim, "Iterative Resonance Self-Shielding Methods Using Resonance Integral Table in Heterogeneous Transport Lattice Calculations," *Annals of Nuclear Energy*, Vol. 38, No. 1, 2011, pp. 32-43.

[4]   D. A. Palma, A. Z. Mesquita, R. M. G. P. Souza and A. S. Martinez, "Real-Time Monitoring of Power and Neutron Capture cross Section of Nuclear Research Reactor," In: *International Conference on Research Reactors: Safe Management and Effective Utilization*, International Atomic Energy Agency, Vienna, 2011.

[5]   W. M. Stacey, "Nuclear Reactor Physics," Wiley, New York, 2001.

[6]   A. C. Gonçalves, A. S. Martinez and F. C. Silva, "Solution of the Doppler Broadening Function Based on the Fourier Cosine Transform," *Annals of Nuclear Energy*, Vol. 35, No. 10, 2008, pp. 1878-1881.

[7]   G. Arfken and H. Weber, "Mathematical Method for Physicists," Academic Press Inc., London, 2001.

[8]   C. M. Amaral and A. S. Martinez, "The Effect of Scattering Interference Term on Pratical Width," *Annals of Nuclear Energy*, Vol. 28, No. 11, 2001, pp. 1133-1143.

[9]   R. S. Keshavamurthy and R. Harish, "Use of Padé Approximations in the Analytical Evaluation of the $J(\xi, \beta)$ Function and Its Temperature Derivative," *Nuclear Science and Engineering*, Vol. 115, No. 1, 1993, pp. 81-88.

[10]   D. A. Palma, A. C. Gonçalves and A. S. Martinez, "An Alternative Analytical Formulation for the Voigt Function Applied to Resonant Effects in Nuclear Processes," *Nuclear Instruments & Methods in Physics Research. Section A, Accelerators, Spectrometers, Detectors and Associated Equipment*, Vol. 654, No. 1, 2011, pp. 406-411.

19

An Approach for Using of Poly Glycolic Acid (PGA) in Reference Standard Dosimetry: PGA/ESR Dosimetry System Response Curve and Post Irradiation Stability

Arbi Mejri[1,2*], Haikel Jelassi[2], Khaled Farah[2], Ahmad Hichem Hamzaoui[3], Hichem Eleuch[4]

[1]Ionizing Radiation Dosimetry Laboratory, National Center for Nuclear Sciences and Technologies,
Tunis, Tunisia

[2]Research Unit: Control and Development of Nuclear Technology for Peaceful Uses, National Center for Nuclear Sciences and Technologies, Tunis, Tunisia

[3]National Institute of Scientific and Technical Research, Tunis, Tunisia

[4]Departments of Physics and Astronomy, Institute for Quantum Science and Engineering, Texas A & M University,
College Station, Texas, USA

ABSTRACT

Reference Standard Dosimeters are used to calibrate radiation environments and routine dosimeters. It can also be used in routine dosimetry applications for radiation processing where higher quality dosimetry measurements are required. Electron Spin Resonance (ESR) is a well-established Reference Standard Dosimetry system in industrial applications of ionising radiation, and its use is also proposed in radiation therapy and accident dosimetry. In the present experimental work, PGA solid state dosimeter (SSD) has been investigated using ESR spectroscopy to study the gamma radiation response of this material and to evaluate its dosimetric characteristics: dose response, room temperature fading, heat treatment effect during post-irradiation storage. Results obtained up to now confirm that PGA seems to be suitable material for ESR dosimetry applications.

Keywords: Dosimetry; Solid State Dosimeter (SSD); Poly Glycolic Acid (PGA); Electron Spin Resonance (ESR); Reference Standard Dosimetery; Post-Irradiation Thermal Treatments; Fading Behaviour

1. Introduction

The ionizing radiation produces electron-hole pairs which individually become trapped at various defect sites (radicalic species) in the polymer structure. The commonly standard spectroscopic technique used for the characterisation of radicalic species, their stabilities and decay kinetics is Electron Paramagnetic Resonance (EPR), also known as Electron Spin Resonance (ESR).

Semiconductors and polymers are examples of materials that have been studied by ESR spectroscopy through the detection of paramagnetic species [1-3]. It has therefore been the main experimental technique in understanding the effect of ionizing radiation on polymers.

Radiation-induced radicals in certain material may be suitable for some applications in radiation processing dosimetry, it provided that radicals having long-term stability can be used in the establishment of relations between radiation dose and radical concentration.

The more used material in Dosimetric applications for

ESR dosimetry is the alanine [4-6] due to its good dosimetric properties, such as sensitivity, stability of the ESR signal with time [6]. Despite the alanine is accepted as reference standard dosimeter by International Atomic Energy Agency (IAEA), its sensitivity is not high for low doses and it has a very high cost. Therefore there are many attempts to characterize a new Reference Standard Dosimeters with better radiation sensitivity to substitute alanine [4,7-11]. The new material should have a stable radical per radiation energy; a sharp linewidth and thermal stability at room temperature.

One of these new materials already studied by X-Band ESR measurements is Poly Glycolic Acid (PGA) used for many years in medical and surgical applications [12-15]. The molecular structure of PGA and the free radicals formed by its interaction with radiation are shown in **Figure 1**. The free radical produced by radiation in PGA which is predominant at room temperature is the molecular species -CH°-COO- Radical I [12]. In ESR Spectrum it's traduced by a doublet signal (**Figure 2**) and it is formed by Hydrogen abstraction from a Methylene

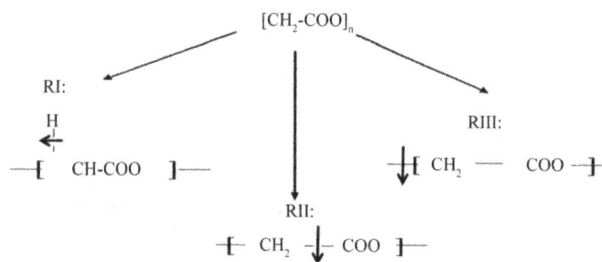

Figure 1. Molecular structure of PGA and the most probable free radicals formed.

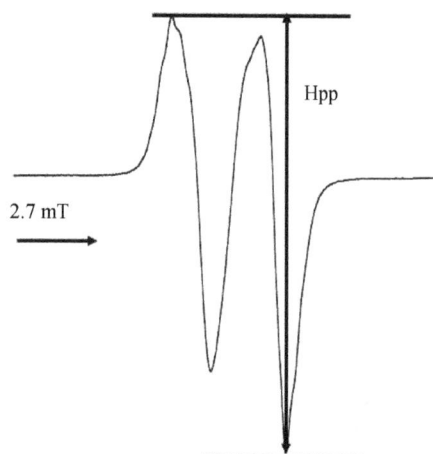

Figure 2. Peak to Peak higher (Hpp) of ESR spectra related to PGA irradiated at a dose of 25 kGy.

group located adjacent to Carbonyl group in the polymer backbone [12]. At room Temperature there is essentially no contribution to the spectrum from radicals II and III.

One of the goals of the present work was to analyze the response of PGA dosimeter for possible application as Reference Standard Dosimeter using ESR technique to ensure quality of high and low dose dosimetry applications.

PGA dosimeter has a good sensitivity for low doses, a wide useful dose range and it does not require special preparation. The main difficulty presented by PGA dosimeter is the undesirable strong initial fading. Many authors overcame this problem (case of other dosimetry system) by taking measurements at the most adequate time intervals after irradiation [16,17]. In order to solve this problem in our work, the PGA dosimeters was submitted to different post-irradiation thermal treatments from 60°C up to 140°C and the effect of these treatments on the fading behaviour of the dosimeters was studied up to

45 days after irradiation at room temperature.

2. Experimental

2.1. Poly Glycolic Acid (PGA) Polymer

Poly Glycolic Acid was obtained from Boehringer Ingelheim of Germany. The characteristics of the materials, according to the supplier, are given in **Table 1**.

2.2. Irradiation

The irradiations have been performed in air at the Tunisian semi-industrial ^{60}Co gamma-irradiation facility [18]. The dose rate was established with the alanine/EPR dosimetry system in term of absorbed dose traceable to the National Physical Laboratory, UK. Before the experiment the dose rate was verified by the standard Fricke dosimeter. Three dosimeters were irradiated for each point of measurement.

2.3. Basic of ESR Technique

The direct detection of paramagnetic species consisting of one or more unpaired electrons in complex samples is ensured by Electron Spin Resonance (ESR) which is the standard spectroscopic technique. The basic principles of ESR consist on the interaction of electromagnetic radiation with magnetic moments caused by electrons (ESR). The magnetic moment of an unpaired electron arises from its "spin" and when placed within an external magnetic field, the electron spin will align parallel or antiparallel in the direction of the magnetic field, which corresponds to a lower ($Ms = -1/2$) or an upper ($Ms = 1/2$) energy state. If electromagnetic radiation corresponding to the energy difference applied to the sample, resonance transition is possible between the lower and the upper energy states [19]. The energy difference (ΔE) between these two states is proportional to the strength of the applied magnetic field (B_0):

$$\Delta E = h\nu = g\mu_B B_0 \qquad (1)$$

where h is Planck's constant, ν is the frequency of the electromagnetic radiation, g is a constant termed g factor (g = 2.0023 for an unpaired electron), and μ_B is the Bohr magneton.

The commonly used ESR frequency in Standard Reference Dosimetry application is in the microwave range at 10 GHz (X-band).

Table 1. Physical and chemical proprieties of PGA.

Polymer	Melting Point (°C)	Glass Transition (°C)	Approximate Strength	Processing Method
Polyesters: Poly (Glycolic Acid)	225 - 230	35 - 40	7.0 GPa (Modulus)	E, IM, CM, SC

E = extrusion; IM = injection moulding; CM = compression moulding; SC = solvent casting.

2.4. ESR Measurement

EPR measurements at 9.837 GHz (X-band) were performed at 25°C using an EMX Bruker spectrometer. The microwave power used, 316 mW, were determined by considering the saturation properties of PGA EPR lines at 25°C. The Sample were accommodated in quartz sample tubes and kept up from the bottom. The different EPR spectra presented in this paper have been normalized to the same receiver gain (10^3) and sample weight (mg) in order to get quantitative comparison of the EPR lines intensities between the different samples. All dosimeters were fixed in the cavity center and flat ones were oriented to lie in the cavity nodal plane. In the present work, it is thus very difficult to compare theses spectra to determine the absolute number of radical spins. Then, in this study, we will only use arbitrary units (arb.u.) to analyze the evolution of Radical concentration in PGA Polymer content as a function of the integrated dose.

2.5. Determination of Keys ESR Parameters

The PGA dosimeter was fixed in the cavity and measured 30 min after switching on the ESR spectrometer. Every spectrum was recorded in about 61 s and delay time was 300 s. Microwave power, modulation amplitude, time constant, conversion time and modulation frequency were also studied, but are not shown here in detail. The operating EPR parameters were set as following: center magnetic field 351.5 mT, sweep width 5 mT, microwave power 316 mW, modulation amplitude 1G, modulation frequency 100 kHz, time constant 163.84 ms and conversion time 60 ms.

3. Results

3.1. ESR Spectra of γ-Irradiated PGA

Before gamma-irradiation, the PGA used in this study has small signal at about 348 mT which can be related to impurity in PGA matrix. The X-Band ESR spectra were subsequently recorded at R.T. (Room Temperature, 25°C). The resulting ESR spectrum of PGA sample after exposure to 25 kGy is shown in **Figure 2**. All PGA samples, regardless of radiation dose produced similar EPR spectra varying only in the intensity of the observed signals.

The radiolytic degradation mechanism of PGA and it's copolymers due to irradiation of samples has been investigated previously [12,20-22] and it's well known that chain scission processes dominate at low temperature. This results in a wide range of relative primary and secondary radical species which are visible by ESR. However at elevated temperatures (R.T.), as in ambient conditions studied here, Hydrogen abstraction reaction by the primary radicals occurs extensively at secondary C-H bonds on the polymer backbone. This produces a nar-

rower range of relatively stable radicals, such as -CH°-COO- which is visible at ambient temperature (25°C) [12].

3.2. Dose Response Curve

In order to find out the useful dose range for this Polymer, the response curve (specific higher Peak to Peak (Hpp) versus dose) was measured in the dose range 0.1 - 200 kGy (**Figure 3**). All data for the dose response curve were immediately acquired after irradiation. As expected in (**Figure 3**) the radical concentration increase linearly as function applied dose in the dose range 0.1 - 200 kGy with the linear regression coefficient better than 0.99 (**Figure 3**). At higher doses the specific Hpp continued to grow slowly up to 200 kGy which was the upper dose level of the present experiments. The PGA response had not yet reached saturation at this dose level. There was linear growth of the Hpp as function of the dose because at room temperature only one radical contribute to the ESR spectrum which is formed by hydrogen abstraction (**Figure 1**) RI -CH°-COO- [12]. In the dose range 0.1 to 10 kGy there was also linear growth of the Hpp as function of the dose see (Inset Curve).

3.3. Post-Irradiation Stability

As a Reference Standard Dosimetry system, the response of PGA/ESR dosimetry system should have a good stability after irradiation because in addition of its use in calibration of radiation environment and routine dosime-

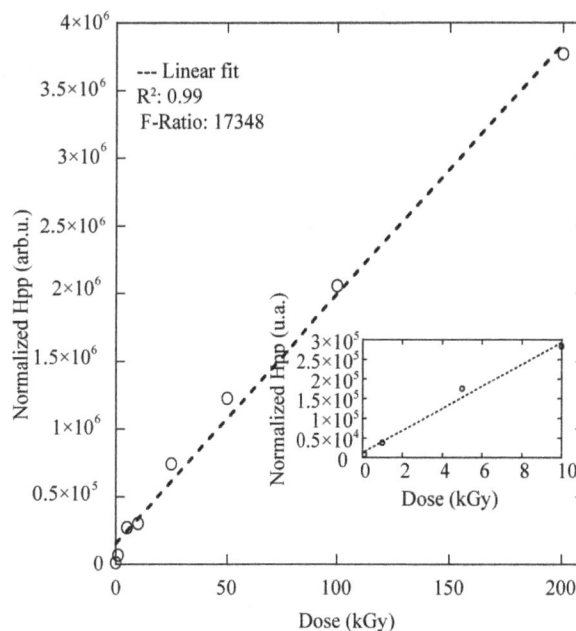

Figure 3. Dose response curve of PGA dosimeter in the dose range of 0.1 - 200 kGy, experimental (°) and calculated by a linear function (dashed line).

ters it can be used as a Transfer Standard Dosimeter for transferring dose information from an accredited or national standards laboratory to an irradiation facility in order to establish traceability for that calibration facility. Hence the time stability of the radiation-induced radical is an important aspect in radiation dose measurement by ESR spectroscopy.

In our work the main difficulty presented by PGA dosimeter is the undesirable strong initial fading. In order to solve these problem PGA dosimeters were submitted to different post-irradiation thermal treatments from 60°C up to 140°C and the effect of these treatments on the fading behaviour of the dosimeters was studied up to 45 days after irradiation at room temperature.

3.3.1. Radical Concentration Fading at Room Temperature

In order to examine the stability of radical, three replicate samples PGA were irradiated with ^{60}Co gamma rays to 25 kGy. After irradiation PGA samples were stored at room Conditions: 25°C, 40% - 60% RH (Relative Humidity). The resulting radical concentration was then followed up to 60 days via ESR technique.

The result concerning long-term room temperature variation in signal intensity was given in **Figure 4**. The signal intensity was found to be non stable at room temperature in the range of storage time. A function was used to fit signal intensity decay as shown in **Figure 4**. The coefficient of correlation (R^2) was 0.999. The stability of radiation-induced radical at room temperature was mainly controlled by the fading process. This fading process seems to follow a simple first-order kinetic.

$$Hpp = 110829,999 + 352107,206e^{(-0.096x)} \quad (2)$$

where k (0.096) is the first order decay rate constant.

3.3.2. Effect of Post-Irradiation Heat Treatments on Radical Concentration Fading

To get more insight in to the decay process of the free radical species, the effect of post-irradiation heat treatments on the PGA response fading was investigated. Six PGA samples were irradiated to a dose of 25 kGy and immediately annealed at six different temperatures (R.T., 60°C, 80°C, 100°C, 120°C and 140°C) for predetermined times (35 min), which was found to be the best heat treatment time. All ESR spectra are recorded at room temperature after cooling the samples to room temperature for predetermined times (1, 2, 5, 10, 15, 30, 45 days).

The radical concentration values were normalized to the first measurements taken 5 min after the heat treatments. The results are presented in **Figure 5**.

For each storage temperature, first-order kinetics model is used to fit the fading of the experimental data. The results show a remarkably difference between fading behaviour of PGA samples treated at various thermal treatments. Theoretical decreases in signal intensity calculated using the rate constant values determined after the fitting procedure were also given in **Figure 5** (dashed lines) with their experimental counterparts. As it can be seen from this figure, the decreasing calculated and experimental decay data agree fairly well.

Post-irradiation heat treatments led to fast fading by accelerating the recombination of defects and the impurities diffusing into the PGA matrix. As it is expected (**Figure 5**), the lower temperature at the same annealing time correspond to the faster decays of the signal intensity.

The standard deviation (*std*) of PGA dosimeters response measurements is about 0.5% (1σ) within the first day after irradiation and heating at 140°C for 35 min and about 3% - 4% between the first and the 45th days. Ac-

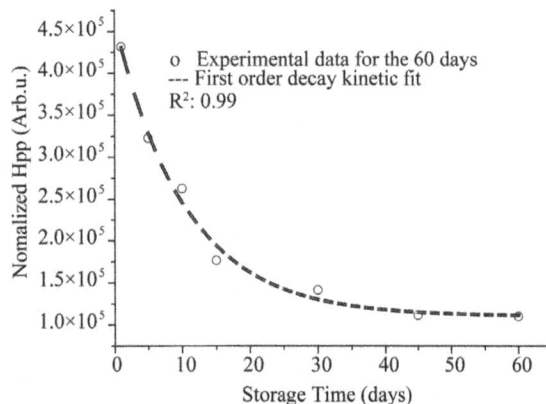

Figure 4. Experimental and calculated ESR signal intensity decay curves for PGA irradiated at a dose of 25 kGy and kept at room temperature. Symbol (experimental) and dashed line (theoretical, calculated by a function describing first-order kinetic).

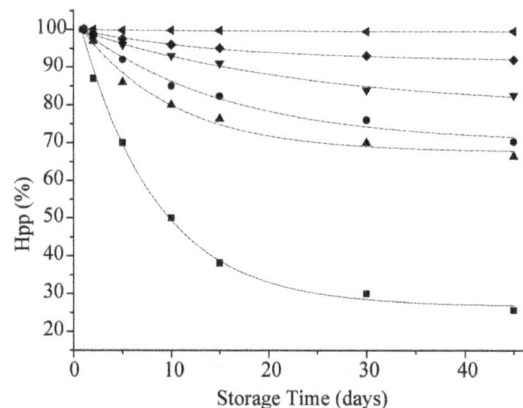

Figure 5. Variations of the ESR signal intensity with storage time for PGA samples irradiated at a dose of 25 kGy and kept at different temperatures: RT (■), 60°C (▲), 80°C (●), 100°C (▼), 120°C (♦), 140°C (◄); calculated by a function describing first-order kinetic (dashed lines).

An Approach for Using of Poly Glycolic Acid (PGA) in Reference Standard Dosimetry: PGA/ESR Dosimetry
System Response Curve and Post Irradiation Stability

129

cording to obtained results, the best results have been obtained with heat treatments at 140°C (35 min).

This procedure is very effective for the removal of unstable entities responsible for the strong fading and it did not affect the metrological properties (reproducibility and useful dose range) of the PGA dosimeter.

The decay behaviours at the higher temperatures have been attributed to the semicrystalline nature of PGA, which will influence the mobility of the polymer chains, particularly in the crystalline and crystalline amorphous boundary regions [12]. The fact that most of the radicals formed in PGA decay well below the melting temperature of the crystallites suggests that they are formed in the amorphous regions of the polymer.

3.3.3. First-Order Kinetic Behaviour of PGA Dosimeter

Signal intensity calculated by second integral of any ESR spectrum is related to the radical populations in the sample and decrease at high temperatures due to the radical recombination transform to non-paramagnetic units. In this case, studying the variation of signal intensity above room temperature would be interesting from kinetic point of view of the free radical species in gamma-irradiated PGA.

Clearly, in **Figure 5**, it can be seen that radiation induced radical concentration (Hpp) is decreasing with time. The curve is much steeper at the beginning and as time progresses; the curve approaches a horizontal line. Thus, this graph evidently shows that the rate is dependent on how much radical concentration is left. How much does the rate depend on the species concentration? To answer this question, we need to find a linear plot and work with the integrated form of the rate law:

$$\ln[\text{Hpp}] = \ln(\text{Hpp}_0) - kt \qquad (3)$$

In the above equations (Hpp_0) is the initial concentration, (Hpp) equals the concentration at time (t) and (k) is the first order constant.

Plotting ln[Hpp] against time creates a linear plot up to a conversion of about 90% with slope –k. As expected from **Figure 6**, a linear relationship is achieved between the natural log of radiation induced radical concentration (Hpp) and storage time (t) for each annealing temperature. According to the plot of ln[A] versus time, one can see that the induced radical concentration (Hpp) fading process, seems to follow a simple first-order kinetic.

3.3.4. Fading Constant Rate Evaluation

The first-order kinetics fading rate constants (k) were calculated for each storage temperature to fit the strong fading of the experimental data. The results show that the fading behaviour of the PGA samples is remarkably different for the various thermal treatments.

The observations indicate that the post irradiation heating (140°C for 35 min) of the PGA samples correspond to the lower constant rate (k) (**Table 2**), witch can traduced by a reducing in the radiation-induced radical concentration (Hpp), and an acceleration in the recombination of the entities responsible for the strong fading (**Figure 5**).

Consequently, the most suitable heating is (140°C for 35 min); it can provide a good stability in radiation-induced radical concentration (3% - 4% for 1σ) between the first and the 45th days. The evaluation of the PGA dosimeter can be performed immediately after irradiation and heat treatment.

3.4. Effect of Post-Irradiation Heat Treatments on Radical Concentration

In order to establish the useful temperature range of heat treatment and to preserve a detectable radiation induced radical response, the radical concentration was plotted against annealing temperature.

As exposed in **Figure 7**. The percentage loss in the to-

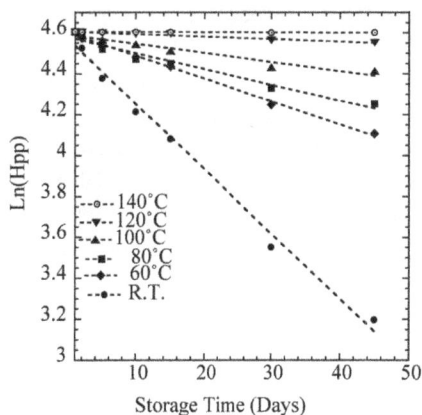

Figure 6. Variations of the natural logarithm of ESR signal intensity (Hpp) with storage time for PGA samples irradiated at a dose of 25 kGy and heat treated for 35 min at different temperatures: RT (●), 60°C (♦), 80°C (■), 100°C (▲), 120°C (▼), 140°C (○).

Table 2. Calculated decay constants value for the proposed radical species contributing to the formation of ESR spectrum of gamma irradiated PGA for temperature range from 25°C to 140°C.

Annealing Temperature (°C)	Decay Constant (k) $\times 10^{-2}$ (day^{-1})
R.T.	9.56
60	6.17
80	3.70
100	3.35
120	1.80
140	1.35

tal concentration of radicals on annealing to a temperature of 80°C was approximately 10%, while annealing to 140°C caused a reduction of the radical concentration by approximately 70%. The annealing experiments also revealed that 3% of the radicals remained at a temperature of 60°C.

According to **Figure 7** data, 140°C is a limit temperature which can not be exceeded caring the disappearance of irradiation induced radical ESR response and respecting the milting point at 225°C, in which major modification in PGA structure may be uncounted.

4. Conclusions

A Poly Glycolic Acid (PGA) Polymer has been investigated with ESR technique in order to evaluate this potential as radiation-sensitive material for dose measurements. The response curve was measured in the dose range 0.1 kGy - 200 kGy. The results demonstrated that the influence of post-irradiation storage on the ESR/PGA response might be very significant. The heating at 140°C for 35 minutes was found to be the most suitable procedure to stabilize the response of irradiated PGA dosimeter.

The linearity of dose response curve and high stability of the radiation-induced radical species in γ-irradiated PGA, it was concluded that PGA presents wanted characteristics of a good dosimetric material and it has the potential to be used as a Reference and/or Transfer Standard Dosimetry System particularly for ^{60}Co γ-ray.

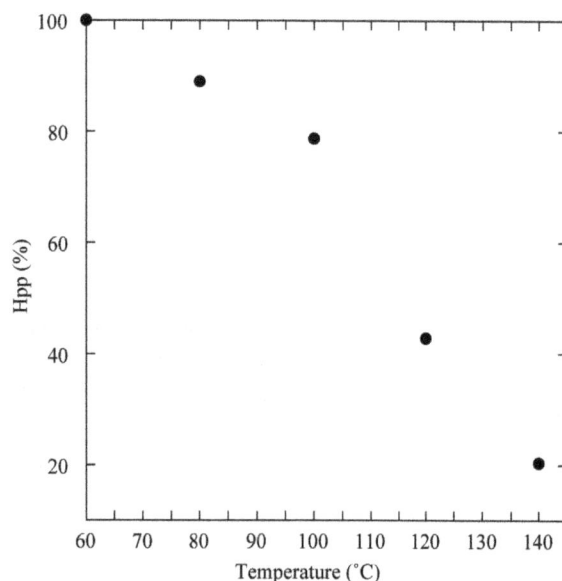

Figure 7. Percentage of radicals remaining in Gamma-irradiated Poly Glycolic Acid (PGA) after annealing at different temperatures for 35 min, measurements were done immediately after irradiation and heating.

REFERENCES

[1] G. D. Watkins, "EPR of Defects in Semiconductors: Past, Present, Future," *Physics of the Solid State*, Vol. 41, No. 5, 1999, pp. 746-750.

[2] A. L. Konkin, H. K. Roth, M. Schroedner, G. A. Nazmutdinova, A. V. Aganov, T. Ida and R. R. Garipov, "Time-Resolved EPR Study of Radicals from 2,2-Dimethoxy-2-Phenylacetophenone in Ethylene Glycol after Flash Photolysis," *Chemical Physics*, Vol. 287, No. 3, 2003, pp. 377-389.

[3] M. Bennati and T. F. Prisner, "New Developments in High Field Electron Paramagnetic Resonance with Applications in Structural Biology," *Reports on Progress in Physics*, Vol. 68, No. 2, 2005, p. 411.

[4] D. F. Regulla and U. Deffner, "Dosimetry by ESR Spectroscopy of Alanine," *The International Journal of Applied Radiation and Isotopes*, Vol. 33, No. 11, 1982, pp. 1101-1114.

[5] M. Ikeya, "New Applications of Electron Spin Resonance: Dating, Dosimetry, and Microscopy," World Scientific Publishing, Singapore, 1993.

[6] T. Kjima and R. Tnaka, "Polymer-Alanine Dosimeter and Compact Reader," *International Journal of Radiation Applications and Instrumentation. Part A. Applied Radiation and Isotopes*, Vol. 40, No. 10-12, 1989, pp. 851-857.

[7] S. P. Dias, A. M. Rossi, R. T. Lopes and E. F. O. de Jsus, "Evaluationof Dosimetric Properties of Paramagnetic Centers Formed in Gamma Irradiated Polymers," *Radiation Protection Dosimetry*, Vol. 85, No. 1-4, 1999, pp. 463-468.

[8] V. Gancheva, E. Sagstuen and N. D. Yordanov, "Study on the EPR/Dosimetric Properties of Some Substituted Alanines," *Radiation Physics and Chemistry*, Vol. 75, No. 2, 2006, pp. 329-335.

[9] G. M. Hassan and M. A. Sharaf, "EPR Dosimetric Properties of Some Biomineral Materials," *Applied Radiation and Isotopes*, Vol. 62, No. 2, 2005, pp. 375-381.

[10] A. Lund, S. Olsson, M. Bonora, E. Lund and H. Gustafsson, "New Materials for ESR Dosimetry," *Spectrochimica Acta Part A: Molecular and Biomolecular Spectroscopy*, Vol. 58, No. 6, 2002, pp. 1301-1311.

[11] E. Lund, H. Gustafsson, M. Danilczuk, M. D. Sastry, A. Lund, T. A. Vestad, E. Malinen, E. O. Hole and E. Sagstuen, "Formates and Dithionates: Sensitive EPR-Dosimeter Materials for Radiation Therapy," *Applied Radiation and Isotopes*, Vol. 62, No. 2, 2005, pp. 317-324.

[12] V. Gancheva, E. Sagstuen and N. D. Yordanov, "Study on the EPR/Dosimetric Properties of Some Substituted Alanines," *Radiation Physics and Chemistry*, Vol. 75, No. 2, 2006, pp. 329-335.

[13] A. Babanalbandi, D. J. T. Hill, J. H. O'Donnell and P. J. Pomery, "An Electron Spin Resonance Analysis on Y-

An Approach for Using of Poly Glycolic Acid (PGA) in Reference Standard Dosimetry: PGA/ESR Dosimetry
System Response Curve and Post Irradiation Stability

131

Irradiated Poly (Glycolic Acid) and Its Copolymers with Lactic Acid," *Polymer Degradation and Stability*, Vol. 52, No. 1, 1996, pp. 59-61.

[14] R. K. Kulkarni, K. C. Pani, C. Neuman and F. Leonard, "Polylactic Acid for Surgical Implants," US National Technical Information Service, AD Report No. 636716, 1966, pp. 1-17.

[15] D. E. Cutright, J. M. Brady, R. A. Miller and M. A. J. Willis, "Systemic Mercury Levels Caused by Inhaling Mist during High-Speed Amalgam Grinding," *Journal of Oral Medicine*, Vol. 28, No. 4, 1973, pp. 100-104.

[16] B. Engin, C. Aydas and H. Demirtas, "ESR Dosimetric Properties of Window Glass," *Nuclear Instruments and Methods in Physics Research Section B: Beam Interactions with Materials and Atoms*, Vol. 243, No. 1, 2006, pp. 149-155.

[17] A. de A. Rodrigues Jr. and L. V. E. Caldas, "Commercial Plate Window Glass Tested as a Routine Dosimeter at a Gamma Irradiation Facility," *Radiation Physics and Chemistry*, Vol. 63, No. 3-6, 2002, pp. 765-767.

[18] K. Farah, T. Jerbi, F. Kuntz and A. Kovacs, "Dose Meas-

urements for Characterization of a Semi-Industrial Co-balt-60 Gamma-Irradiation Facility," *Radiation Measurements*, Vol. 41, No. 2, 2006, pp. 201-208.

[19] J. A. Weil and J. R. Bolton, "Electron Paramagnetic Resonance: Elementary Theory and Practical Applications," 2nd Edition, John Wiley & Sons, Hoboken, 2007.

[20] A. G. Hausberger, R. A. Kenley and P. P. DeLuca, "Gamma Irradiation Effects on Molecular Weight and *in Vitro* Degradation of Poly (D, L-Lactide-Co-Glycolide) Microparticles," *Pharmaceutical Research*, Vol. 12, No. 6, 1995, pp. 851-856.

[21] M. B. Sintzel, A. Merkli, C. Tabatabay and R. Gurney, "Influence of Irradiation Sterilization on Polymers Used as Drug Carriers—A Review," *Drug Development and Industrial Pharmacy*, Vol. 23, No. 9, 1997, pp. 857-879.

[22] L. Montanari, M. Constantani, E. Ciranni-Signoretti, L. Valvo, M. Santucci, M. Barolomei, P. Fattibene, S. Onori, A. Faucitano, B. Conti and I. Genta, "Gamma Irradiation Effects on Poly (DL-Lactide-Co-Glycolide) Microspheres," *Journal of Controlled Release*, Vol. 56, No. 1-3, 1998, pp. 219-229.

20

The Radiation Degradation of Neutral Red Solution by γ-Ray

Xiu-Hua Liu, Yi Deng, Yin-Hang Zhou, Liang Xia, Lan-Lan Ding, Yu-Chuan Zhang
China Academy of Engineering Physics, Mianyang, China

ABSTRACT

Neutral red is kind of biologic colourant and acidity-basicity indicator. Radiation degradation of neutral red in aqueous solution was done by γ-ray. The removal rate of chemical oxygen demand, total organic carbon, chroma and the changing of pH value were studied under various conditions. With the increase of absorbed doses, the chemical oxygen demand and chroma decreased conspicuously. The absorbed dose rate has little effect on the degradation of neutral red. When the absorbed doses are the same, the chemical oxygen demand and chroma decreased more obviously with the increase of neutral red concentration. Weak basic condition and proper H_2O_2 addiation are propitious to removal of chemical oxygen demand of neutral red.

Keywords: γ-Ray; Irradiation; Neutral Red; Degradation

1. Introduction

For the rapid development of dyestuff industry, the dyeing wastewater is recognized as one of the intractable industrial organic wastewater for its large amount discharge, complex composition with toxicity and deep color [1]. Neutral red is an important coloring agent for its aqueous solution in deep red. Neutral red is often used as linsey-woolsey coloring agent, biological stain and acid-base indicator. Therefore, neutral red is also an important composition in dyeing wastewater.

Ionizing radiation seems to be an effective technology for the degradation of organic pollutants. Relative researches show that irradiation can achieve the effective treatment of organophosphorus compounds, halogenated hydrocarbon, carboxymethylcellulose, etc. [2-8]. Radiation degradation can be performed at ambient temperature and for large-scale treatment. Since the γ-ray has higher penetration, radiation degradation of neutral red in aqueous solution was done by γ-ray. The removal rate of COD, TOC, Chroma and the changing of pH value were studied under various conditions. The effect of absorbed dose rate, absorbed dose and H_2O_2 content on degradation efficiency of neutral red has been obtained.

2. Experiment Method

2.1. Irradiation Method of γ-Ray

The radiation degradation of neutral red was done by a ^{60}Co-γ-source with an activity of 230 kCi (average energy of 1.25 MeV). Neutral red solution was irradiated in 100 mL colorimetric glass vessels at certain dose rates. Absorbed dose was controlled by irradiated time. Every colorimetric vessel contained 55 mL neutral red solution. The absorption dose was calibrated by silver dichromate stoichiometric method (Chinese National Standard: JJG 1028-91). All experiments were performed at ambient temperatures.

2.2. Analytical Method

Several main indices of neutral red solution were quantitatively measured before and after irradiation. Chemical oxygen demand (COD) was measured by bichromate method (Chinese National Standard: GB 11914-89). Chroma was determined by diluted multiple method (Chinese National Standard: GB 11903-89). Total organic carbon (Chinese National Standard: TOC) and total inorganic carbon (TIC) were measured by OI Analytical 1030 C Aurora Combustion Total Organic Carbon Analyzer according to nondispersive infrared absorption method (Chinese National Standard: HJ 501-2009). The pH value was determined by Mettler Toledo SG2 pH meter according to glass electrode method (Chinese National Standard: GB 6920-86). UV-Vis spectra of neutral red solution was carried out by a PE Lamda 12 UV-Vis spectrophotometer.

3. Result and Discussion

3.1. The Effect of Absorbed Dose

Neutral red solutions with concentration at 10 mg/L, 28.9

mg/L, 60 mg/L were irradiated for different doses from 0 to 20 kGy at 74.68 Gy/min, in accordance with different concentrations. A little precipitation produced after radiation. The evolution of pH values, COD contents and Chroma of superstratum limpid liquid are showed in **Figure 1**. After irradiation, COD and pH value decreased rapidly. pH value decreased more obviously with the increase of absorbed doses, indicating that there is at least an acidic compound produced during the radiation degradation of neutral red.

COD of neutral red solutions decreased gradually with the increase of absorbed dose. For the solutions with the concentrations at 10 mg/L, 28.9 mg/L, 60 mg/L, the COD decreased more than 57.6% at a dose of 5 kGy. When the absorbed doses are the same, the COD and

chroma of the neutral red solutions decreased more obviously with concentration increase. The chroma of neutral red solution decreased dramatically after irradiation. The color removed completely with an initial neutral red concentration of 10 mg/L at a dose of 1 kGy, an initial neutral red concentration of 28.9 mg/L at a dose of 5 kGy, an initial neutral red concentration of 60mg/L at a dose of 10 kGy.

Figure 2 is the UV-Vis spectra of neutral red solutions. The absorbency data are normalized to the highest peak of 0 kGy spectrum. The UV-Vis spectra illustrate that the main characteristic absorption peaks both at 266 nm and

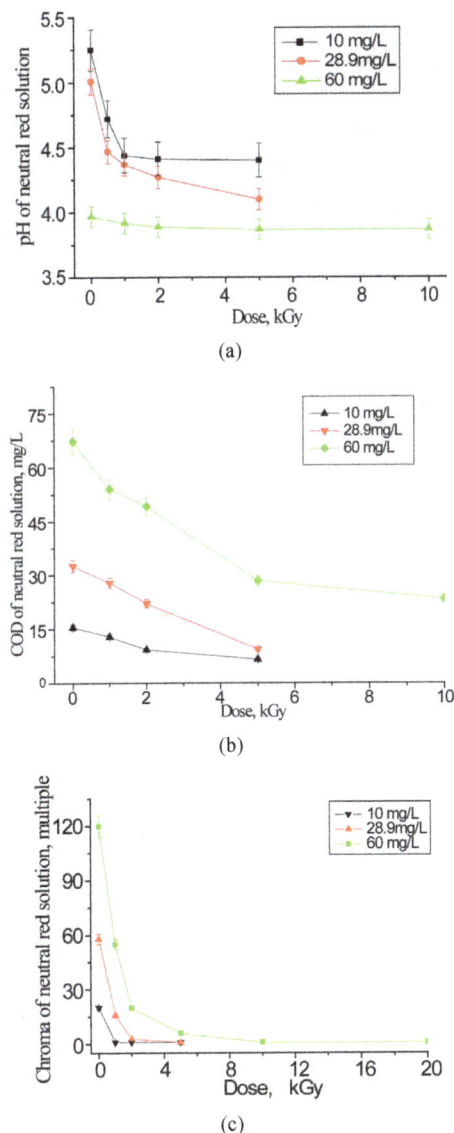

(a)

(b)

(c)

Figure 1. pH values (a), COD contents (b) and chroma (c) of the neutral red solutions under different absorbed doses.

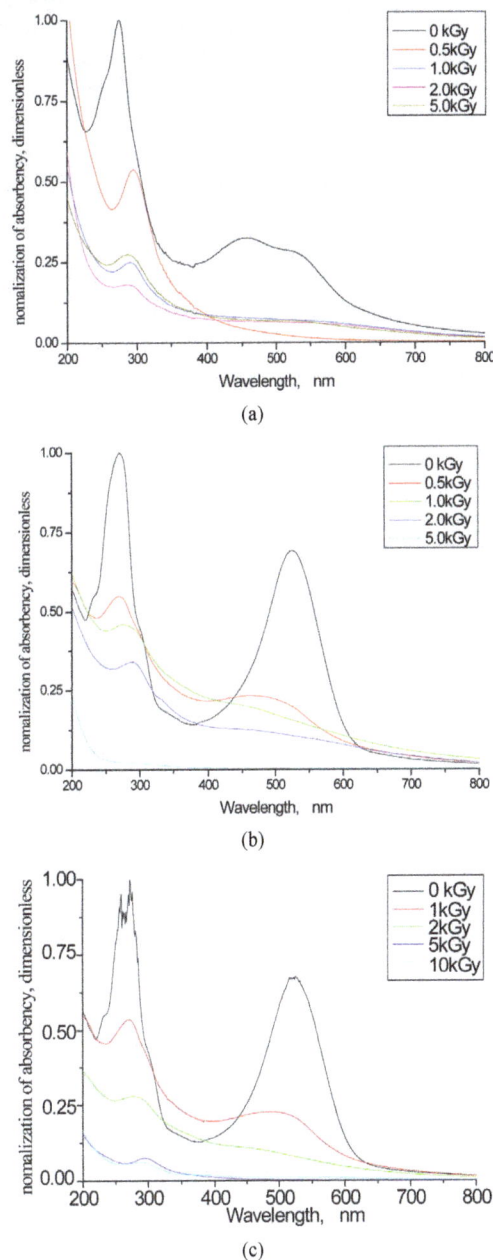

(a)

(b)

(c)

Figure 2. UV-Vis spectra of the neutral red solutions under different absorbed doses.

523 nm of neutral red disappeared basically for the an initial concentration of 10 mg/L at a dose of 1 kGy, initial concentration of 28.9 mg/L and 60 mg/L at a dose of 5 kGy. The absorbencies are close to zero.

3.2. The Effect of Absorbed Dose Rate

The neutral red solutions with concentration at 28.9 mg/L were irradiated for 5 kGy at 4.51 Gy/min, 8.90 Gy/min, 30.86 Gy/min, 61.73 Gy/min and 86.91 Gy/min respecttively. The pH values drop to from 5.86 before irradiation to 4.25 ~ 4.70 after irradiation, which is shown in **Figure 3(a)**. Although the acidity increased greatly, the pH values of the solutions irradiated at different dose rate are close to each other. The UV-vis absorptionspectra (in **Figure 3(b)**) of the same solutions irradiated at different dose rate are also identical. Both the pH values and UV-vis absorptionspectra indicates that the absorbed doserate has little effect on the degradation efficiency of neutral red.

3.3. The Effect of pH Value

The neutral red solutions at pH values of 36.98, 7.73, 9.06 respectively were irradiated for 5 kGy at 61.73 Gy/min. Their pH value changing is showed in **Figure 4(a)**. The pH value changing is not obvious for acidic solutions but distinct for basic solutions. The studies in section 2.1 shows that there is at least an acidic compound produced during the radiation degradation of neutral red, the pH value of basic solutions decreased rapidly might because of acid-base neutralization.

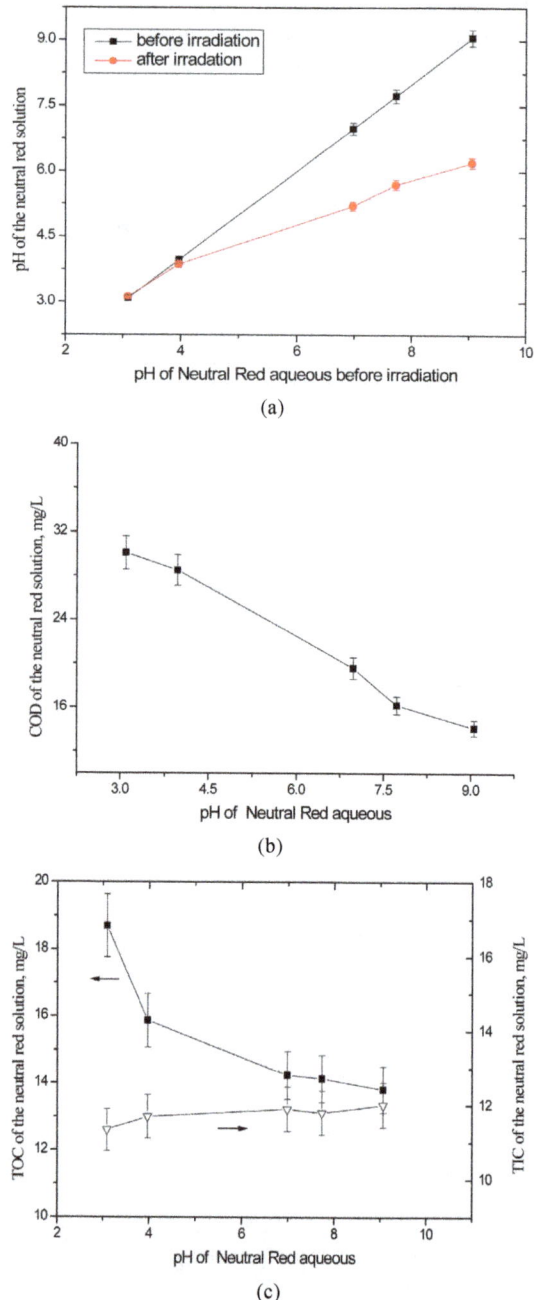

Figure 3. pH values (a) and UV-Vis spectra (b) of the neutral red solutions under different absorbed dose rates.

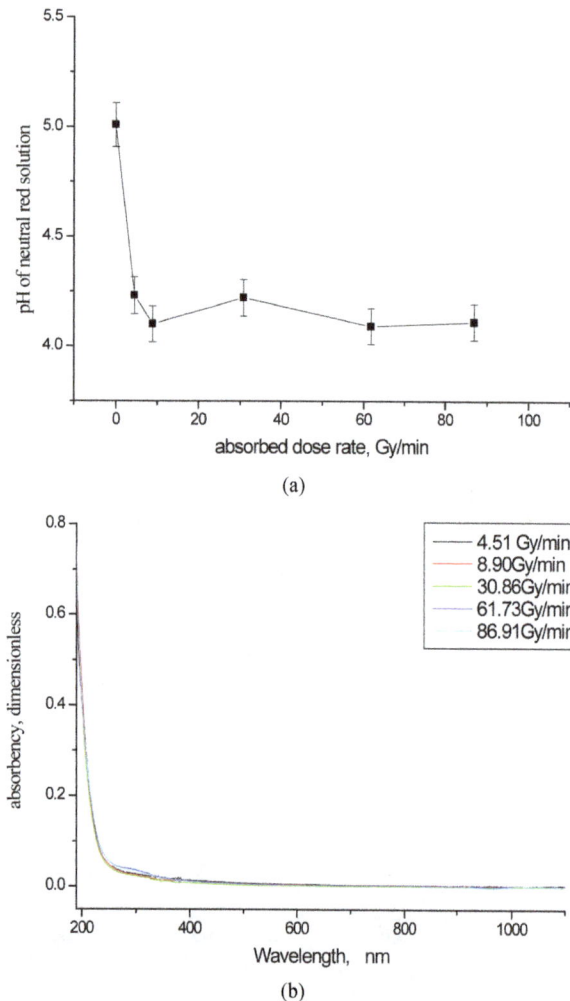

Figure 4. pH values (a), COD (b), TOC and TIC (c) of the neutral red solutions under different absorbed doses.

The COD, TOC and TIC of neutral red solutions at various pH values were determined before and after irradiation, which are showed in **Figures 4(b)** and **(c)**. The COD content of basic solution was less, indicating that alkalescency condition is propitious to the removal of COD. With the increase of alkalescence, TOC decreased obviously while TIC increase appreciably. The decrease of TOC reflects the mineralization of neutral red. While the increase of TIC is less than the decrease of TOC, and the TC content decreases gradually, indicating that some neutral red were degraded to carbon dioxide.

3.4. The Effect of Hydrogen Peroxide

Effect of H_2O_2 on the degradation of neutral red solution is evaluated by COD, pH determination (**Figure 5**) with the initial concentration of 0 mmol/L, 5.3 mmol/L, 8.9 mmol/L, 17.8 mmol/L, 35.5 mmol/L, 88.2 mmol/L and 174.8 mmol/L. The concentration of neutral red solution is 60 mg/L. Whether adding H_2O_2 in the neutral red solution or not, the pH value decreases after irradiation. And the pH value decreases more obviously after the addition of H_2O_2. The COD get lower with the H_2O_2 concentration of 5.3 mmol/L - 35.5 mmol/L. The addition of H_2O_2 enhanced the removal of COD. However, a further increase of H_2O_2 does not lead to further increase of COD removal when the concentration up to 88.2 mmol/L. On the premise of the concentration of H_2O_2 greater than or equal to up to 88.2 mmol/L, the COD of the neutral red solution increases with the increase of H_2O_2, but the chroma of the neutral red solutions does not increase. In the actual determination, the COD content of 49.0 mmol/L H_2O_2 prapared with ultrapure water is 68 mg/L, indicating that the H_2O_2 has positive interference with the determination of COD. The H_2O_2 can produce oxygen under irradiation, and the oxygen can decompose organic compounds. Radiolysis occurs in the first place to produce some reactive primary species such as hydroxyl radicals ($\cdot OH$), hydrogen atoms ($H\square$), hydrated electrons (e_{eq}^-) and hydrogen ions when the neutral red solutions are exposed to γ-ray [9]. Hydrated electrons can react with H_2O_2 to form $\cdot OH$ and OH

$$\left(K = 1.1 \times 10^{10} \, \text{L} \cdot \text{mol}^{-1} \cdot \text{s}^{-1} \right).$$

$H\square$ can react with H_2O_2 to form $\cdot OH$ and H_2O $K = \left(9 \times 10^7 \, \text{L} \cdot \text{mol}^{-1} \cdot \text{s}^{-1} \right)$ [10]. The $\cdot OH$ has strong oxidation and can accelerate the decomposition of neutral red. But higher concentration H_2O_2 could not react completely. The remaining H_2O_2 cause interference with the determination of COD and make the COD larger.

4. Conclusion

The studies on the radiation degradation of neutral red solutions show that gamma-ray irradiation is an effective

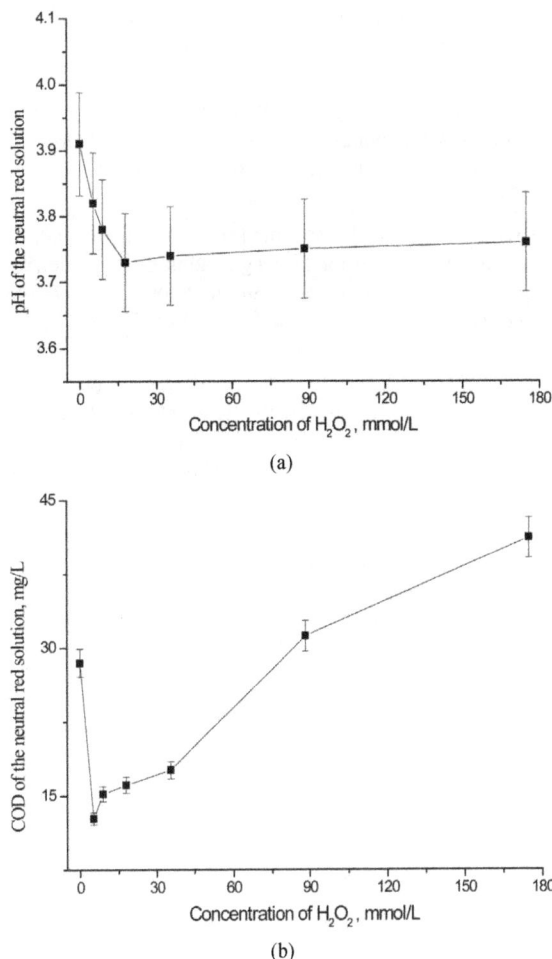

(a)

(b)

Figure 5. pH values (a) and COD contents (b) of the neutral red solutions with H_2O_2 existing.

method. The characteristic absorption peaks of neutral red disappeared basically and the absorbency are close to zero for the an initial concentration of 10 mg/L at a dose of 1 kGy, initial concentration of 28.9 mg/L and 60 mg/L at a dose of 5 kGy. With the increase of absorbed doses, the COD and chroma decreased conspicuously. The absorbed dose rate has little effect on the degradation of neutral red. When the absorbed doses are the same, with the increase of the concentration of neutral red between 0 mg/L and 60 mg/L, the COD and chroma decreased more obviously. Weak basic condition and proper H_2O_2 addiation are propitious to removal of COD of neutral red.

5. Acknowledgements

This work was financially supported by the Sichuan Provincial Science and technology program of China (Grant No. 2009GZ0037).

REFERENCES

[1] I. H. Faisal, Y. Kazuo and F. Kensuke, "Hybrid Treat-

ment Systems for Dye Wastewater," *Critical Reviews in Environmental Science and Technology*, Vol. 37, No. 4, 2007, pp. 315-377.

[2]   T. Polonca and I. Arcon, "Degradation of Organophosphorus Compounds by X-Ray Irradiation," *Radiation Physics and Chemistry*, Vol. 67, No. 3-4, 2003, pp. 527-530.

[3]   A. A. Basfar, K. A. Mohamed, A. J. Al-Abduly and A. A. Al-Shahrani, "Radiolytic Degradation of Atrazine Aqueous Solution Containing Humic Substances," *Ecotoxicology and Environmental Safety*, Vol. 72, No. 3, 2009, pp. 948-953.

[4]   J. Choi, H. S. Lee, J. H. Kim, K. W. Lee, J. W. Lee, S. J. Seo, K. W. Kang and M. W. Byun, "Controlling the Radiation Degradation of Carboxymethylcellulose Solution," *Polymer Degradation and Stability*, Vol. 93, No. 1, 2008, pp. 310-315.

[5]   R. Zona, S. Solar and P. Gehringer, "Degradation of 2,4-Dichlorophenoxyacetic Acid by Ionizing Radiation: Influence of Oxygen Concentration," *Water Research*, Vol. 36, No. 5, 2002, pp. 1369-1374.

[6]   A. A. Basfar, H. M. Khan and A. A. Al-Shahrani, "Trihalomethane Treatment Using Gamma Irradiation: Kinetic Modeling of Single Solute and Mixture," *Radiation Physics and Chemistry*, Vol. 72, No. 5, 2005, pp. 555-563.

[7]   E. A. Arbra [Former Soviet Union], "Industrial Electron Accelerators and Their Applications in Radiation Processing," Atomic Energy Press, Beijing, 1990 (in Chinese).

[8]   S. K. Huang, L. L. Hsieh, C. C. Chen, P. H. Lee and B. T. Hsieh, "A Study on Radiation Technological Degradation of Organic Chloride Wastewater—Exemplified by TCE and PCE," *Applied Radiation and Isotopes*, Vol. 67, No. 7-8, 2009, pp. 1493-1498.

[9]   J. W. T. Spinks and R. J. Woods, "Introduction to Radiation Chemistry," Wiley Press, New York, 1990.

[10] M. Wang, R. Y. Yang, W. F. Wang, S. W. Bian and Z. Q. Shen, "γ-Ray Induced Degradation of Reactive Blue KNR in Aqueous Solution," *Journal of Radiation Research and Radiation Processing*, Vol. 22, No. 2, 2004, pp. 92-96 (in Chinese).

21

Dose-Dependence of Trap Parameters of OSL Decay from Al₂O₃:C

Ayse Güneş Tanır*, **Mustafa Hicabi Bölükdemir, Rasoul Ghomi**

Faculty of Sciences, Gazi University, Ankara, Turkey

ABSTRACT

Optically Stimulated Luminescence (OSL) trap parameters can only be reliably determined through the detailed analysis of OSL decay curves. In this study the kinetic parameters of a blue-light stimulated luminescence (BLS) decay curve from Al₂O₃:C sample irradiated at 0.1, 0.15, 0.2, 0.4 and 0.6 Gy beta doses were obtained using the same basic methods with some modifications applied and also by using our suggestion: Active-OSL Approximation (AOSL). The results were compared with those of other studies on the trap parameters of Al₂O₃:C material.

Keywords: Optically Stimulated Luminescence; Kinetic Parameters; Carbon Doped Aluminum Oxide; Deconvolution

1. Introduction

The Al₂O₃:C sample has become an important material as an OSL dosimeter because of its highly sensitive response to ionization radiation. Its use as TL dosimeters was first suggested by Akselrod *et al.*, [1]. Subsequently, many researcher groups [2-10] have investigated the OSL properties of Al₂O₃:C because of its applications in the field of space, medical therapy and medical diagnostic.

Conventional OSL signals were measured using the illumination source with a constant intensity and luminescence was plotted over time. In many cases there is no mode-based analytical function that is able to describe the shape of the CW-OSL decay curve. In general the experimental data is described as the sum of simple exponential curves and it is often difficult to determine either the number of exponential curves or the number of contributing traps. The evaluation of OSL trap parameters is one of the fundamental requirements in understanding the luminescence mechanism.

Many approaches for determining the trap parameters through the application of various models exist in the literature. The simplest model attempted to explain the luminescence kinetic as consisting of single trap/single recombination centers. Several authors have reported that the decay curve of OSL can be described as the sum of multiple exponentials. Also, it is reported in many works that the OSL decay curves from Al₂O₃:C samples suggest an overlap of several peaks or even a distribution of traps having different activation energy and frequency factors [11-15].

In the present work the kinetic parameters of OSL decay curves from the Al₂O₃:C given at different doses were obtained using various known methods to which some modifications were applied. The methods used were: Curve-Fitting of Thermal Analysis, Linear Modulation Technique [16], General Order (GO) [17] Model and Active-OSL (AOSL) Approximation [18,19].

2. Methodology

The methods used in this study for decay curve analysis and their applications are as follows:

2.1. The Curve-Fitting Method

The experimental OSL intensity as the sample preheats temperature increased was fitted to [2,20]:

$$\eta = 1/\left(1 + c\exp\left(-E/kT\right)\right) \qquad (1)$$

Equation (1) describes the thermal quenching. Where η is luminescence efficiency; c is a dimensionless constant; E, is the activation energy and k is the Boltzman's constant. In this work, T is the preheat temperature. In first the Al₂O₃:C aliquot was tested to observe the background luminescence: ~130 - 150 counts/s. The integrated blue-light stimulated luminescence experimental data (0 - 20 s) from Al₂O₃:C in the range of 295 - 673 K was fitted to the Equation (1). The aliquot was irradiated with 0.1Gy β-dose and measured for 20 s at room temperature (RT): ~295 K. It was bleached by being heated at 400°C and again exposed to 0.1Gy β-dose/heated at 346 K for 5 minutes, left 30 minutes then measured at RT. These steps were repeated for 373, 398, 423, 448, 473, 498, 523,

---
*Corresponding author.

548, 573 and 673 K. Then the normalized BLS intensity was plotted against the temperature. These procedures were also repeated for 0.1, 0.15, 0.2, 0.4 and 0.6 Gy beta doses.

2.2. General Order (GO) Technique

According to General Order (GO) technique suggested by Rasheedy [17]

$$I(t) = -\frac{dn}{dt} = \frac{n^b}{N^{b-1}} s \exp(-E/kT)$$

where N is the concentration of the total traps and n is the concentration of filled traps at temperature T. In this model the expression for calculating the order of kinetics b is given as,

$$b = \frac{\dfrac{\ln\left(I_1^\gamma/I_3\right)}{\gamma-1} - \dfrac{\ln\left(I_2^\alpha/I_3\right)}{\alpha-1}}{\dfrac{\ln\left(n_{1e}^\gamma/n_{3e}\right)}{\gamma-1} - \dfrac{\ln\left(n_{2e}^\alpha/n_{3e}\right)}{\alpha-1}} \qquad (2)$$

and an expression for calculating E is given as [17],

$$E = \frac{k \ln\left\{(n_{2e}/n_{3e})^b I_3/I_2\right\}}{(1/T_2 - 1/T_3)} \qquad (3)$$

$$S = \frac{I_i \exp\left(\dfrac{E}{KT_i}\right)}{n_e\left(n_{ie}^b\right)} \qquad (4)$$

where, n_{ie} is the area under the glow peak from T_i to the end of the glow peak; n_e is the area under the whole glow peak; $\gamma = T_1/T_3$ and $\alpha = T_2/T_3$. As far as we know, this method of analysis has only been applied for TL traps not for OSL up to now. The peak-shape form of the decay curve is required to apply this method. To obtain the peak-shape curve, the reduction rate of BSL signal was calculated as percentage of the original signal using the experimental data and plotted versus T [21,22]. In this study three temperature points were selected on PS curve, one of them was on the left side of the peak and the others were on the right side of it. These selected temperature positions have critical importance unlike with Ogunxdare et al., [23,24].

2.3. Linear Modulation Technique (LM-OSL)

LM-OSL technique suggested by Bulur [16] transforms CW-OSL signal to a peak-shape form. The details of the transformation process are given in the study by Bulur [25].

CW-OSL decay curve for the first-order kinetic of a simple solid material that consisting of one trap and one recombination center is described as follow [12]:

$$L(t) = n_0 B \exp(-Bt) \qquad (5)$$

where n_0 is the initial trap concentration; B is a constant describing the decay of luminescence curve and is proportional to the photoionization cross section and the intensity of light I_0 ($B = \alpha I_0$) Bulur [25]. A new independent variable, u, is described to transform the CW-OSL curve to LM-OSL.

$$\mu = \sqrt{2tP}$$

where u is time in second and $P \cong 2t$.

$$I(\mu) = n_0 \frac{B}{P} \mu \exp\left\{-\frac{B}{2P}\mu^2\right\} \qquad (6)$$

Equation (6) is the first order LM-OSL curve.

2.4. AOSL Approximation

The process is similar to the successive decay of a radioactive element but not identical [18,19]. Accordingly, the equations describing the OSL counts and the activity (or OSL intensity) are proposed as follows:

$$N = N_1 + N_2 = N_{10} \exp(-\lambda_1 t) + \frac{\lambda_1 \lambda_2}{\lambda_2 - \lambda_1}$$
$$\times N_{10}\left[\exp(-\lambda_1 t) - \exp(-\lambda_2 t)\right] \qquad (7)$$
$$+ N_{20} \exp(-\lambda_2 t)$$

$$I = I_1 + I_2 = \lambda_1 N_{10} \exp(-\lambda_1 t) + \frac{\lambda_1 \lambda_2}{\lambda_2 - \lambda_1}$$
$$\times N_{10}\left[\exp(-\lambda_1 t) - \exp(-\lambda_2 t)\right] \qquad (8)$$
$$+ N_{20} \exp(-\lambda_2 t)$$

Equation (7) describes the luminescence photons counted experimentally over finite time intervals. N_1 is the number of atoms in the parent element which decays at λ_1 into its daughter element; and N_2 is the number of atoms in the daughter element, which decays at a constant decay rate λ_2, into a stable element having N_3 stable atoms. Assume that at time $t = 0$, $N_1 = N_{10}$ and $N_2 = N_{20}$. While Equation (7) has been used to plot the OSL decay curves, Equation (8) was found to be applicable to find the kinetic parameters of IRSL. It is important to note here, for the compensation of theoretical and experimental data the term λ_2 should be inserted in the numerator of the second term in Equation (7). This situation is different from radioactive decay law. Radioactive decay law of successive disintegration can be found in the book written by Krane [26].

3. Experimental

In this work the apparatus developed by Spooner et al., [27] was used. All measurements were carried out using an automated ELSEC 9010 OSL reader system with a ring of 24 blue-light OSL attachments. These LEDs with blue-light ($\sim$470Δ30 nm) from WENRUN were settled to system by us. They have a power output of about 6 cd at 20 mA current and an emission angle of 25°. A green long-pass Schott GG-420 filter was mounted in front of blue LEDs to minimize the amount of directly scattered blue light reaching the PM photocathode. In 24 diodes, the total power delivered to the sample was measured as 21.6 mW/cm^2 at distance of 16 mm. Detection was made through 3 Hoya U-340 filters (3 mm).

A ^{90}Sr-^{90}Y β-source was used for irradiation. The dose rate given to sample was 0.028 Gy·s^{-1}. Bleaching was carried out by exposing to daylight and checked by measuring the signals from the sample. All the luminescence measurements were made at room temperature (RT). Luminescence was detected using a Thorn-EMI 9235QA PM Tube (with a Schott BG-39 filter) having a dark count rate of about 130 - 150 cps at room temperature. Al$_2$O$_3$:C single crystal discs (diameter 5 mm and thickness 1 mm) were used in the measurements.

4. Results and Discussion

Figure 1 shows the pulse annealing experimental curves for the integrated BSL (0 - 20 s) in the range of 295 - 673 K from α-Al$_2$O$_3$:C. The experimental data was fitted to the Equation (1) and the activation energy values were obtained from this fitted curve:

$\eta = 1/(1 + 4.25 \times 10^{10}\exp(-1.15/(0.000086174 \times T)))$ for 0.1 Gy;

$\eta = 1/(1 + 3.33 \times 10^{10}\exp(-1.05/(0.000086174 \times T)))$ for 0.2 Gy;

$\eta = 1/(1 + 5.60 \times 10^{11}\exp(-1.07/(0.000086174 \times T)))$ for 0.4 Gy;

$\eta = 1/(1 + 2.63 \times 10^{9}\exp(-1.066/(0.000086174 \times T)))$ for 0.6 Gy.

The activation energies and dimensionless constants can be seen from these equations. It can be seen that the activation energy of α-Al$_2$O$_3$:C does not vary with beta-dose in the range 0.1 - 0.6 Gy and that its mean value is 1.084 eV.

Figure 2 was plotted using the fitted data in **Figure 1**. They are peak-shape (PS) curves for α-Al$_2$O$_3$:C at the doses given. The percentage of reduction in the BSL signals was plotted against annealing temperature. As seen from **Figure 2**, peak temperatures are different a relationship between dose and peak temperature was not observed.

To apply the GO technique suggested by Rasheedy [17] the three temperature points were selected on the peak-shape (PS) plots. For example the data for 0.1 Gy are $T_1 = 503$ K; $T_2 = 563$ K and $T_3 = 593$ K. The value of b, that is the order of kinetics of OSL decay mechanism, was calculated as 1.71 using the Equation (2). The order of kinetics that describes the retrapping number of charges should be the integral number so it is possible to assume as 1.71 $\cong$ 2. The results were listed at **Table 1** for other doses values. The activation energy was found to be 1.09 $\pm$ 0.035 eV using the GO Model for 0.1 Gy beta-dose. In this study, although the GO model was applied for the analysis of the OSL traps, the selected temperature points on the PS curve should not be random; one of them should be selected before the maximum temperature and the others should be selected after the peak temperature. The comparison of trap parameters determined by using GO model and the pulse annealing curve fitting are shown in **Table 1**. The discrepancy between activation energies at 0.1 Gy and 0.6 Gy was found to be 92% using curve fitting and 99.09% using the GO model. Therefore, it can be seen that the activation energy of α-Al$_2$O$_3$:C does not change with beta dose in the range 0.1 - 0.6 Gy. Bulur *et al.*, [6] reported that the activation energy of Al$_2$O$_3$:C as $E = 1.15$ eV and Akselrod *et al.*, [28] found to be about 1.08 eV.

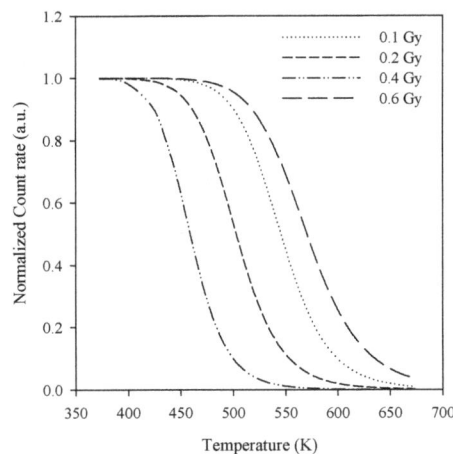

Figure 1. The experimental pulse annealing curves at different beta doses.

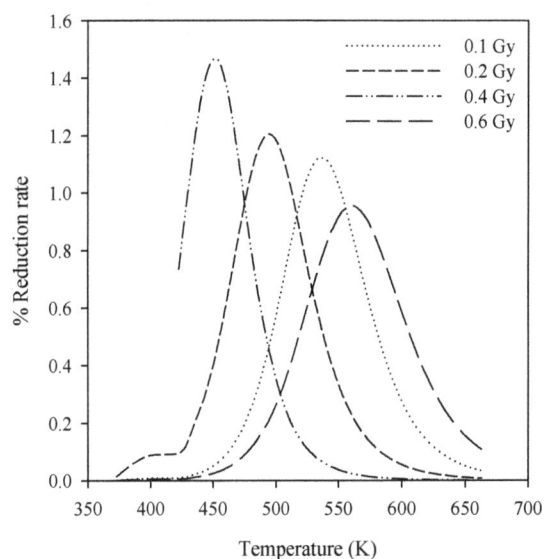

Figure 2. Percentage of the BSL signal reduction plotted against annealing temperature for α-Al$_2$O$_3$:C given different beta-doses.

Table 1. Trap parameters of α-Al$_2$O$_3$:C using GO model and pulse annealing curve fitting.

Dose (Gy)	b	E (eV)	$s\,(1/s)$ $\times 10^{10}$	E (Puls-anneling curve fitting) (eV)	r^2
0.1	1.71	1.09	0.6	1.15	0.99
0.2	1.87	0.91	0.7	1.05	0.99
0.4	1.85	0.94	1.3	1.07	0.99
0.6	1.60	1.10	0.3	1.066	0.99

The experimentally measured CW-OSL decay curves from the α-Al$_2$O$_3$:C is shown in **Figure 3** for different beta-doses. Their dose dependence can clearly be seen. The total luminescence counts increase as dose increases as expected in the luminescence technique and the increasing dose distorts the shape of the peak. An initial increase was observed in the BSL intensity from Al$_2$O$_3$:C. McKeever and Chen [29] reported that the reason for the initial increase in luminescence intensity was thermally metastable traps. So, in this study the measurements used to calculate the parameters were taken after the maximum intensity. These decay curves were fitted to the sum of two simple exponential functions and the decay constants are shown in the **Table 2**. These curves were transformed to LM-OSL curves using the transformation equation suggested by Bulur [16] (**Figure 4**). Although the measured decay curves at 0.1, 0.15 and 0.2Gy of Al$_2$O$_3$:C can be fitted to the sum of two simple exponential functions, the LM-OSL curve cannot clearly display the different two peaks (**Figure 4**). This situation was also seen in the work reported by Bulur *et al.*, [30].

Figure 5 shows the comparison between the experimental decay curve and the analysis model's decay curve known as AOSL Approximation. The peak-shaped forms obtained using the AOSL approximations are shown in **Figure 6**. The two peaks having different decay constants can be clearly seen for a 0.2 Gy dose. The decay curves at 0.4 and 0.6 Gy were fitted to only one simple exponential function. In the AOSL approximation the graph of the logarithm of the luminescence intensity versus the time clearly shows the two decay constants (**Figure 7**). The decay constants determined from these graphs are also shown in **Table 2**. However the decay constant, λ_2, decreases as dose increases but λ_1 does not change. The line equations for λ_1 at **Figure 7** are as follows:

$y = -0.197x + 9.359$ ($r^2 = 0.999$) for 0.1 Gy;
$y = -0.196x + 9.740$ ($r^2 = 0.999$) for 0.15 Gy;
$y = -0.192x + 10.046$ ($r^2 = 0.999$) for 0.2 Gy.

They are as follows for λ_2:

$y = -0.0699x + 8.6$ ($r^2 = 1$) for 0.1 Gy;
$y = -0.049x + 8.42$ ($r^2 = 0.999$) for 0.15 Gy;
$y = -0.0346x + 8.40$ ($r^2 = 0.999$) for 0.2 Gy.

Yukihara *et al.*, [13] also observed that the rate of de-

cay increases as dose increases.

5. Conclusions

The determination of OSL trap parameters is one of the fundamental requirements in understanding the lumines-

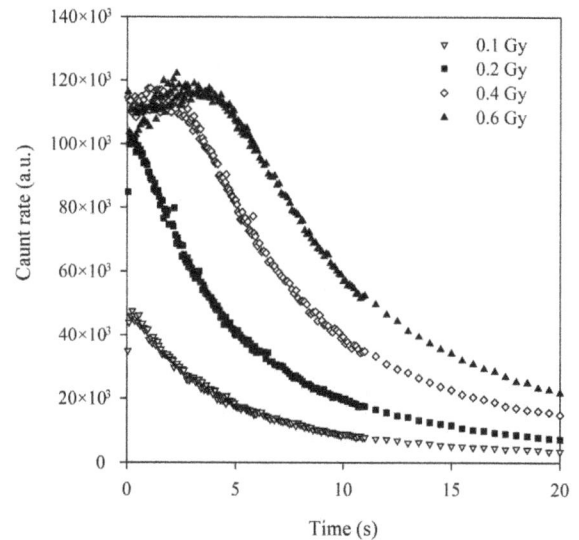

Figure 3. The decay curves measured experimentally.

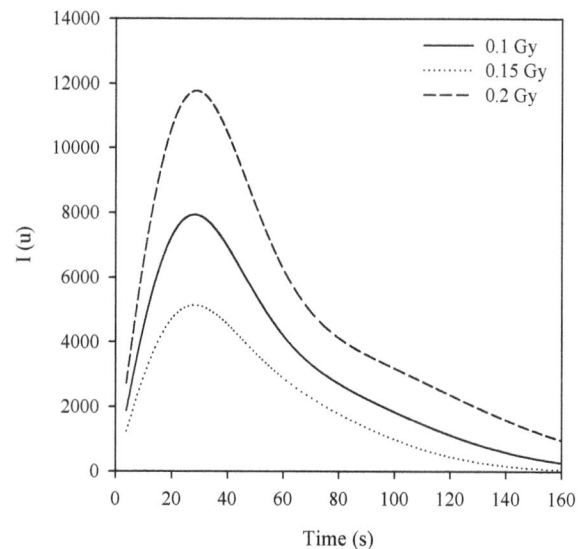

Figure 4. LM-OSL curves.

Table 2. The decay constants using CW-OSL curve-fitting and AOSL approximation.

Dose (Gy)	$\lambda_1\,(1/s)$		$\lambda_2\,(1/s)$	
	CW-OSL	AOSL	CW-OSL	AOSL
0.1	0.296	0.197	0.070	0.0699
0.15	0.260	0.196	0.049	0.0493
0.2	0.240	0.192	0.034	0.0346

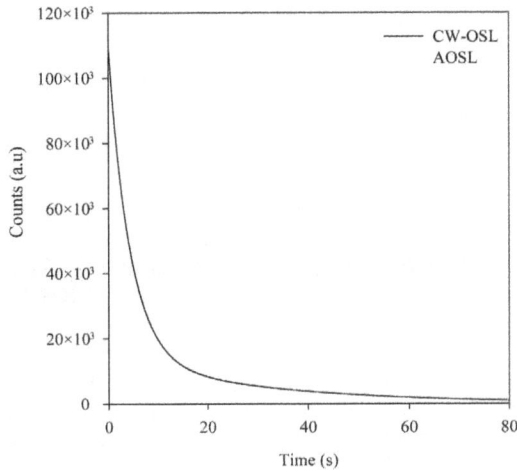

Figure 5. Comparison of decay curves between CW-OSL and AOSL.

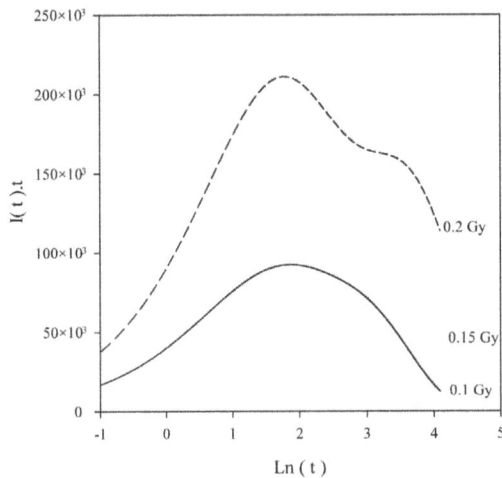

Figure 6. Peak-shape form of luminescence intensity using AOSL approximation.

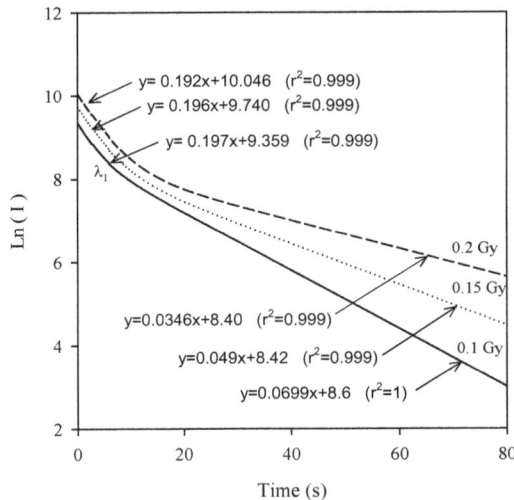

Figure 7. The graph of the logarithm of luminescence intensity versus time using AOSL approximation: The difference between the decay constants was clearly determined.

cence mechanism. Trap parameters of Al$_2$O$_3$:C sample are also important for dating and dosimetric studies using the OSL technique. This study shows that the activation energy of Al$_2$O$_3$:C does not change with beta radiation dose but that its decay constants do change. It may be said that the results of activation energies obtained from different experimental conditions are nearly the same.

The decay curve of Al$_2$O$_3$:C nears a single simple exponential function as the dose increases. AOSL approximation is reliably valid in determining the decay constants.

6. Acknowledgements

We are grateful to Dr. Enver Bulur for his helpful discussions, especially in giving the α-Al$_2$O$_3$:C sample.

REFERENCES

[1] M. S. Akselrod, V. S. Kortov, D. J. Kravetsky and V. I. Gotlib, "Highly Sensitive Thermoluminecent Anion-Defective α-Al$_2$O$_3$:C Single Crystal Detector," *Radiation Protection Dosimetry*, Vol. 32, 1990, pp. 15-29.

[2] A. E. Akselrod and M. S. Akselrod, "Correlation between OSL and the Distribution of TL Traps in Al$_2$O$_3$:C," *Radiation Protection Dosimetry*, Vol. 100, No. 1-4, 2002, pp. 217-220.

[3] L. Bøtter-Jensen, L. N. Agersnap, B. G. Markey and S. W. S. McKeever, "Al$_2$O$_3$:C as a Sensitive OSL Dosemeter for Rapid Assessment of Environmental Photon Dose Rates," *Radiation Measurements*, Vol. 27, No. 2, 1997, pp. 295-298.

[4] L. Bøtter-Jensen, D. Banarjee, H. Jungner and A. S. Murray, "Retrospective Assessment of Environmental Dose Rates Using Optically Stimulated Luminescence from Al$_2$O$_3$:C and Quartz," *Radiation Protection Dosimetry*, Vol. 84, No. 1-4, 1999, pp. 537-542.

[5] E. Bulur and H. Y. Göksu, "Pulsed Optically Stimulated Luminescence from α-Al$_2$O$_3$:C Using Green Light Emitting Diodes," *Radiation Measurements*, Vol. 27, No. 3, 1997, pp. 479-488.

[6] E. Bulur, H. Y. Göksu and W. Wahl, "Infrared (IR) Stimulated Luminescence from α-Al$_2$O$_3$:C," *Radiation Measurements*, Vol. 29, No. 6, 1998, pp. 625-638.

[7] J. M. Edmund and C. E. Andersen, "Temperature Dependence of the Al$_2$O$_3$:C Response in Medical Luminescence Dosimetry," *Radiation Measurements*, Vol. 42, No. 2, 2007, pp. 177-189.

[8] E. G. Yukihara and S. W. S. McKeever, "Ionization Density Dependence of the Optically And Thermally Stimulated Luminescence from Al$_2$O$_3$:C," *Radiation Protection Dosimetry*, Vol. 119, No. 1-4, 2006, pp. 206-217.

[9] V. H. Whitley and S. W. S. McKeever, "Linear Modulation Optically Stimulated Luminescence and Thermolu-

mines- cence Techniques in Al$_2$O$_3$:C," *Radiation Protection Do- simetry*, Vol. 100, No. 1-4, 2002, pp. 61-66.

[10] R. H. Biswas, M. K. Murari and A. K. Singhv, "Dose-Dependent Change in the Optically Stimulated Luminescence Decay of Al$_2$O$_3$:C," *Radiation Measurements*, Vol. 44, No. 5-6, 2009, pp. 543-547.

[11] B. G. Markey, S. W. S McKeever, M. S. Akselrod, L. Bøtter-Jensen, L. N. Agersnap and L. E. Colyott, "The Temperature Dependence of Optically Stimulated Luminescence from α-Al$_2$O$_3$:C," *Radiation Protection Dosimetry*, Vol. 65, No. 1-4, 1996, pp. 185-189.

[12] S. W. McKeever, "Optically Stimulated Luminescence Dosimetry," *Nuclear Instruments and Methods in Physics Research Section B*, Vol. 184, No. 1-2, 2001, pp. 29-54.

[13] E. G. Yukihara, V. H. Whitley, S. W. S. McKeever, A. E. Akselrod and M. S. Akselrod, "Effect of High-Dose Irradiation on the Optically Stimulated Luminescence of Al$_2$O$_3$:C," *Radiation Measurements*, Vol. 38, No. 3, 2004, pp. 317-330.

[14] V. Pagonis, R. Chen and J. L. Lawless, "A Quantitative Kinetic Model for Al$_2$O$_3$:C TL Response to Ionizing Radiation," *Radiation Measurements*, Vol. 42, No. 2, 2007, pp. 198-204.

[15] D. R. Mishra, S. Anuj, N. S. Rawat, M. S. Kulkarni, B. C. Bhatt and D. N. Sharma, "Method of Measuring Thermal Assistance Energy Associated with OSL Traps in α-Al$_2$O$_3$:C Phosphor," *Radiation Measurements*, Vol. 46, No. 8, 2011, pp. 635-642.

[16] E. Bulur, "An Alternative Technique for Optically Stimulated Luminescence (OSL) Experiment," *Radiation Measurements*, Vol. 26, No. 5, 1996, pp. 701-709.

[17] M. S. A. Rasheedy, "New Evaluation Technique for Analyzing the Thermoluminescence Glow Curve and Calculating the Trap Parameters," *Thermochimica Acta*, Vol. 429, No. 2, 2005, pp. 143-147.

[18] G. Tanır and M. H. Bolukdemir, "An Alternative View on the Kinetics of Optical Stimulated Luminescence Decay," *Journal of Radioanalytical and Nuclear Chemistry*, Vol. 285, No. 3, 2010, pp. 563-568.

[19] G. A. Tanır and M. H. Bolukdemir, "Application of Active-OSL Approximation to Some Experimental Opti-

cal Stimulated Luminescence Decay," *Turkish Journal of Physics*, Vol. 35, 2011, pp. 265-272.

[20] D. Curie, "Luminescence in Crystals," Wiley, New York, 1963.

[21] M. S. Akselrod, A. C. Lucas, J. C. Polf and S. W. S. McKeever, "Optically Stimulated Luminescence of Al$_2$O$_3$," *Radiation Measurements*, Vol. 29, No. 3-4, 1998, pp. 391-400.

[22] G. Kitis, I. Liritzis and A. Vafeiadou, "Deconvolution of Optical Stimulated Luminescence Decay Curves," *Journal of Radioanalytical and Nuclear Chemistry*, Vol. 254, No. 1, 2002, pp. 143-149.

[23] Y. Kirsh and R. Chen, "Analysis of the Blue Phosphorescence of X-Irradiated Albite Using a TL-Like Presentation," *Nuclear Track Radiation Measurements*, Vol. 18, No. 1-2, 1991, pp. 37-40.

[24] F. O. Ogundare, F. A. Balogun and L. A. Hussain, "Evaluation of Kinetic Parameters of Traps in Thermoluminescence Phosphors," *Radiation Measurements*, Vol. 41, No. 7-8, 2006, pp. 892-896.

[25] E. Bulur, "A Simple Transformation for Converting CW-OSL Curves to LM-OSL Curves," *Radiation Measurements*, Vol. 32, No. 2, 2000, pp. 141-145.

[26] I. K. S. Krane, "Introductory Nuclear Physics," John Wiley and Sons, New York, 1988, p. 161.

[27] N. A. Spooner, M. A. Aitken, B. W. Franks and C. McElroy, "Archaeological Dating by Infrared Stimulated Luminescence Using a Diode Array," *Radiation Protection Dosimetry*, Vol. 34, No. 1-4, 1990, pp. 83-86.

[28] M. S. Akselrod, L. N. Agersnap, V. Whitley and S. W. S. McKeever, "Thermal Quenching of F Centre Luminescence in Al$_2$O$_3$:C," *Radiation Protection Dosimetry*, Vol. 84, No. 1-4, 1999, pp. 39-42.

[29] S. W. S. McKeever and R. Chen, "Luminescence Models," *Radiation Measurements*, Vol. 27, No. 5-6, 2007, pp. 625-661.

[30] E. Bulur, L. Botter-Jensen and A. S. Murray, "LM-OSL Signals from Some Insulators: An Analysis of the Detrapping Probabilty on Stimulation Light Intensity," *Radiation Measurements*, Vol. 33, No. 5, 2001, pp. 715-719.

22

New Method for Diagnostics of Ion Implantation Induced Charge Carrier Traps in Micro- and Nanoelectronic Devices

Mukhtar Ahmed Rana[1,2]

[1]Physics Department, International Islamic University, Kashmir Highways, Islamabad, Pakistan.
[2]Physics Division, Directorate of Science, Institute of Nuclear Science and Technology, Islamabad, Pakistan

ABSTRACT

An important problem of defect charging in electron-hole plasma in a semiconductor electronic device is investigated using the analogy of dust charging in dusty plasmas. This investigation yielded physical picture of the problem along with the mathematical model. Charging and discharging mechanism of charge carrier traps in a semiconductor electronic device is also given. Potential applications of the study in semiconductor device technology are discussed. It would be interesting to find out how dust acoustic waves in electron-hole plasma in micro and nanoelectronic devices can be useful in finding out charge carrier trap properties of impurities or defects which serve as dust particles in electron-hole (*e-h*) plasma. A new method based on an established technique "deep level transient spectroscopy" (DLTS) is described here suggesting the determination of properties of charge carrier traps in present and future semiconductor devices by measuring the frequency of dust acoustic waves (DAW). Relationship between frequency of DAW and properties of traps is described mathematically proposing the basis of a technique, called here, dust mode frequency deep level transient spectroscopy (DMF-DLTS).

Keywords: Semiconductors; Ion Implantation; Defects; Dust Acoustic Waves; Deep Level Transient Spectroscopy; Electron-Hole Plasma; Nanoelectronics

1. Introduction

During the fabrication, defect or particulate contamination of microelectronic device structures is a serious concern. It can be due to unwanted growth on growth reactor walls or impurities in supply of atomic species in the reactor. Extended defect structures are also formed in different semiconductors used in electronic devices during ion implantation. Extended structures in ion implanted Sb doped Ge (Ge: Sb) form extended defect structures which show high thermal stability and also transform to other defect phases rather than repair. These defects have fatal effects on transport of charge carriers in semiconductor thin films used in electronic devices, which makes knowledge about them important [1-4]. Due to limitation of carrier mobility in Si, Ge has emerged as a key ingredient material for fabrication of a new generation of ultrafast devices in the regime of tens of gigahertz. Ion implantation can be used to write a network of devices on a semiconductor wafer due to possibility of controlled implantation beam size and energy, and masking of the target wafer.

Elemental implantation defects, especially charge carrier traps, and their electronic implications in Ge and similar materials have not been understood with sufficient details, so a comprehensive understanding of properties of defects (vacancies and self-interstitials) in Ge and related materials, and their interactions with impurity atoms is still required [4-7]. These elemental defects join impurities in implanted/doped semiconductors to form defect clusters which act as charge carrier traps. A generalized model for charging and discharging of defects in semiconductor electronic devices is presented here along with its perspective applications.

2. A Model of Defect/Impurity Charging

2.1. Physical Picture

In a perfect crystal of, say, Ge or Si, all the atoms are at their lattice sites and do not have ability to trap additional charge carriers. If impurities are introduced in a perfect crystal, they get ability to trap charge carriers. It may be noted that self interstitial and vacancies are also defects and can trap charge carriers. In real crystalline structure grown for use in devices, contain a variety of defects and sometimes form defect clusters. These clusters are normally composed of vacancy-dopant or interstitial-dopant clusters. These extended clusters can trap a considerable fraction of electrons or holes present in the active volume of a device.

It is thought that these types of extended defect clusters are present in N-type Ge and prevent the feasibility of high speed electronic devices. Depending upon crystal and dopant, preset defect clusters can trap electrons or holes depending on the coulomb potential around defect, impurities or a defect cluster. Electron-hole plasmas are found in electronic devices crystals exposed to optical excitations etc. Defect clusters are exposed to electron and hole currents due to which they are charged. Defect charging in semiconductors is quite complex. After charging processes end, charged defects start discharging naturally. Charging/discharging of defect clusters in semiconductor devices are important and affect performance of devices. Charging and discharging of defects can be used for characterization of charge carrier traps in micro and nanoelectronic devices. Behaviour of deep level traps, which forms the basis for DLTS technique, is described by a schematic in **Figure 1** showing electron and hole traps in a semiconductor. Capture & emission by deep-level traps in a semiconductor. DLTS makes use of the measurement of the depletion region of a simple semiconductor electronic device, normally, a Schottky diode. In such a measurement, the regular reverse polarization of the diode is modified by the probing voltage pulse, resulting in the flow of carriers from the bulk to the measurement region, charging the defects or traps. After the pulse ends, thermal emission of carriers from the traps produces the capacitance transient in the device, called the DLTS signal.

2.2. Mathematical Model

Suppose an electron-hole plasma is generated in an electronic device with electron density as n_e and hole density n_h For simplicity, we assume here extended defect clusters capable of trapping a considerable number of electrons. We consider a case of net electron trapping or negatively charged defects in a semiconductor device. The model situation resembles the typical case of ion implanted Ge:Sb. For the simplicity, we assume that at the time of generation of electron-hole plasma, $n_e - n_h$. In implanted Ge:Sb, highly stable vacancy-antimony-Ge clusters ($V_l Sb_m Ge_n$) are present which are thought to be electron traps in this n-type Ge. These extended defect clusters get charged like dust particles and grains in electron-ion plasma. Let the density of defect cluster or dust grain in e-h plasma, generated in the above mentioned device, is n_d.

We further assume here that densities of electrons and holes are below the limit at which inter particle distance

becomes lower than de Broglie wavelength of any specific specie which assures that that the case e-h plasma being discussed here is of classical nature. If we take charge of electron as $-e$, hole as $+e$ and defect or dust grain $s - Z_d \cdot e$, the well-known condition of charge neutrality can be written as below:

$$-n_e e + n_h e - n_d Z_d e = 0 \qquad (1)$$

Using simple algebra,

$$\frac{n_e}{n_h} = 1 - \frac{n_d}{n_h} Z_d \qquad (2)$$

For any practical device, defect grain density is much lower than densities of electrons and holes, so the number

$$\frac{n_d}{n_h} Z_d \ll 1$$

which is the case of isolated dust or defect grains [8-11]. It may also be noted that in present devices defect grains can be only up to a few nanometer size. Maximum charge on defect depends upon defect structure and its size. A isolated defect or dust grain in e-h plasma attains a floating potential φ_p which is determined by electron repulsion and hole collection in case of net negative charging of defects. The electron current responsible for depositing electrons on the defect is given by,

$$I_e = -4\pi a^2 e n_e \left(\frac{k_B T_e}{2\pi m_e}\right)^{1/2} \exp\left(\frac{e\varphi_p}{k_B T_e}\right) \qquad (3)$$

where m_e and T_e are electron effective mass and temperature, and a is the radius of the equivalent sphere of defect grain. The hole current compensating electron charging of a defect is given by,

$$I_h = -4\pi a^2 e n_h \left(\frac{k_B T_h}{2\pi m_h}\right)^{1/2} \exp\left(\frac{e\varphi_p}{k_B T_h}\right) \qquad (4)$$

where m_h and T_h are hole effective mass and temperature. For further details of Equations (3), (4), please see Refs. [8-12]. The charging equation of a negatively charged grain can be written as below, [8-10].

$$\frac{dZ_d}{dt} = -\frac{I_e + I_h}{e} \qquad (5)$$

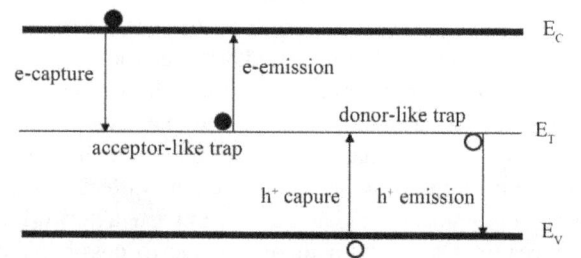

Figure 1. Energy band diagram for a semiconductor with deep-level traps. E_C is lower edge of the conduction band while E_V is the upper edge of the valence band.

Case of hole or positive charging of defects can be described using a simple analogy and comparison between electrons and holes.

2.3. Mechanism of Charging and Discharging of Electron Traps in an Electronic Device

Electron traps are centers in a semiconductor capable of trapping electrons with a certain binding energy. Traps cause power dissipation and reduce conductivity of the device. When electrons and holes are present in the form of a plasma in an electronic device, dust or defect grains are exposed to both electron (T_e) and hole (T_h) currents. These currents simultaneously charge and discharge defect grains. When e-h plasma is first generated ($t = 0$) in an electronic device, a charging transient is produced after which charging and discharging perturbation continues until the device operation ends (end of e-h pair generation). After this, a discharging transient is produced in which electrons are released due to thermal agitation and thermally generated holes. UV light can also produce discharging through photoelectron emission.

3. New Method for the Diagnostics of Charge Carrier Traps in a Micro- and a Nanoelectronic Device

The frequency ω of dust acoustic mode in electron ion plasma, with dust charged as negative, is given by [8-10],

$$\omega = kZ_d \left(\frac{n_{d0}}{n_{i0}}\right)^{1/2} \left(\frac{k_B T_i}{m_d}\right)^{1/2} \left[1 + \frac{T_i}{T_e}\left(1 - \frac{Z_d n_{d0}}{n_{i0}}\right)\right]^{-1/2} \quad (6)$$

where k is the wavenumber, k_B is the Boltzman constan, Z_d is charge on the dust, n_{d0} and n_{i0} are, respectively, equilibrium dust and ion densities, m_i and m_d are, respectively, ion and dust masses, T_e and T_i are electron and ion temperatures respectively and e is the electronic charge. With take the case of semiconductor device working at room temperature and having low electron and hole densities (due to low doping level) so that plasma behavior is classical. With charge traps or dust in the device, the above equation the following form by replacing ion parameters with hole parameters,

$$\omega = kZ_d \left(\frac{n_{d0}}{n_{h0}}\right)^{1/2} \left(\frac{k_B T_h}{m_d}\right)^{1/2} \left[1 + \frac{T_h}{T_e}\left(1 - \frac{Z_d n_{d0}}{n_{h0}}\right)\right]^{-1/2} \quad (7)$$

Now, m_h and n_{h0} are, respectively, hole mass and equilibrium density. Dust particles in e-h plasma are impurities or defect clusters which can be positively charged, negatively charged or neutral. Dust particles in e-h plasma can have a range of mass and mobility coefficient as they can be free or loosely or strongly bound to the host lattice structure. So, dust model in e-h plasma in electronic devices can be of quite complex nature.

Here, following questions about e-h-dust plasma in semiconductor devices are investigated. What can be the nature of dust modes in e-h-dust plasma in semiconductor devices? Can these dust modes be useful in diagnostics of present or future micro or nanoelectronic devices? What can be the practical method of utilizing dust modes in diagnostics of semiconductor devices?

Dust particles (impurities or defects) in a semiconductor device act as electron or hole traps depending upon its electronic structure. If a pulse of electrons or holes is injected into a semiconductor device, as in a charge carrier trap characterization technique DLTS [13-15], these dust particles will be charged with electrons are holes. After the charge carrier feeding pulse ends, charged defects or dust particles will start discharging and it will continue until dust particles or defects reach their equilibrium charge. Dust mode frequency will also change as dust charge changes with time. This changing mode frequency with time is given by the following,

$$\omega(t) = kZ_d(t)\left(\frac{n_{d0}}{n_{h0}}\right)^{1/2}\left(\frac{k_B T_h}{m_d}\right)^{1/2}\left[1 + \frac{T_h}{T_e}\left(1 - \frac{Z_d(t)n_{d0}}{n_{h0}}\right)\right]^{-1/2} \quad (8)$$

In DLTS technique, the device is biased at a reverse bias V_R at which the bias is pulsed by a trap filling pulse V_P for a time t_p. Filled defect states start emptying, when bias returns back to V_R, and that results in a capacitance transient. Filling pulse is applied in a series and the varying capacitance of the diode $C(t,T)$ is given by [16],

$$C(t,T) = C_\infty - \Delta C_0 \exp(-e_p t) \quad (9)$$

where C_∞ is the equilibrium capacitance value at reverse voltage, ΔC_0 is the lowering of C_∞ at $t = 0$ and e_p is the hole emission coefficient. The above equation will be valid for electron traps with e_p replaced with e_n (the electron emission coefficient) and ΔC_0 is the rise of C_∞ at $t = 0$. DLTS spectrum is generated by applying a sine-function correlation with period P to $C(t,T)$ as,

$$S(T) = -\frac{1}{P}\int_0^P \sin\left(\frac{2\pi t}{P}\right) \cdot C(t,T)\,\mathrm{d}t \quad (10)$$

The above procedure is applied to determine fundamental Fourier component of the transient, $C(t,T)$. For exponential transients, the period P, the emission coefficient e_p and T_{max} for which $S(T)$ shows a maximum, are related as [17]

$$1/e_p(T_{\max}) = 0.42435 \times P \quad (11)$$

By choosing different appropriate values of period P, corresponding values of e_p and T_{max} can be determined for a specific transient $C(t,T)$ corresponding to a particular hole trap.

As clear from Equation (3), the frequency of dust

would depend on varying charge Z_d of dust particles in the same way as $C(t,T)$ during dust or trap discharge after the application of the filling pulse. So, dust mode frequency transient $\omega(t,T)$ for the same hole trap as in Equation (4), an analogue to capacitance transient $C(t,T)$, can be defined as,

$$\omega(t,T) = \omega_\infty - \Delta\omega_0 \exp(-e_p t) \qquad (12)$$

where ω_∞ and $\Delta\omega_0$ are, respectively, equilibrium dust mode frequency and full magnitude of the frequency transient at $t = 0$. In the same way as in Equation (5), dust mode frequency DLTS or DMF-DLTS signal, $S_\omega(T)$, can be generated by applying the correlation sine-function as in Equation (5),

$$S_\omega(T) = -\frac{1}{P}\int_0^P \sin(\frac{2\pi t}{P}) \cdot \omega(t,T)\, dt \qquad (13)$$

DMF-DLTS will show a set of peaks of $S_\omega(T)$ signal, each corresponding to a particular dust specie. This spectrum of peaks will be similar to a DLTS spectrum (**Figure 2**) with different DLTS peaks, with peak parameters carrying finger prints of dust species. Once dust mode frequency ω_∞ and change $\Delta\omega_0$ are measured, further details of experimental and analysis aspects of implementation of this new DLTS can be seen by a selected studies available in the common literature [13-19].

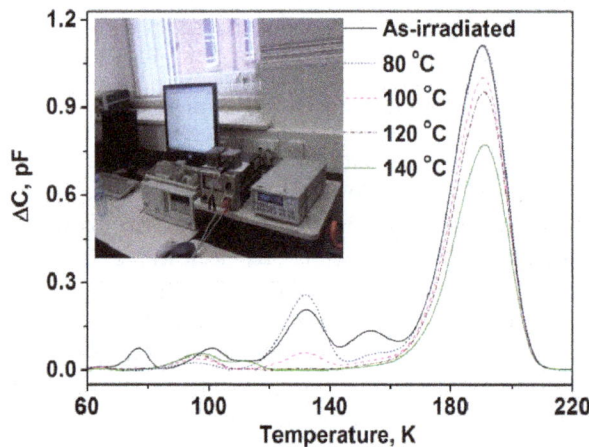

Figure 2. DLTS signal showing traps caused by the electron irradiation induced in Sb-doped ($\rho \approx 20$ Ω·cm) Ge sample for the filling pulse of −0.5 V. Reverse bias −5 V and filling pulse duration of 10ms were used. As a simple rule, each peak represents an electronically active trap. Measurements of the same sample annealed for 30 m at temperatures of 100°C - 220°C are also shown which reduction or evolution of traps due to thermal treatment. The inset shows the measurement system at Microelectronics & Nanostructures Group, Electrical and Electronic Engineering, UMIST, University of Manchester, Manchester M60 1QD, UK University of Manchester, UK. These results were collected with the help of A. R. Peaker, V. P. Markevich and I. D. Hawkins.

4. Potential Applications and Discussion

The present investigation provides the essential information for understanding charging/discharging of defects in semiconductor electronic devices using transient and steady state conditions of operation. It is well known that switching on and off of a range of electronic devices affects their life. Charging and discharging currents produced, respectively, in switching on and off of an electronic devices affect its quality life.

Deep level transient spectroscopy [13-15] makes use of charging and discharging capacitance transients for characterization of charge carrier traps in a semiconductor electronic device. According to this study small scale charging and discharging perturbations continue even during steady state operation of devices. These perturbations would have higher magnitude and intensity if an electronic devices operation different operational stages. These perturbations produce a fluctuating filed in the device. Fluctuating fields can have more deleterious effects on the operating device, especially if the fluctuation frequency is high.

Here, we classify defects in a semiconductor electronic device into three classes which are electronically inactive defects, electronically static defects and electronically dynamic defects. Electronically dynamic defects continue producing charging and discharging transients in an operating device and are most undesirable. Electronically inactive defects only decrease active volume of the device and have minimum undesirable effects whereas electronically static defects trap charge carriers but do not produce negligible charging and discharging transients after getting charged at start up of the device. The work presented here in conjunction with the fundamental plasma work [20,21] may lead to applications in the field of Nanotechnology [22,23] and nuclear instruments and methods [24-27].

Figure 3 is a pictorial view of findings of the paper. Physical significance of charge carrier traps in a semiconductor is shown graphically at the right bottom of the figure. Immediately above it shown is a physical process of bending of a crystal (as a pictorial view) which can introduce charge carrier traps in a crystal.

5. Conclusion

Defects are charged due to electron and hole currents. Meaningful charging/discharging continues until $I_e + I_h = 0$. Start up of a device and in some conditions the condition $I_e + I_h = 0$ is not fulfilled, so charge fluctuations continue casing energy dissipation which is undesirable due to the problem of decreased efficiency. Energy dissipation also limits life of the device. Using this investigation, defects in semiconductor electronic devices are classifieds into three groups, electronically inactive defects,

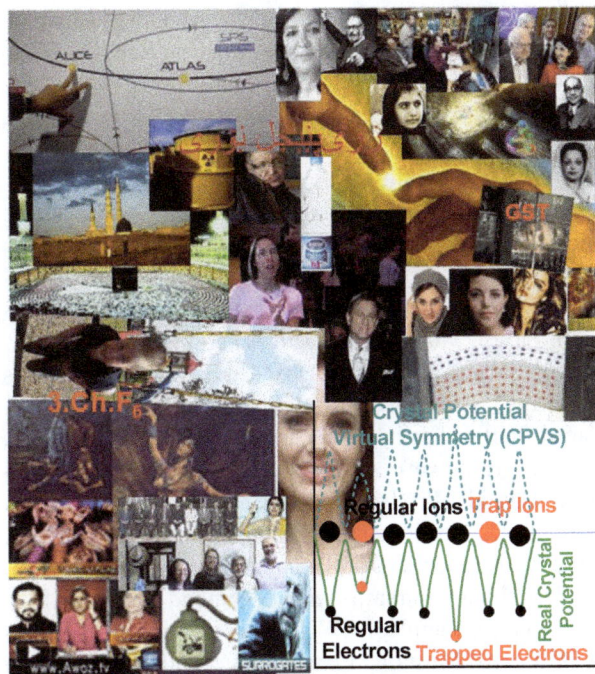

Figure 3. This is a pictorial view of the findings reported in the paper. For the better apprehension of the reader, some aesthetic/geometric Visualization of traps are shown. Right bottom (RB) is geometric visualization of charge carrier traps (here electrons shown) in the form of a plot. These traps have an analogy with massive dust particles in traditional and fullerene plasma [9 ,10]. Above RB plot, a pictorial view of bending crystal model is show. Bending of a crystal can possibly create traps in it. Other than RB plot, typical mounty or mounted pictures are taken freely available on the web. Leading sources of the pictures are webpages of FermiLab Today (US), CERN (Switzerland), Dawn News (Pakistan), OGRA (Pakistan), PINSTECH/SID Library (PAEC, Islamabad, Pakistan).

electronically static defects and electronically dynamic defects. Electronically dynamic defects are most harmful for semiconductor electronic devices. A new technique DMF-DLTS is proposed here by describing measurement principle of the technique. This new technique combined with conventional and Laplace DLTS techniques can serve as a combination of carrier trap characterization techniques capable of extracting further advanced information about carrier traps in semiconductor devices.

6. Acknowledgements

Generous help regarding "deep level transient spectroscopy" (DLTS) and discussions with Prof. Tony Peaker, Dr. Vladimir Markevich, Dr. Ian Hawkins, Prof. Aimin Song, Miss Sam Aldean, Miss Kerry Plant, Dr. Leszek Majewski, Prof. Bruce Hamilton, Ms. Marie Davies, Dr. Huda El mubarek (Royal Academy of Engineering and EPSRC Research Fellow), Dr. Ahmed El-makadema, Ms. Katy Woolfenden (the Central library) and Prof. Mo-hamed Missous at the University of Manchester, UK, are gratefully acknowledged. Revealing discussion with Dr. Waqas Masood, Prof. Pervez Hoodbhoy, Mr. Haidar Rizvi, Sughra, Mother Teresa, Haseeb Hasnain, Qamarul Haq and Dr. Nawab Ali about geomagnetic field, solar winds, and Saturn's rings in links with the space plasma ought to be thanked. Thanks are also for Mr. Asif Osman for morality motivated and philosophical healthy discussions. Helpful and dedicated thoughts sharing of Dr. Shoaib Ahmad (presently at CASP, Church Road, GCU, Lahore) about micro and nanoelectronic devices are respectfully appreciated. This work started as composing of knowledge from distant fields of plasma physics and nanoelectronics which later achieved to be a mature draft/manuscript. The financial help from the HEC, Islamabad (Pakistan) is heart-feelingly acknowledged. The help/discussion with Ms. Ee Jin Teo, Ms. Halen (Philpine), Markus Zmeck, Andrew Bettiol (Australia), Ong Yiew Leung, Abdullah Al Mamun, Abdullah Al Mamun, Shazib Pervaiz, at the National University of Singapore, are appreciated. Thanks are also due for *G.* Giacomelli, *L.* Patrizii, *V. Togo* from the Bologna University; and Vincenzo Guidi, Andrea Mazzolari, the University of Ferrara; Sara Maria Carturan, INFN—Laboratori Nazionali di Legnaro (Italy); G. Schiwietz (Helmholtz-Zentrum Berlin f. Materialien u. Energie, Hahn-Meitner-Platz 1, 14109 Berlin, Germany) and P. L. Grande (Uni. Federal do Rio Grande do Sul, Av. Bento Gonçalves 9500, 91501-970 Porto Alegre, RS, Brazil), Sandro Guedes, Julio Hadler and Eduardo Curvo (Instituto de Fisica Gleb Wataghin, Universidade Estadual de Campinas, UNI-CAMP, 13083-970 Campinas, SP, Brazil) have been helpful; Nidal Dwaikata, Ghassan Safarinib, Mousa El-hasanb (An-najah National University, Nablus, Jerusalem, Palestine); Shoji Hashimoto, Shin Ugawa, Kazuki Nanko & Koji Shichi (Soil Resources Laboratory, Department of Forest Site Environment, Forestry and Forest Products Research Institute, 1 Matsunosato, Tsukuba, Ibaraki, 305-8687, Japan); K. Stübner, R. C. Jonckheere, L. Ratschbacher (Institut für Geowissenschaften, Technische Universität Bergakademie Freiberg, Sachsen, Germany); Xiangdong Ji (Institute of Nuclear, Particle, Astronomy and Cosmology Offic: Physics Building Rm. 606 (Shanghai Jiatong University, Shanghai, PRC; MIT, USA); Fatih Gerçekcioğlu (Kirikkale University, Turkey); Laureline Bourcier (Institute for Reference Materials and Measurements, Joint Research Center, European Commission, B-2440 Geel, Belgium); Prof. Man Gyun Na (Chosun University, Korea (South)), S. Mostafa Ghiaasiaan Editor, Editor ANE at G. W. Woodruff School of Mechanical Engineering Erskine Love Manufacturing Building Georgia Institute of Technology, Atlanta.

This paper marks the concurrent occurrence of 4 events

including Islamic Conference 2012 at the Makkah, Saudi Arabia, the Pakistan's Independence Day Celebrations (Mubarek), the 2012 Olympics, London, UK, and NASA's Mars Mission which benefitted me in further shaping my mind. Participation in 2 workshops: 1) Pakistan's Strategic Needs; 2) The Role of China in Nuclear Power in Pakistan), at NAFDEP, the Pitfalls Waterlanes, the Neelam Silk Route, Islamabad, were beneficial.

Discussion/help/inspiration from individuals/places/concepts/institutions (local or at places mentioned above, mentioned otherwise) Ahmed Raza Brailvi, Shah Ahmed Noorani, Mufti Mahmood, Hussain Ahmad, Liaqat Balauch, Tahir-ul-Qadri, Haji Abdul Wahab, Pir Pagaro, Ray Wind, Shaukat-Khanum, Naseer-ud-Din Shah, Qari Khushi Muhammad Al-Azahari, Data Nagari, Haroon Rashid, Adnan Kayani, Naheed, Saad Sajid Bashir, Shahid Bilal Butt, Ishtiaq Kashmiry, Razia Butt, Atta-ul-Haq Qasmi, Skill Development Initiative (KINPOE & KANUPP, Karachi), Zafar Shuttle Service, Faiz-ul-Hassan, Khalid Minhas, Nafeesa Nazali, Farhat, Ayesha, Shazia, Serwat, Kanwal, Rahat Bakers, Kalabagh, Nadra, Ada Kidman, Military College Jhelum, Sarai Alamghir; Pakistan Naval Academy, Karachi; Pakistan Air Force Academy, Risalpur: Iqbal, Karsaz, Rashid Minhas, M. M. Alam, Mehran, Bahadur, Maj Aurangzeb, Brig (R) Muhammad Afzal Malik, Lt Col Abdul Qayyum Satti, Capt Shahbaz Anjum, Miss Shazia Tabassum, Miss Aneela Malik, Miss Amber Amin, Mr. Sajjad; Balqis Kaur, Ameer Chak, Bashir, Bajwa, Gujjar, University of Balauchistan, Quetta: Abdul Manan Bazai, Zahid Ali Marwat, Mr. Muhammad Arif, Ms. Faiqa Abdul Hayee, Ms. Faiza Mir. Fida Muhammad Bazai; Faisal, Hina Jeelani, McDonald, KFC, NIM, Asma Jahangir, Veena Malik, Mahreen A. Raja, Naseeb (Narowal), Matiullah/Shafiq, Babar, Rashid, Arshi, Maulvi Malik Mushtaq (Master Teddi, Shakargarh), Aziz, Sorruyya Deendar, Prof. Zaka Ullah, Belal Baaquie, Murray Barrett, Meng Hau Kuok, Nidhi Sharma, Kuldip Singh, Edward Teo, Xuesen Wang, Artur Ekert, B. V. R. Chowdary, Ms. Teo Hoon Hwee, Ms. Zhou Weiqian, Faiza, Fariha Pervez, Sajjad Ali, Komal Rizvi, Abrar-ul-Haq. Ali Haider, Alamgir, Aman, Yasin Malik, Zaid Bin Annehan, Hamza Tzortzis, Asma Jahangir, Aslam Masih, Muhammad Bashir, Rawla Kote, Azad Kashmir (24 Hours), Aziz Qureshi, Nasreen Qureshi, R.M. Qureshi; Sy Hong Kok (Singapore), Ghazzala, Ashiq Hussain Dogar, Shagufta, Rana Rab Nawaz, Muhammad Saeed, Maria Sultan Böhmer Wasti, Sharmila Tagore, Zulfiqar, Saleem, Shikh Zaid Bin Sultan Al Nahian (UAE), Maimoona Murtaza Malik, Nida, Safdar, Sohail, Nichia (Japan), Sajjad Ali, Afnaan, Hasnain, Kainat, Nauman, Sameena, Shafqat, Shaukat Ali Bhatti, Imran, Awais, Bilal, Maaz, Masrur, Shakeel, Syed Noor, Tahira Sayed, Syed Hameed Qaiser, Fareeha Pervez, Shamoon Maseeh, Raheem Shah, Zain, Abrar-ul-Haq, the proposed Mireeza Foundation, PIMS, Islamabad, Urba, Abdullah, Ghannia, Asad, Iqra, Dar-ul-Amman, Naat Khawan Rabeya, Ashiq Anjum, Sana, Tandoori, Lavish Dine, Tariq Dogar, Pak Continental, Gourmet Cola (Lahore), Mukhtar Maseeh, Adnan, Saima, Mohsin, Lubna, A. R. Peaker, University of Pretoria, Republic of South Africa: M Diale, T Hlatswayo, J. Janse van Rensburg, D. Langa, W. Meyer, C. Moji, J. Nel, Claudia Zander, E. A. Meyburgh, P. Chakraborty, J. van der Merwe; Fauzi, Manzoor Hussain, Nuclear Decommissioning Authority (NDA): John Mathiesen, Easton Murray, Elizabeth Atherton; Shahid, Azhar, Dar-us-Salaam (Singapore), Al-Huda Quran & Hadees Academy, Islamabad, Khalid, Jameel, Hafiz, Faisal, Saeed, Riffat, Fasih, Mano, Umar, Farah, Saima, Sidra, Showkat (Sahiwal), Aftab Iqbal (Okara). Javed Chaudhary, Hamid Mir, Irfan Siddiqi, Asma Shirazi, Saleem Safi, R. U. Yusufzai, Shah Rukh Khan, Shehzad Roy, Kareena Kapoor, Baba Nanak, Kabeer, K. K. Khattak, Dr. Karishma (Gujraat), Azhar Mehmood Kayani (AFIC, Rawalpindi), Kamila, LCWU, Lahore:Yasmin Ali, Shafaq Arif, Lubna Mustafa, Rehana Zia; Dalda and Meezan Cook-books, Sardar Abdul Qayyum, Azad Kashmir, PTCL, OGDCL, Daak Khana, Madina Cash & Carry, Makkah Traders, RanaAcademy, Saqib Osman, Muhammad Akhtar, Mushtaq Ahmad. Inspiration from the mind awakening Hamd & Naat writers & reciters/movies/characters/dramas/books/singers/newpapers/channels/websites, Altaaf Hussain Hali, A Beautiful Mind, The Ugly Truth, Pearl Harbor, Black Rain (Hiroschima, Nagasaki), The Message, Un-Kahe, Ghazi Ilam Din Shaheed, Tere Naam, Shabana, Ram Chand Pakistani, Tere bin Laden, Banjaran, Silsila, Kabhi Kabhi, Heer, Cleopatra, Vizontele, Tears of Gaza, Downfall, Mr. Bean, My Fair Lady, Das Leben der Anderen, Tokio Hotel, Half Plate, Devils on the Doorstep, Dess, Waqt, Time, BBC (Hard talk), Newweek (Fareed Zakaria), Fatmire Bajramaj (Kosovo & Germany), Marco Soggetto (Eni, Milano, Italy), CNN (Christiane Amanpour), Google, Wikipedia, Youtube, Dutsche Welle, VoA, Beijing Radio, Radio Masco, Ranghun, Bol, Radio Pakistan, FM 92.5, ARY Qur'an TV, Geo, Dunya, Dawn, Express News, PTV.

REFERENCES

[1]   F. D. Auret, A. R. Peaker, V. P. Markevich, L. Dobaczewski and R. M.Gwilliam, "High-Resolution DLTS of Vacancy-Donor Pairs in P-, As- and Sb-Doped Silicon," *Physica B: Condensed Matter*, Vol. 376-377, 2006, pp. 73-76.

[2]   V. P. Markevich, I. D. Hawkins, A. R. Peaker, K. V. Emtsev, V. V. Litenov, L. I. Murin and L. Dobaczewski, "Vacancy-Group-V-Impurity Atom Pairs in Ge Crystals Doped with P, As, Sb, and Bi," *Physical Review B*, Vol. 70, No. 23, 2004, Article ID: 235213.

[3]   H. J. Queisser and E. E. Haller, "Defects in Semiconduc-

tors: Some Fatal, Some Vital," *Science*, Vol. 281, No. 5379, 1998, pp. 945-950.

[4] P. Muret, J. Pernot, T. Teraji and T. Itoh, "Near-Surface Defects in Boron-Doped Diamond Schottky Diodes Studied from Capacitance Transients," *Applied Physics Express*, Vol. 1, No. 3, 2008, Article ID: 035003.

[5] J. Coutinho, S. Oeberg, V. J. B. Torres, M. Barroso, R. Jones and P. R. Briddon, "Donor-Vacancy Complexes in Ge: Cluster and Supercell Calculations," *Physical Review B*, Vol. 73, No. 23, 2006, Article ID: 235213.

[6] S. Roy and A. Asenov, "Where Do the Dopants Go?" *Science*, Vol. 309, No. 5733, 2005, pp. 388-390.

[7] V. P. Markevich, "A Comparative Study of Ion Implantation and Irradiation-Induced Defects in Ge Crystals," *Materials Science in Semiconductor Processing*, Vol. 9, No. 4-5, 2006, pp. 589-596.

[8] P. K. Shukla and A. A. Mamun, "Introduction to Dusty Plasma Physics," Institute of Physics Publishing, Bristol and Philadelphia, 2002.

[9] W. Masood, H. Rizvi, H. Hasnain and Q. Haque, "Rotation Induced Nonlinear Dispersive Dust Drift Waves Can be the Progenitors of Spokes," *Physics of Plasmas*, Vol. 19, No. 3, 2012, Article ID: 032112.

[10] W. Masood, A. M. Mirza and S. Nargis, "Revisiting Coupled Shukla-Varma and Convective Cell Mode in Classical and Quantum Dusty Magnetoplasmas," *Journal of Plasma Physics*, Vol. 76, No. 3-4, 2010, pp. 547-552

[11] M. S. Barnes, J. H. Keller, J. C. Forster, J. A. O'Neill and D. K. Coultas, "Transport of Dust Particles in Glow-Discharge Plasmas," *Physical Review Letters*, Vol. 68, No. 3, 1992, pp. 313-316.

[12] A. Piel and A. Melzer, "Dusty Plasmas—The State of Understanding from an Experimentalist's View," *Advances in Space Research*, Vol. 29, No. 9, 2002, pp. 1255-1264.

[13] D. V. Lang, "Recalling the Origins of DLTS," *Physica B*, Vol. 401-402, No. 1, 2007, pp. 7-9.

[14] D. V. Lang, "Deep-Level Transient Spectroscopy: A New Method to Characterize Traps in Semiconductors," *Journal of Applied Physics*, Vol. 45, No. 7, 1974, pp. 3023-3032.

[15] D. V. Lang and L. C. Kimerling, "Observation of Recombination-Enhanced Defect Reactions in Semiconductors," *Physical Review Letters*, Vol. 33, No. 8, 1974, pp. 489-492.

[16] S. Voss, N. A. Stolwijk, H. Bracht, A. N. Larsen and H. Overhof, "Substitutional Zn in SiGe: Deep-Level Transient Spectroscopy and Electron Density Calculations," *Physical Review B*, Vol. 68, No. 3, 2003, Article ID: 035208.

[17] D. S. Day, M. Y. Tsai, B. G. Streetman and D. V. Lang, "Deep-Level-Transient Spectroscopy: System Effects and Data Analysis," *Journal of Applied Physics*, Vol. 50, No. 8, 1979, pp. 5093-5098.

[18] S. Q. Wang, F. Lu, Z. Q. Zhu, T. Sekiguchi, H. Okushi, K. Kimura and T. Yao, "Compensating Levels in p-Type ZnSe:N Studied by Optical Deep-Level Transient Spectroscopy," *Physical Review B*, Vol. 58, No. 16, pp. 10502-10509.

[19] J.-U. Sachse, J. Weber and H. Lemke, "Deep-Level Transient Spectroscopy of Pd-H Complexes in Silicon," *Physical Review B*, Vol. 61, No. 3, 2000, pp. 1924-1934.

[20] M. Bonitz, C. Henning1 and D. Block, "Complex Plasmas: A Laboratory for Strong Correlations," *Reports on Progress in Physics*, Vol. 73, No. 6, 2010, Article ID: 066501.

[21] E. K. El-Shewy, M. A. Zahran, K. Schoepf and S. A. Elwakil, "Contribution of Higher Order Dispersion to Nonlinear Dust-Acoustic Solitary Waves in Dusty Plasma with Different Sized Dust Grains and Nonthermal Ions," *Physica Scripta*, Vol. 78, No. 2, 2008, Article ID: 025501.

[22] L. Boufendi, M. Ch Jouanny, E. Kovacevic, J. Berndt and M. Mikikian, "Dusty plasma for Nanotechnology," *Journal of Physics D: Applied Physics*, Vol. 44, No. 17, 2011, Article ID: 174035.

[23] M. Z. Iqbal and N. Zafar, "Study of Alpha-Radiation-Induced Deep Levels in p-Type Silicon," *Journal of Applied Physics*, Vol. 73, No. 9, 1993, pp. 4240-4247.

[24] E. Gaubas, G. Juška, J. Vaitkus and E. Fretwurst, "Characterization of the Radiation-Induced Defects in Si Detectors by Carrier Transport and Decay Transients," *Nuclear Instruments and Methods in Physics Research A*, Vol. 583, No. 1, 2007, pp. 185-188.

[25] R. M. Keyser and T. W. Raudorf, "Germanium Radiation Detector Manufacturing: Process and Advances," *Nuclear Instruments and Methods in Physics Research A*, Vol. 286, No. 3, 1990, pp. 357-363.

[26] C.-X. Liu, S. Cheng, H.-T. Guo, W.-N. Li, X.-H. Liu, W. Wei and B. Peng, "Proton-Implanted Optical Planar Waveguides in Yb3+-Doped Silicate Glasses," *Nuclear Instruments and Methods in Physics Research B*, 2012, in Press.

[27] S. M. C. Miranda, N. Franco, E. Alves and K. Lorenz, "Cd Ion Implantation in AlN," *Nuclear Instruments and Methods in Physics Research B*, 2012, in Press.

23

Differential Cross Section of Electron Scattering from ^{3}He and ^{3}H Nuclei with Considering Pionic Contribution

Nazli Hamdolahi[1], Farhad Zolfagharpour[2*], Negin Sattary Nikkhoo[2]
[1]Department of Physics, University of Azad Tehran Markaz, Tehran, Iran
[2]Department of Physics, University of Mohaghegh Ardabili, Ardabil, Iran

ABSTRACT

The excess pion inside the nucleus could change not only nucleons structure function, but also electron scattering cross section from nuclei. In this paper, we calculate pionic contribution in nuclear structure functions of ^{3}He and ^{3}H nuclei and the differential cross sections of electron scattering from these nuclei. At first, we calculate the Fermi motion and binding energy contribution as an important nuclear medium effect in scattering cross sections and then we add pionic contribution to structure functions and differential cross sections of electron scattering from these nuclei that the Fermi motion and binding energy are considered.

Keywords: Pionic Contribution; Structure Function; Differential Cross Section

1. Introduction

One of the important questions in nuclear physics is difference between a free nucleon structure function with a bound nucleon structure function. In this paper, we have investigated one of the main reasons of this difference in nucleus structure function in $0.1 \leq x \leq 0.3$ range by considering pion sea around nucleons inside nucleus. A nucleus consists of nucleons, mesons, and objects like Δ particle [1]. Nucleons and mesons made of quarks and sea quarks. Quarks inside both of nuclei and mesons could not be marked and distinguished, so the electron scattering from bound nucleons are different from free nucleons [2-4]. Therefore, the free nucleon structure function is different from bound nucleon structure function. Pions are the lightest mesons, and presences of them in a nucleus are more than other mesons. So, we investigate presence of pions in nucleus. Since, a free nucleon is a particle that virtual pions embedded around nucleons and creating a pion cloud around nucleons. When a nucleon is inside nuclei, pion cloud is increasing around the nucleon. When electrons scatter from nuclei, the probability of the interaction between a virtual photon of incident electrons with a bound nucleon is increasing, because interaction between virtual photon with pions is increasing. The virtual photon scatters from not only a quark insides nucleons, but also a quark or sea quark from pion cloud [5]. This effect is obvious in range $x \leq 0.2$. Therefore, pion cloud plays a role in nuclei structure functions. In this paper, first, we calculate pionic contri-

butions in ^{3}He and ^{3}H nuclei, and then we calculate cross sections by considering these contributions.

2. Pion Structure Function

The nucleus structure function, in nuclear conventional theory, with respecting the pion cloud effect is [5]:

$$
F_2^A\left(x,Q^2\right) = \int_x^A f_\pi\left(z\right) F_2^\pi\left(\frac{x}{z},Q^2\right)dz \\
+ \int_x^A f_N\left(z\right) F_2^N\left(\frac{x}{z},Q^2\right)dz, \; z \geq x
\tag{1}
$$

First term indicates the pionic contribution and the next term indicates the nucleon contribution in the nucleus. First term contribution is noticeable around $x \cong 0.2$. x is Bjorken variable. z is the light-cone momentum fraction of the nucleus carried by nucleon, and its distribution is inside nucleus $f_N\left(z\right)$ taken from [6]. $F_2^N\left(\frac{x}{z},Q^2\right)$ is the nucleon structure function [7]. The momentum distribution $f_\pi\left(z\right)$ in a free nucleon case is given as follow [8]:

$$
f_\pi^A\left(z\right) = \frac{3g^2}{16\pi^2}\Delta\lambda z\left(\frac{1}{\lambda}\exp\left(-2\lambda\frac{t_0+m_\pi^2}{m_\pi^2}\right)\right. \\
\left. + \frac{1}{2}Ei\left(-2\lambda\frac{t_0+m_\pi^2}{m_\pi^2}\right)\right)
\tag{2}
$$

where

$$Ei(-z) = -\int_{z}^{\infty} dt \frac{e^{-t}}{t} \qquad (3)$$

which $g = 13.5$ is the coupling constant. The cut off parameter λ plays the most substantial role. When the nucleon is embedded in a nucleus, several modifications such as the polarization of nuclear medium occur. They may be expressed by an effective change of λ. We have taken $\lambda = 0.026$ [8] and pion mass $m_\pi = 139.570$ MeV.

We take $\Delta\lambda > 0$ to have

$$\eta_\pi = \int_{0}^{M_\pi/m} dz \, z f_\pi^A(z) \qquad (4)$$

where η_π is the momentum fraction of pion. We have taken $\eta_\pi = 0.130$ for ^{2}H and $\eta_\pi = 0.155$ for ^{3}He and ^{3}H nuclei from [8], also $\Delta\lambda = 0.0031$ for ^{2}H and $\Delta\lambda = 0.00367$ for ^{3}He and ^{3}H nuclei.

For pion structure function, we used follow parameterization [9,10]:

$$F_2^\pi(x) = \frac{5}{9} x V_\pi(x) + \frac{4}{3} x s_\pi(x) \qquad (5)$$

$$x V_\pi(x) = \frac{\Gamma(\alpha+\beta+1)}{\Gamma(\alpha)\Gamma(\beta+1)} x^\alpha (1-x)^\beta \qquad (6)$$

$$x S_\pi(x) = \frac{1}{6} A(p+1)(1-x)^p \qquad (7)$$

$$\alpha = 0.36 - 0.074\overline{S}, \quad \beta = 0.99 + 0.60\overline{S} \qquad (8)$$

$$\overline{S} = \ln\left(\frac{\ln(Q^2/\Lambda^2)}{\ln(Q_0^2/\Lambda^2)}\right) \qquad (9)$$

$$Q_0^2 = 25(\text{GeVc}^{-1})^2, \Lambda = 0.2\text{GeVc}^{-1}, p = 8.7,$$
$$A = 0.51 - 2\alpha/(\alpha+\beta+1) \qquad (10)$$

In **Figure 1**, we plotted the momentum distribution of pions inside ^{2}H, ^{3}He and ^{3}H nuclei.

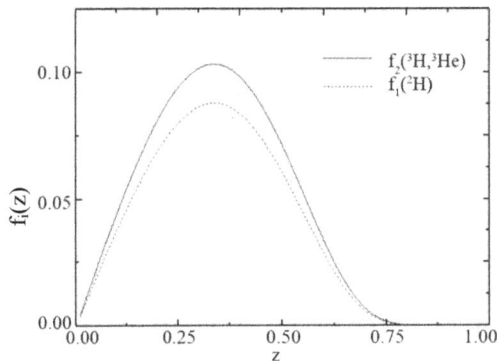

Figure 1. The momentum distribution of pions for ^{2}H, *i.e. i* = 1 in figure, shown with dotted curve, ^{3}He and ^{3}H, *i.e. i* = 2 in figure, shown with full curve.

3. Calculating Differential Cross Section of Electron Scattering from Nucleus

The structure functions for charged lepton scattering from a nucleon are related to cross section by [11]:

$$\frac{d^2\sigma}{d\Omega dE'} = \frac{4\alpha^2(E'')^2 \cos^2\left(\frac{\theta}{2}\right)}{Q^4}$$
$$\times \left[\frac{F_2(x,Q^2)}{\upsilon} + \frac{2F_1(x,Q^2)\tan^2\left(\frac{\theta}{2}\right)}{M}\right] \qquad (11)$$

$$Q^2 = -q^2 = -4EE'\sin^2\left(\frac{\theta}{2}\right), \nu = E - E'$$

where

$$\alpha = \frac{e^2}{4\pi} \sim \frac{1}{137}$$

is the fine structure constant, four-momentum transfer squared is Q^2. Initial and scattered lepton energies are E and E', respectively. Energy of the virtual photon is $\nu = E - E'$, and

$$x = \frac{Q^2}{2M\nu}$$

is Bjorken scaling variable. M is the nucleon rest mass. θ is the detected lepton scattering angle. F_1 and F_2 are the deep inelastic structure functions.

4. Results and Discussion

In **Figures 2** and **3**, nucleus structure functions for ^{3}He

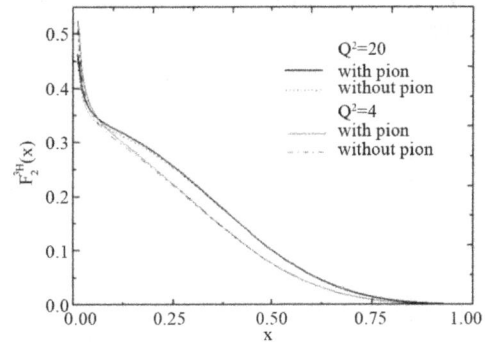

Figure 2. This figure shows ^{3}H structure functions. The full curve is plotted for $Q^2 = 4$ G$_e$V^2 with considering the Fermi motion, the binding energy, and the pionic contribution effect, the dotted curve is plotted with considering the Fermi motion, and binding energy effect. The full curve is plotted for $Q^2 = 20$ G$_e$V^2 with considering the Fermi motion, the binding energy, and the pionic contribution effect, the line-dash is plotted with considering the Fermi motion, and the binding energy effect.

and ^{3}H nuclei have been plotted by considering the Fermi motion, the binding energy and the pionic contribution. In this calculation we do not consider any difference between pionic cloud around the protons and neutrons and the contribution is equal ^{3}H and ^{3}He nuclear structure functions. This figure shows that the pionic contribution has maximum around $x \approx 0.2$ that one expected.

In addition, the full curve in **Figures 4** and **5** show differential cross sections of electron scattering from ^{3}He and ^{3}H nuclei with considering the Fermi motion, the binding energy and the pionic contribution. The dotted curves in **Figures 4** and **5** show differential cross sections of electron scattering from ^{3}He and ^{3}H nuclei with considering only the Fermi motion and the binding energy effects. These figures show the pionic contribution

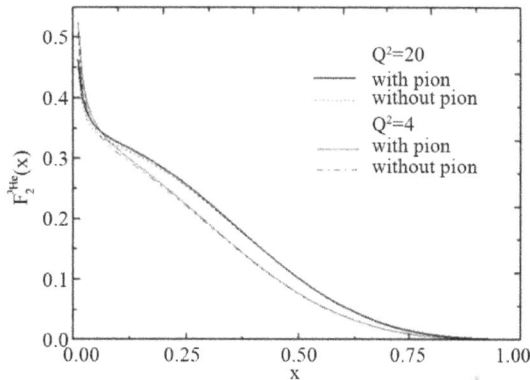

Figure 5. Differential cross section of electron scattering from ^{3}He at $\theta = 8°$, $Q^2 = 4$ GeV^2, and E = 7.26 GeV. The full curve is plotted with considering the Fermi motion, the binding energy, and the pionic contribution. The dotted curve is plotted with considering only the Fermi motion and the binding energy.

cloud improve cross section up to %1 or %2, which is perceptible in small x. In these figures pionic contribution increases when v goes up because pion cloud inside nuclei plays a role like sea quark inside nucleons.

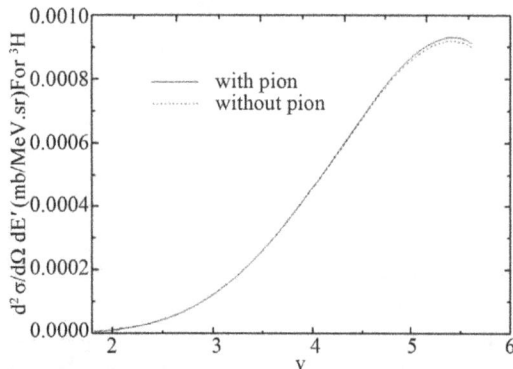

Figure 3. This figure shows ^{3}He structure functions. The full curve is plotted for $Q^2 = 4$ GeV^2 with considering the Fermi motion, the binding energy, and the pionic contribution effect, the dotted curve is plotted with considering the Fermi motion, and binding energy effect. The full curve is plotted for $Q^2 = 20$ G$_e$V^2 with considering the Fermi motion, the binding energy, and the pionic contribution effect, the line-dash is plotted with considering the Fermi motion, and the binding energy effect.

REFERENCES

[1]   G. Cattapan and L. Ferreira, "The Role of the Δ in Nuclear Physics," *Physics Report*, Vol. 362, No. 5-6, 2002, pp. 303-407.

[2]   J. Ashman, B. Badelek, G. Baum, J. Beaufays, C. P. Bee, C. Benchouk, *et al.*, "Measurement of the Ratios of Deep Inelastic Muon-Nucleus Cross Sections on Various Nuclei Compared to Deuterium," *Physics Letter B*, Vol. 202, No. 4, 1988, pp. 603-610.

[3]   P. Amaudruz, M. Arneodo, A. Arvidson, B. Badelek and G. Baum, *et al.*, "Precision Measurement of the Structure Function Ratios F_2^{He}/F_2^D, F_2^C/F_2^D and F_2^{Ca}/F_2^D," *Zeitschrift fur Physik C*, Vol. 51, No. 3, 1991, pp. 387-393.

[4]   The European Muon Collaboration, M. Arneodo, A. Arvidson, J. J. Aubert, B. Badelek, J. Beaufays, C. P. Bee, C. Benchouk, *et al.*, "Measurements of the Nucleon Structure Function in the Range $0.002 < x < 0.17$ and $00.2 < Q^2 < 8$ GeV^2 in Deuterium, Carbon and Calcium," *Nuclear Physics B*, Vol. 333, No. 1, 1990, pp. 1-47.

[5]   M. Arneodo, "Nuclear Effects in Structure Functions," *Physics Report*, Vol. 240, No. 5-6, 1990, pp. 301-393.

[6]   S. V. Akulinichev, S. Shlomo, S. A. Kulagin and G. M. Vagradov, "Lepton-Nucleus Deep Inelastic Scattering," *Physical Review Letters*, Vol. 55, No. 21, 1985, pp. 2239-2241.

[7]   M. Gluck, E. Reya and A. Vogt, "Dynamical Parton Distributions of Parton and Small-x Physics," *Zeitschrift fur Physik C*, Vol. 67, No. 3, 1995, pp. 433-447.

[8]   T. Uchiyama and K. Saito, "European Muon Collaboration Effect in Deuteron and in Three-Body Nuclei,"

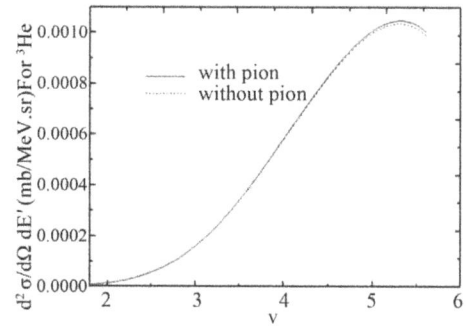

Figure 4. Differential cross section of electron scattering from ^{3}H at $\theta = 8°$, $Q^2 = 4$ G$_e$V^2, and $E = 7.26$ G$_e$V. The full curve is plotted with considering the Fermi motion, the binding energy, and the pionic contribution. The dotted curve is plotted with considering only the Fermi motion and the binding energy.

Physical Review, C. Vol. 38, No. 5, 1988, pp. 2245-2250.

[9] E. L. Berger and F. Coester, "Nuclear Effects in Deep-Inelastic Lepton Scattering," *Physical Review D*, Vol. 32, No. 5, 1985, pp. 1071-1083.

[10] K. Nakano, "Remarks on Pionic Constraints in the EMC

Effect," *Journal of Physics G, Nuclear Part*, Vol. 17, No. 11, 1991, pp. L201-L207.

[11] J. Gomez, R. G. Arnold, P. E. Bosted, C. C. Chang, A. T. Katramatou, G. G. Petratos, *et al.*, "Measurement of the a Dependence of Deep-Inelastic Electron Scattering," *Physical Review D*, Vol. 49, No. 9, 1994, pp. 4348-4372.

24

Radon Awareness among Saudi People in Riyadh, Saudi Arabia

Abdulaziz S. Alaamer
Department of Physics, College of Science, Al-Imam Mohammad Ibn Saud Islamic University, Riyadh, KSA

ABSTRACT

People should be aware of sources of health hazards, such as radon gas; and efforts should be made to educate them. Radon awareness among people is important for monitoring its level in their residential houses to reduce the risk of adverse health effects. Furthermore, radon awareness among public would support and facilitate researchers working for such surveys during the field work. In the present study, a public survey was conducted to investigate radon awareness level among Saudi people in Riyadh. A questionnaire was designed and distributed among 2297 persons of various educational background. It was found that about 82% of the Saudi public in Riyadh were ignorant of radon and its associated health risks. It was also found that only ~18% of educated public knew about radon. It was concluded that Saudi public needs to be educated in this respect properly.

Keywords: Radon; Awareness; Riyadh; Saudi Arabia

1. Introduction

Radon is a noble radioactive gas produced by decay of ^{226}Ra, a member of ^{238}U-decay series. Uranium is present in almost all types of rocks, soils, plants and ground waters. Radon (^{222}Rn) is a significant indoor exposure source of natural background. Exposure to radon from soil, building materials or water may be through ingestion and/ or by inhalation of ^{222}Rn gas released from these sources, which pose potential health hazards [1-3]. In the last two decades, there has been a great deal of awareness about the health risks associated with radon and its decay products. ^{222}Rn and its progeny are one of the most significant natural sources of radiation exposure to the population [4]. Sustained exposure to indoor radon and its progeny is believed to be associated with a potential health risk of lung cancer [5]. The Surgeon General of the United States has assessed that radon is the second leading cause of lung cancer in the United States [6]. It is estimated that 50% - 55% of the average annual dose from natural background radiation sources is contributed by ^{222}Rn [7].

Radon has been surveyed on a large-scale mainly in the USA and Europe [8-10]. The study of radon has been carried out in air, soils, building materials and water. Production rates from the soil depend upon the geological characteristics of the soil itself and its underlying geological strata [11]. The measurement of indoor radon levels have also been carried out in main cities of Saudi Arabia [12-16].

Although radon gas level in Saudi Arabia is generally low [16,17], people should know the present levels of radon in their dwellings to avoid its health hazards. This present study gives results of a survey conducted in Riyadh city, the capital of Saudi Arabia regarding level of radon awareness among the general public. It is expected that this study will play an important role to educate Saudi people about radon and get them interested in assessing radon level in their houses. Furthermore, when uneducated Saudi people are aware of radon, they will not object the installation of radon do- simeters in their houses [18].

2. Materials and Methods

A questionnaire was designed containing questions relating to radon awareness among people in Riyadh of various educational backgrounds. The questionnaire was distributed randomly among university students, school teachers, government employees, municipality engineers, house contractors, real state offices and academics. Uneducated people were excluded from the questionnaire. Both genders, male and female participated in this survey. A total number of 2297 participants responded to the questionnaires. The questionnaire was divided into three sections as detailed below:

Section 1:

1) What is your qualification? [High Secondary School, Undergraduate, Graduate, Postgraduate]

2) What is your study specialization? [Humanities, Medical, Scientific, Other]

3) Do you know Radon or have heard about it? [Yes, No]

Section 2:

4) What is Radon (1)? [Electronic device, Natural material, Artificial material, I don't know]

5) What is Radon (2)? [Gas, Liquid, Solid, I don't know]

6) Is Radon: [Active, Inactive, Radioactive, I don't know]

7) Is Radon: [Colorless, Odorless, Tasteless, I don't know]

8) Where does Radon come from? [Underground, Water, Air, I don't know]

Section 3:

9) Is Radon harmful? [Yes, No, may be, I don't know]

10) Is Radon useful? [Yes, No, may be, I don't know]

11) If Radon is harmful or useful, where in body? [Head, Chest, Abdomen, I don't know]

3. Results and Discussion

Table 1 represents the results of survey conducted. It is shown that among the participants, 4.1% participants were high secondary school students, 30.2% college undergraduates, 64.6% college graduates, and 1.1% master and PhD holders. None of the school students had heard

about radon at all. However, they were excluded from further analysis of the study. About 82.4% of participants had never heard about radon. Only 17.6% knew or had heard of radon. Those who had not heard about radon included undergraduates (91.2%), graduates (77.7%) and postgraduates (53.8%). In general, almost all humanity-studied people (96.1%) had not heard of radon. Only 3.9% people who had heard of radon among them were probably highly educated people. Obviously, about 50.7%, 27.0% and 27.6% among specialists of science, medical and others, e.g., engineers respectively, had heard about radon.

The fourth question, which was "what is radon?", examined the truth and certainty of those who answered that they knew radon. Those who answered that radon was an electron device meant that they had not really heard about radon. About 95.5% participants from this section answered radon was a natural material as shown in **Table 2**. Section 2 of the questioner investigated the knowledge of radon characteristics as shown from **Table 3**. The majority of participants (~97%) knew that radon is a radioactive, colorless, odorless, tasteless gas and underground material. Only 2.0% participants were unaware of radon. As reflected from **Table 4**, Section 3 of questioner investigated the awareness of health hazards of radon. Almost all participants (~99%) knew that radon is a harmful to the body.

Table 1. Survey results of the participants who knew or had heard of radon.

	High S. School		Undergraduate		Graduate		Postgraduate	
What is your qualification?	94		694		1483		26	
% of total participants	4.1%		30.2%		64.6%		1.1%	
	Yes	No	Yes	No	Yes	No	Yes	No
	0	94	61	633	331	1152	12	14
	0.0%	100.0%	8.8%	91.2%	22.3%	77.7%	46.2%	53.8%
	Humanities		Medical		Scientific		Other	
Study specialization	1421		37		546		199	
% of total educated participants	64.5%		1.7%		24.8%		9.0%	
	Yes	No	Yes	No	Yes	No	Yes	No
	56	1365	10	27	277	269	55	144
	3.9%	96.1%	27.0%	73.0%	50.7%	49.3%	27.6%	72.4%

Table 2. Testing whether participants truly had heard about radon.

Question	Electronic device	Natural material	Artificial material	I don't know
What is radon (1)	3	386	14	1
% of total participants	0.4%	16.8%	0.6%	0.0%
% of total participants who know radon	0.7%	95.5%	3.5%	0.2%

Table 3. Awareness of radon characteristics.

Question	Gas	Liquid	Solid	I don't know
What is radon (2)	385	6	5	8
% of total participants	16.8%	0.3%	0.2%	0.3%
% of total participants who know radon	95.3%	1.5%	1.2%	2.0%
	Active	Inactive	Radioactive	I don't know
Is radon	1	0	392	11
% of total participants	0.0%	0.0%	17.1%	0.5%
% of total participants who know radon	0.2%	0.0%	97.0%	2.7%
	Colorless	Odorless	Tasteless	I don't know
Is radon	392	392	392	7
% of total participants	17.1%	17.1%	17.1%	0.3%
% of total participants who know radon	97.0%	97.0%	97.0%	1.7%
	Underground	Water	Air	I don't know
Where does radon come from?	400	255	311	0
% of total participants	17.4%	11.1%	13.5%	0.0%
% of total participants who know radon	99.0%	63.1%	77.0%	0.0%

Table 4. Awareness of health hazards of radon.

	Yes	No	May be	I don't know
Is radon harmful?	401	0	3	0
% of total participants	17.5%	0.0%	0.1%	0.0%
% of total participants who know radon	99.3%	0.0%	0.7%	0.0%
	Yes	No	May be	I don't know
Is radon useful?	0	404	0	0
% of total participants	0.0%	17.6%	0.0%	0.0%
% of total participants who know radon	0.0%	100.0%	0.0%	0.0%
	Head	Chest	Abdomen	I don't know
If radon is harmful or useful, where in body?	2	386	2	14
% of total participants	0.1%	16.8%	0.1%	0.6%
% of total participants who know radon	0.5%	95.5%	0.5%	3.5%

From the above results it shows that in general, radon awareness increases among educated people with an increasing educational background level and with science education in particular. The radon awareness level among the general public in the USA is relatively higher than those of educated people in Saudi Arabia. Internet and other information media may be used to enhance radon awareness level. However, this survey showed that the radon awareness of general public is scarce. This suggests that efforts by the concern authorities should be made to improve radon awareness level via the press, media and lectures.

4. Conclusion

In general, Saudi people have poor radon awareness level. Among educated people, only ~18% are aware of radon and its health hazards. Radon awareness level increases among educated people with increasing scientific education level. Efforts should be made by the concern authorities to enhance radon awareness level in the general public using tools of the press, media and lectures.

REFERENCES

[1] K. N. Yu, Z. J. Guan, M. J. Stokes and E. C. M. Young, "A Preliminary Study on the Radon Concentrations in Water in Hong Kong and the Associated Health Effects," *Applied Radiation and Isotopes*, Vol. 45, No. 7, 1994, pp. 809-810.

[2] M. J. Barnett, K. E. Holbert, B. D. Stewart and W. K. Hood, "Lung Dose Estimates from Radon in Arizona Ground Water Based on Liquid Scintillation Measurements," *Health Physics*, Vol. 68, No. 5, 1995, pp. 699-703.

[3] Z. A. Tayyeb, A. R. Kinsara and S. M. Farid, "A Study on the Radon Concentration in Water in Jeddah and Associated Health Effects," *Journal of Environmental Radioactivity*, Vol. 38, No. 1, 1998, pp. 97-104.

[4] S. Oikawa, N. Kanno, T. Sanada, N. Ohashi, M. Uesugi, K. Sato, J. Abukawa and H. Higuchi, "A Nationwide Survey of Outdoor Radon Concentration in Japan," *Journal of Environmental Radioactivity*, Vol. 65, No. 2, 2003, pp. 203- 213.

[5] M. Galan Lopez, A. Martin Sanchez and V. Gomez Escobar, "Estimates of the Dose Due to Radon-222 Concentrations in Water," *Radiation Protection Dosimetry*, Vol. 111, No. 1, 2004, pp. 3-7.

[6] International Commission on Radiological Protection, "Protection against 222Rn at Home and at Work. International Commission on Radiological Protection Publication 65," *Annals of the ICRP*, Vol. 23, No. 2, 1993.

[7] United Nations Scientific Committee on the Effects of Atomic Radiation, "Sources and Effects of Ionizing Radiation," *UNSCEAR* Report to the General Assembly, New York, 1993.

[8] E. S. Ford, C. R. Eheman, P. Z. Siegal and P. L. Garbe, "Radon Awareness and Testing Behavior: Findings from the Behavioral Risk Factor Surveillance System, 1989-1992," *Health Physics*, Vol. 70, No. 3, 1996, pp. 363-366.

[9] Y. Wang, C. Ju, A. D. Stark and N. Teresi, "Radon Awareness, Testing, and Remediation Survey among New York State Residents," *Health Physics*, Vol. 78, No. 6, 2000, pp. 641-647.

[10] B. Gregory and P. P. Jalbert, "National Radon Results: 1985 to 2003," US Environmental Protection Agency, Washington DC, 2003,

[11] B. Lévesque, D. Gauvin, R. G. McGregor, R. Martel, S. Gingras, A. Dontigny, W. B. Walker, P. Lajoie and E. Letourneau, "Radon in Residences: Influences of Geological and Housing Characteristics," *Health Physics*, Vol. 72, No. 6, 1997, pp. 907-914.

[12] F. Abu-Jarad and M. I. Al-Jarallah, "Radon Activity in Saudi Houses," *Radiation Protection Dosimetry*, Vol. 14, 1986, pp. 243-249.

[13] F. Abu-Jarad, M. I. Fazal-ur-Rehman, Al-Jarallah and A. Al-Shukri, "Indoor Radon Survey in Dwellings of Nine Cities of the Eastern and the Western Provinces of Saudi Arabia," *Radiation Protection Dosimetry*, Vol. 106, No. 3, 2003, pp. 227-232.

[14] M. S. Garawi, "Measurement of Radon Concentration in Private Houses in the Eastern Part of AL-Qaseem Province of Saudi Arabia," *Radiation Protection Dosimetry*, Vol. 63, No. 3, 1996, pp. 227-230.

[15] M. I. Al-Jarallah and Fazal-ur-Rehman, "Indoor Radon Concentration Measurement in the Dwellings of Al-Jauf Region of Saudi Arabia," *Radiation Protection Dosimetry*, Vol. 121, No. 3, 2006, pp. 293-296.

[16] F. S. Al-Saleh, K. A. Zarie, K. Al-Magran and M.S. Sharaf, "Measurements of Radon Concentration and Exhalation Rate from Some Construction Materials Used in Riyadh Region (KSA)," *Isotopes and Radiation Research*, Vol. 33, 2001, pp. 261-265.

[17] F. S. Al-Saleh, "Measurements of Indoor Gamma Radiation and Radon Concentrations in Dwellings of Riyadh City, Saudi Arabia," *Applied Radiation and Isotopes*, Vol. 65, No. 7, 2007, pp. 843-848.

[18] S. Rahman, M. Faheem, S. Rehman and Matiullah, "Radon Awareness in Pakistan," *Radiation Protection Dosimetry*, Vol. 121, No. 3, 2006, pp. 333-336.

25

Selected Chemical Aspects of Nuclear Power Development

Andrzej G. Chmielewski, Marcin Brykala, Tomasz Smolinski
Institute of Nuclear Chemistry and Technology, Warsaw, Poland

ABSTRACT

The Fukushima nuclear power plant (NPP) accident consequences are a new challenge for nuclear power development; however the sequence of the event has illustrated importance of radiation- and radiochemistry processes on the safe operation and shut down of nuclear reactor and decontamination of formed liquid and solid wastes. A chemistry program is essential for the safe operation of a nuclear power plant. It ensures the integrity, reliability and availability of the main plant structures, systems and components important to safety, in accordance with the assumptions and intent of the design. The proper implementation of these procedures minimizes the harmful effects of chemical impurities and corrosion on plant structures, systems and components. It supports the minimization of buildup of radioactive material and occupational radiation exposure as well as limiting of the release of chemicals and radioactive material to the environment [1].

Keywords: Chemistry; Nuclear Safety; Water Radiolysis; Radioactive Waste Treatment

1. Introduction

The history of radiochemistry has started over hundred years ago. The first separation scheme was developed by Maria Skłodowska-Curie and Pierre Curie [2]. The biggest development of the science was related to military and peaceful applications of fissile and radioactive elements in nuclear power, industry and medicine. The role of chemistry in the present safe and effective operation of all stages of nuclear power related technology is unquestionable.

The Polish Government decided to lunch a Polish Nuclear Energy Program which assumes the construction NPP by the year 2023. The Institute of Nuclear Chemistry and Technology (INCT, known in Poland as IChTJ) have elaborated own programme considering a role of their expertise in radiochemistry, radiation chemistry, radiobiology and radiometric methods to support of this programme. The construction of the Center of Radiochemistry and Nuclear Chemistry for Nuclear Power and Nuclear Medicine has been completed this year.

2. Chemistry and Safety

2.1. Radiomonitoring, Radioecology and Emergency Preparedness

Monitoring of fuel clad integrity, the first and the most important barrier against radioactivity release into the environment, is the most important line of process radiation monitoring at nuclear power installations (NPIs).

Current radioanalytical methods for detection of such radionuclides are usually difficult and time consuming, they require a great effort for sample separation and long-time measurements. Inductively coupled plasma mass spectrometry (ICP-MS) can be a complementary method to radiometric ones. ICP-MS methods of radionuclide determination have started to be developed since the last 15 years. In the INCT works on ICP-MS methods for U, Pu, ^{90}Sr, ^{241}Am are being carried out. The procedures are developed mainly for application to environmental analysis and preparation of test materials for proficiency tests on radionuclide determination which the INCT is a provider [3].

2.2. Coolant Chemistry, Reactor Elements Corrosion and Decontamination of the Primary Circuit

Coolant chemistry and reactor elements corrosion plays a very important role in nuclear engineering, covering all aspects of water radiolysis in boiling water reactor (BWR) and pressurized water reactor (PWR) units, effects of radiolytic phenomena on corrosion, etc.

There are three main tools used to handle reactor coolant system chemistry: pH control, zinc injection and dissolved hydrogen [4].

Decrease in pH of the water causes corrosion of metal equipment and destruction of water supply. Optimal pH of primary water approximately 7.2 +/– 0.2 is recommended, because that minimizing the solubility and corrosion of the surfaces of the main components in the primary system, the difference of solubility and thus of transport of radioactivated isotopes in the various parts of the system at different temperatures, typically in the 270°C - 325°C range. In practice, if the duration of a fuel cycle is extended 18 months or more is extending, it meant that the boron concentration at the beginning of cycle had to be raised from 1200 to 1800 mg/kg (ppm). This requires a lithium concentration greater than 2.2 mg/kg to maintain optimum pH.

The correct pH of water is very important for minimum solubility of nickel ferrite, which is the major component of CRUD. It is product of corrosion, mainly R_3O_4 (R = Fe, Ni or Cr), and a ratio of these elements depends on the type of construction material and type of reactor. Crud circulates inside the cooling system and in the vicinity of the core is activated, then settles on poorly secured locations of pipes, cooling system components and etc., which is a major threat.

The next tool used to handle reactor coolant system chemistry is addition of zinc, where is used zinc's affinity for the tetrahedral sites of spinel oxide minerals' structure. Thus, it can displace other cations such as iron, or nickel from oxide films, which activates into ^{58}Co, a major radioelement. Zinc is added at a concentration of about 5 µg/kg (ppb), which is sufficient to reduce the dose rates by 15% during the first fuel cycle and up to 30% to 40% after a few cycles with zinc injection. Relatively high cost of zinc depleted (^{64}Zn) salts causes searching new additives like aluminum salts. A study by gel chromatography in combination with radioisotope and elemental analyses showed that the Co(II) ions incorporated in polymeric hydrolysis products in the step of their formation were practically fully replaced by aluminum ions in the course of the polymer aging. Because such polymers are precursors of loose corrosion product deposits formed in primary circuits of NPIs and incorporating the most radiation-hazardous radionuclide ^{60}Co, addition of aluminum salts to primary coolants improves the radiation situation in the maintenance area of NPIs and, therefore, can be appropriate solution [5].

The third important tool to handle reactor coolant system chemistry is to optimize hydrogen concentration, applied to mitigate water radiolysis. For a long time, a value of 25 - 35 ml/kg (within a range of 25 - 50) has been used. The US Electric Power Research Institute (EPRI) is considering increasing the hydrogen concentration towards 50 ml/kg and possibly beyond, depending on the results of the ongoing qualification work on the risk of decreasing the time to crack initiation. The intention is to minimize crack growth rates. However, Japanese utilities are evaluating the action of decreasing hydrogen concentration to delay crack initiation.

On the other hand, the knowledge of radiolytic proc-

esses governed by physics and chemistry plays a very important role in the development of fuel reprocessing technologies, waste reprocessing and spent fuel and conditioned waste storage.

One of the challenges of chemistry is decontamination coolant water. It is necessary to establish precise chemistry to achieve optimum results. There are a lot of chemical and physical processes used to decontaminate coolant. It is used ion exchangers, solvent extraction and membrane methods. Water in primary coolant circle might be contaminated by fission products for example cesium or strontium. It might be removed from coolant by using special sorbents. In INCT by J. Narbutt groups synthetized sorbent to composite ion exchanger for selective removal radiocesium [6]. The sorbent allow to remove cesium ions from aqueous solutions containing cations of other metals.

2.3. Fuel Reprocessing

Generation IV reactor concepts and accompanying reprocessing technologies aqueous-based and pyrometallurgical, is a new challenge from the chemistry point of view [7]. Mainly, how to reduce the cost and to improve the safety of the management of such high-level nuclear waste are being considered.

An advanced closed fuel cycle relies on the possibility to maximise the energy usage of nuclear spent fuel and to provide improved waste forms for long-term storage: the removal of minor actinides (i.e.: Np, Am, Cm, Cf) with U and Pu from the waste and in some cases long-lived fission products (LLFP: ^{99}Tc, ^{129}I and ^{135}Cs) and heat generation fission products (^{90}Sr and ^{137}Cs), are partitioned by chemical separation. Separated all elements of significant Trans-Uranium (TRU) elements are immobilized or transmutated in either a thermal or in fast neutron reactors [8] into short-lived or stable isotopes prior to disposal.

Various technologies of partitioning, based on hydrometallurgical processes, were developed to extract the minor actinides from nuclear waste—an acidic (HNO_3) solution as a rule, the aqueous raffinate, which remains after recovering uranium and plutonium in the solvent extraction process PUREX [9]. The technique can also be extended for the recovery of neptunium, but separation of trivalent americium and curium from trivalent lanthanides (Ln) cannot be separated directly in this process. Studies on solving of this issue are carried out for decades by many research teams.

Therefore, two steps of separations processes are at present considered. In the first step, at high acidity, a group separation of MA and lanthanides is carried out, followed by a selective separation of MA from lanthanides at lower acidity.

In this technology developed by the EU consortium

coordinated by C. Madic, in the first step the minor actinides are directly extracted from the PUREX raffinate
together with fission lanthanides as nitrates, using diamide extractants—the malonamides or diglycolamides
(DIAMide EXtraction process) [10]. Various extraction
systems were studied for the second step. The MA (III) +
Ln (III) mixture generated after this first step is low-
acidic to facilitate the second process, the SANEX process (Selective ActiNide EXtraction), which the goal is to
separate the trivalent An from the Ln directly from the
DIAMEX product. This process is based on the BTP,
which belongs to a new family of extractants, the Bis-
Triazinyl-Pyridine developed by Z. Kolarik et al., and is
very efficient for a selective extraction of MA (III) at
high acidity [11] and shows good capabilities in cen-
trifugal extractors.

The EU ACSEPT project in which INCT is active
plays a very important role to achieve the progress in the
chemical separations field [12]. One of the concept of
ACSEPT project is optimize of GANEX (Group Acti-
Nide EXtraction) In the second cycle of separation of
actinides from lanthanides is possible, as well as from
other fission products. Objective is achieved by using as
the organic phase of extraction mixture of bis-triazine-
bipyridine (BTBP) and tri-n-butyl phosphate (TBP) in
cyclohexanone. The aqueous phase contains 4 M nitric
acid, actinides, and lanthanides, and FP. BTBP molecules
containing soft-donor nitrogen atoms favoring complexa-
tion of actinide 5f orbitals have the ability to separate
actinides (III, IV, V and VI) from lanthanides (III). The
TBP allows the extraction of uranium and plutonium
[13].

One promising concept after the partitioning step is
embedding the minor actinides (MA: Am, Cm, Cf) in
uranium-based nuclear fuel by sol-gel process. This will
allow the MAs to be destroyed by fast-neutron reactions
in the upcoming generation-IV reactors.

Even a manufacturing of uranium oxide fuels is well
developed process, synthesis of mixed uranium-MA fuels
is a process under development. At the INCT a new
variant of a sol-gel method called—Complex Sol-Gel
Process (CSGP) has been elaborated to obtain uranium
dioxides (Patent Pending—[14]). This method has been
used to synthesis of uranium dioxides doped by neodym-
ium, as surrogates of trivalent plutonium and americium.
The main modification step is the formation of uranyl-
neodymium-ascorbate sols from components alkalized by
aqueous ammonia. Those sols were gelled into: 1) irregularly agglomerates by evaporation of water; 2) medium
sized microspheres (diameter < 100 μm) by INCT variant
of sol-gel processes by water extraction from drops of
emulsion sols in 2-ethylhexanol-1 by this solvent (**Figure
1**). Uranium dioxide was obtained by a reduction of gels
with gas mixture of argon and hydrogen at temperatures

900°C [15]. The work has been carried out within the
collaborative Project ACSEPT (Actinide Recycling by
Separation and Transmutation), contract No. FP7-CP-
2007-211267. Continuations of this work carry out
within the collaborative Project ASGARD (Advanced for
Generation IV reactors: Reprocessing and Dissolution,
contract No. FP7-CP-2011-295825).

In frame of Strategic Project "Technologies Supporting Development of Safe Nuclear Power Engineering"
Domain 4 "Development of spent nuclear fuel and radioactive waste management techniques and technologies" by the National Centre for Research and Development, in INCT, works for elaborated of method synthesis
of uranium carbides and nitrides doped by surrogates of
MA are carried out. By using of CSGP Process, finally
carbides and nitrides will gelled into irregularly agglomerates and medium sized microspheres (diameter <100
μm). The physical properties of carbide and nitride fuel
make them attractive because they are conducive to high
specific rod powers with relatively low fuel centre temperatures: start of life power capability is increased,
power-to-melt margin is increased and fatter (more economic) pins are facilitated. Also, the very important advantages in case of transmutation is a higher actinide
density to oxides. Because fuel temperatures are low, the
fuel suffers little or no restructuring and the release of
fission gases and volatile fission products is low [16,17].

Other developed option of fuel to the incineration of
plutonium and minor actinides in thermal reactors, fast
reactors, and advanced systems are Inert Matrix Fuels
(IMF). The desired properties of the material(s) are a

**Figure 1. SEM analysis of uranium dioxide in the shape of
irregular powders (top) and spherical particles (diameter
below 100 μm) obtained by CSGP.**

high melting point, good thermal conductivity, good compatibility with the cladding, low solubility in the coolant, good mechanical properties, and high density. In all cases, the IMF candidate is compared to standard UO_2 fuel as the reference [18,19]. A critical assessment of selection of materials as diluents for burning of plutonium fuels in nuclear reactors based on their physical and chemical properties are discussed in a recent review. This review appraises the out-of pile properties of Al_2O_3, MgO, $MgAl_2O_4$, CeO_2, Ce_2O_3, Y_2O_3, ZrO_2, $CePO_4$, $ZrSiO_4$, B4C, SiC, AlN, Mg_3N_2, Si_3N_4, CeN, YN and ZrN include, e.g. melting point, vapour pressure and thermal conductivity; the chemical behavior towards steel, liquid sodium, water, air and nitric acid in the context of suitability of diluents [20].

In the task of the Strategic Project there was elaborated method of synthesis of zirconium dioxide doped by surrogates of MA (finally plutonium and MA). This method will be used to study of nuclear transmutation of long lived actinides to short-lived nuclides.

2.4. Radioactive Waste Treatment

All fuel cycles generate some types of long-lived radioactive wastes; thus the ultimate need for disposal of those wastes. The disposal of SNF and HLW has been a major technical and institutional challenge. The modern concept of safe management of high-level liquid radioactive wastes (HLRW) is strongly connected with reprocessing of nuclear spent fuel. It provides for the necessity in fractional separation of highly active components of these wastes including [235]U and [239]Pu, minor actinides ([241]Am, [244]Cm, [237]Np), and fission products (^{137}Cs and ^{90}Sr) followed by their immobilization or transmutation into short-lived or stable isotopes prior to disposal [21]. Selective removal of radionuclides from the bottoms residue of evaporation equipment used in NPPs has an enormous advantage over the conventional methods currently being used to condition liquid radwastes (cementing, bituminization). The advantage is primarily due to the decrease in the volume of the conditioned wastes put into solid radwaste repositories.

The nuclear industry produces large volumes of radioactive solution waste, which requires treatment prior to final disposal or storage. A highly efficient treatment concept is the removal of harmful radionuclides from the bulk waste solution. Application of such a technique will result in considerable reductions in the volumes of waste that require solidification prior to final disposal, as well as in radioactive discharges from storage containers into the environment. Number of methods are used to treat aqueous radioactive wastes, including chemical precipitation, evaporation and ion exchange, as well as less developed solvent extraction, biotechnological processes and membrane methods.

Before final storage of HLW in geological repositories, wastes have to be immobilized and encapsulated. The possible water penetration and leaching of radionuclides from stored radioactive waste makes this problem very important issue. The standard material used to encapsulate fission products and minor actinides is the vitreous matrix. The main advantages of this type of containment matrix are that it is very leach-resistant (dissolution in water) and radiation-resistant material. It is relatively easy to produce on a technical level, and it can accommodate a wide range of radioelements in compact packaging.

The most significant disadvantage of glasses for such use is their high processing temperature of ~2000°C. Sintering of sol-gel-derived glasses can be accomplished at much lower temperatures, and sol-gel techniques have been successfully used for preparation of porous glass hosts for nuclear wastes [22,23]. Appropriately prepared, sintered ceramic bodies can have higher stabilities and be more resistant to leaching than are many melt-processed glasses [24]. In one frame of Strategic Project, researches in INCT are focused on elaborating of "cold" vitrification process and immobilization HLW in ceramic matrixes. Mentioned CSGP process [25] has been adapted to synthesize silica glasses capable of incorporating significant concentrations of high-level nuclear wastes. An example flow-chart for the preparation silica gels is shown in **Figure 2**.

Gels in the form of powders or monoliths were prepared by hydrolysis and subsequent polycondensation of tetraethoxide/me nitrate solutions containing ascorbic acid (ASC) as a catalyst, instead of the HCl or NH_4OH that are routinely used for catalysis in glass synthesis.

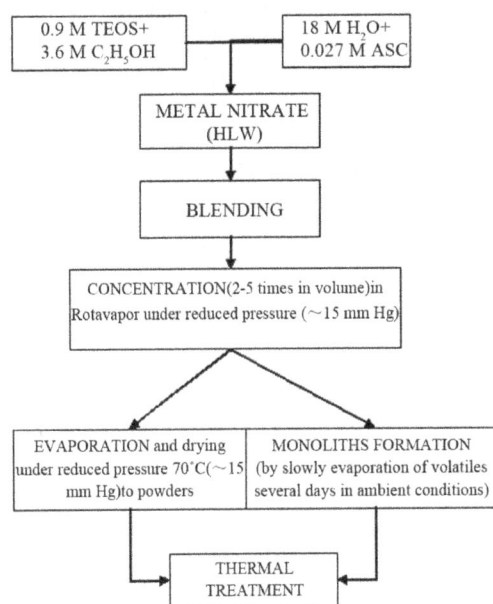

Figure 2. Flow-chart of CSGP for preparation of silica glass.

The second task of the project is immobilization of radionuclides SYNROC materials [26]. The INCT sol-gel method of synthesis titanates are used to synthesis titanate matrixes for encapsulation nuclear wastes [27].

2.5. Radio- and Radiation Chemistry in INCT Poland

A chemistry program is essential for the safe operation of a nuclear power plant. Moreover in Polish situation, when Polish Government has lunched: "Program of development of nuclear energy in Poland" R & D background has a key role. The only R & D Institute in Poland which covers most of the fields of nuclear chemistry, radiochemistry, nuclear chemical engineering and radiation chemistry is the Institute of Nuclear Chemistry and Technology. The Institute runs the PhD studies in this field. Some researches of the Institute were presented earlier in the thematic chapters. Another R & D unit the National Centre for Nuclear Research (NCBJ), Świerk was established in September 2011. NCBJ is strongly involved in developing nuclear reactor engineering and promoting practical applications of nuclear physics methods. Concerning the radiochemistry major market products manufactured in the NCBJ include radiopharmaceuticals and other radioisotopes. Other activities regarding radio- and radiation chemistry are ongoing at some universities. Regarding the radiochemistry, the most academic centers are working in the radioecology field Department of Radiochemistry and Colloid Chemistry, Faculty of Chemistry, Maria Curie-Skłodowska University. Lublin after Chernobyl accident focused its activities on defining a radiological state of environment —determination of the contamination level of soil, river and lake sediments, as well as ground level air with the anthropogenic isotopes such as: ^{137}Cs, ^{90}Sr and Plutonium. General research interest of the Analytics and Environmental Radiochemistry Chair at Faculty of Chemistry, University of Gdańsk, is focused on analytical chemistry and radiochemistry [28,29]; application of activity disequilibrium between ^{210}Po/^{210}Pb, ^{234}U/^{238}U, ^{238}Pu/$^{239+240}$Pu and ^{241}Pu/$^{239+240}$Pu as well as isotopic ratio ^{240}Pu/^{239}Pu for research on polonium, uranium and plutonium sources in the natural environment; impact of the Chernobyl accident on radioactive pollution; radiological risk of radionuclides intake with air, water and food consumption as well as cigarette smoking by consumers. The group working in the Department of Chemistry of the University of Warsaw has been engaged for years in the study of isotope effects on various physicochemical properties of chemical substances [30]. There are three research groups active in the field of radiation chemistry in Poland. One exists at the INCT and uses as a main experimental method nanosecond pulse radiolysis with UV/VIS detection based on the electron linear accelera-

tor 10 MeV. The second is located at the Institute of Applied Radiation Chemistry (IARC), Technical University of Łodź equipped in PR1—pulse radiolysis system using electron beam pulses of duration variable from 2.5 ns to 4.5 µs, dose per pulse from 2 Gy to 1 kGy, wavelength spectroscopic range 250 - 2000 nm at room temperature, recorded data time range from 500 ns to 2 s FS (full scale) with 1 ns resolution time. IARC is working on some aspects of radiochemistry as well. The third radiation chemistry group exists in the Chemistry Department of the University of Podlasie, Siedlce. Some aspects of radiochemistry and radiation chemistry developments are discussed by Narbutt [13] and Chmie- lewski [31].

INCT in collaboration with other institutions was responsible for chemistry program preparation, personnel training and pilot studies for NPP Żarnowiec. These are well known facts that this kind of knowledge is not being delivered in the frame of the most well elaborated contract. Technology provider will not guarantee full responsibility for the operations which has to be updated to the state of art available in the next 5 - 10 years. Moreover foreign suppliers will not able to provide an adequate training of a personnel to be employed in the given country conditions.

3. Conclusion

This paper has been prepared to underline the role of chemistry and its contribution to the nuclear energy power development. It shows only few aspects which connect chemistry with nuclear engineering, but it is just "the tip of the iceberg". Mostly, it is referring to the INCT and author's works. However, its content illustrate well the role of chemistry in the present safe and effective operation of all stages of nuclear power related technology. Hopefully, it will pay attention on this issue and stimulate researches in this field.

4. Acknowledgements

This paper has been prepared in frame of Strategic Project "Technologies Supporting Development of Safe Nuclear Power Engineering" Domain 4 "Development of spent nuclear fuel and radioactive waste management techniques and technologies" by the National Centre for Research and Development (SP/J/4/143 321/11).

REFERENCES

[1]   International Atomic Energy Agency, "Chemistry Programme for Water Cooled Nuclear Power Plants," Specific Safety Guide No. SSG-13, 2011.

[2]   J. Hurwic, "Maria Skłodowska-Curie and Radioactivity," Galant Edition, Warsaw, 2011.

[3]   H. Polkowska-Motrenko and L. Fuks, "Proficiency Test-

ing Schemes on Determination Of Radioactivity in Food and Environmental Samples Organized by the NAEA," *Nukleonika*, Vol. 55, No. 2, 2010, pp. 149-154.

[4] F. Nordmann, "PWR and BWR Chemistry Optimization," *Nuclear Engineering International*, Vol. 56, No. 689, 2011, pp. 24-29.

[5] T. V. Epimakhov, L. N. Moskvin, A. A. Efimov and O. Y. Pykhteev, "Positive Effect of Adding Aluminum Salts to Primary Coolants of Nuclear Power Installations," *Radiochemistry*, Vol. 52, No. 6, 2010, pp. 581- 584.

[6] J. Narbutt, "Inorganic Ion Exchangers as Selective Adsorbents and Potential Primary Barriers for Radionuclides," In: *The Environmental Challenges of Nuclear Disarment*, Kluver Academic Publishers, Netherlands, 2000, pp. 237-243.

[7] D. Olander, "Nuclear Fuels—Present and Future," *Journal of Nuclear Materials*, Vol. 389, No. 1, 2009, pp. 1-22.

[8] Nuclear Energy Agency, "National Programmes in Chemical Partitioning. A Status Report," NEA No. 5425, Nuclear Energy Agency, Organization for Economic Co-Operation and Development, 2010.

[9] J. N. Mathur, M. S. Murali and K. L. Nash, "Actinide Partitioning—A Review," *Solvent Extraction and Ion Exchange*, Vol. 19, No. 3, 2001, pp. 357-390.

[10] C. Madic, F. Testard, M. J. Hudson, *et al.*, "PARTNEW—New Solvent Extraction Processes for Minor Actinides," Final Report, CEA-R-6066, 2004.

[11] Z. Kolarik, U. Müllich and F. Gasner, "Selective Extraction of Am(III) over Eu(III) by 2,6-Ditriazolyl- and 2,6-Ditriazinylpyridines," *Solvent Extraction and Ion Exchange*, Vol. 17, No. 1, 1999, p. 23.

[12] J. Narbutt, "Hydrometallurgic Separation of Minor Actinides from High Active Nuclear Waste for Their Transmutation—Collaborative Pro*ject ACSEPT, 7. Framework Programme, EU, Euratom,*" *V Polish Conference on Radiochemistry and Nuclear Chemistry*, Kraków, 24-27 May 2009, p. 2.

[13] J. Narbutt, "Trends in Radiochemistry at the Beginning of the 21st Century," *Nukleonika*, Vol. 50, Suppl. 3, 2005, pp. S77-S81.

[14] A. Deptuła, M. Brykała, W. Łada, D. Wawszczak, T. Olczak and A. G. Chmielewski, "Method for Preparing of Uranium Dioxide in the Form of Spherical and Irregular Grains," Polish Patent Pending No. P-389385 (27-10-2009), European Patent Application No. 10188438.5—1218 (2010), Russian Federation Patent Application 2010136670 (2010), Belarus Patent Application 20101305 (2010), Ukraine Patent Application 201010756 (6-09-2010).

[15] A. Deptuła, M. Brykała, W. Łada, D. Wawszczak, T. Olczak, G. Modolo, H. Daniels and A. G. Chmielewski, "Synthesis of Uranium and Thorium Dioxides by Complex Sol-Gel Processes (CSGP)," *Proceedings of the 3rd International Conference on Uranium, 40th Annual Hydrometallurgy Meeting*, Vol. II, Saskatoon, 2010, pp. 145-154.

[16] K. Minato, M. Akabori, M. Takano, Y. Arai, K. Nakajima,

A. Itoh and T. Ogawa, "Fabrication of Nitride Fuels for Transmutation of Minor Actinides," *Journal of Nuclear Materials*, Vol. 320, No. 1-2, 2003, pp. 18-24.

[17] International Atomic Energy Agency, "High Temperature Gas Cooled Reactor Fuels and Materials," International Atomic Energy Agency, Vienna, 2010.

[18] C. Degueldre, *et al.*, "Plutonium Incineration in LWRS by a Once through Cycle with a Rock-Like Fuel," *Materials Research Society Symposium Proceedings*, Vol. 412, 1996, p. 15.

[19] H. Kleykamp, "Selection of Materials as Diluents for Burning of Plutonium Fuels in Nuclear Reactors," *Journal of Nuclear Materials*, Vol. 275, No. 1, 1999, pp. 1-11.

[20] International Atomic Energy Agency, "Viability of Inert Matrix Fuel in Reducing Plutonium Amounts in Reactors," International Atomic Energy Agency, Vienna, 2006.

[21] T. A. Maryutina, M. N. Litvina, D. A. Malikov, *et al.*, "Multistage Extraction Separation of Am(III) and Cm(III) in Planet Centrifuges," *Radiochemistry*, Vol. 46, No. 6, 2004, pp. 596-602.

[22] T. Woignier, J. Reynes, J. Phalippou and J. L. Dussossoy, "Nuclear Waste Storage in Gel-Derived Materials," *Journal of Sol-Gel Science and Technology*, Vol. 19, No. 1-3, 2000, pp. 833-837.

[23] T. Woignier, J. Reynes, J. Phalippou and J. L. Dussossoy, "Sintered Silica Aerogel: A Host Matrix for Long Life Nuclear Wastes," *Journal of Non-Crystalline Solids*, Vol. 225, 1998, pp. 353-357.

[24] D. R. Clarke, "Ceramic Materials for the Immobilization of Nuclear Waste," *Annual Review of Materials Research*, Vol. 13, 1983, pp. 191-218.

[25] A. Deptula, W. Lada, T. Olczak, M. T. Lanagan, S. E. Dorris, K. C. Goretta and R. B. Poeppel, "Method for Preparing High-Temperature Superconductors," Polish Patent No. 172618, 1997.

[26] A. E. Ringwood, "Immobilization of Radioactive Wastes in SYNROC," *American Scientist*, Vol. 70, No. 2, 1982, pp. 201-207.

[27] A. Deptula, K. C. Goretta, T. Olczak, W. Lada, A. G. Chmielewski, U. Jakubaszek, B. Sartowska, C. Alvani, S. Casadio and V. Contini, "Preparation of Titanium Oxide and Metal Titanates as Powders, Thin Films, and Microspheres by Novel Inorganic Sol-Gel Process," *Materials Research Society Symposium Proceedings*, Vol. 900E, 2005.

[28] B. Skwarzec, K. Kabat and A. Astel, "Seasonal and Spatial Variability of 210Po, 238U and 239 + 240Pu Levels in the River Catchment Area Assessed by Application of Neural-Network Based Classification," *Journal of Environmental Radioactivity*, Vol. 100, No. 2, 2009, pp. 167-175.

[29] D. Strumińska-Parulska and B. Skwarzec, "Plutonium Isotopes ^{238}Pu, $^{239+240}Pu$, ^{241}Pu and $^{240}Pu/239Pu$ Atomic Ratios in Southern Baltic Sea Ecosystem," *Oceanologia*,

Vol. 52, No. 3, 2010, pp. 499-512.

[30] A. Makowska, A. Siporska and J. Szydłowski, "Isotope Effects on Miscibility of 1-Alkyl-3-Methylimidazolium Bis(Trifluoromethyl)Sulfonyl Imides with Aromatic Hy-

drocarbons," *Fluid Phase Equilibria*, Vol. 282, No. 2, 2009, pp. 108-112.

[31] A. G. Chmielewski, "Chemistry for the Nuclear Energy of the Future," *Nukleonika*, Vol. 56, No. 3, 2011, pp. 241-249.

26

Charge Fluctuations in *pp* and *AA* Collisions at RHIC and LHC Energies

Shakeel Ahmad[1], M. Zafar[1], M. Irfan[1], A. Ahmad[2]
[1]Department of Physics, Aligarh Muslim University, Aligarh, India
[2]Department of Applied Physics, Aligarh Muslim University, Aligarh, India

ABSTRACT

Various measures of event-by-event net charge and charge ratio fluctuations in *pp* and *AA* collisions at RHIC and LHC energies are studied using the different Monte Carlo generators: URQMD, HIJING and HIJING/ $B\bar{B}$ and the results are compared with the predictions for the independent emission, hadron gas and QGP phase. Values of the D-measures are observed to exhibit significant energy dependence for both *pp* and *AA* data. Furthermore, there is essentially no significant difference in the values of the D-measures predicted by the various Monte Carlo codes used in the present study. A slight centrality dependence of the D-measures in terms of net charge fluctuations is observed in the case of *Au-Au* data at 200 A GeV/c. These findings, thus, suggest that a difference in the D-measures for *pp* and *AA* collisions either the re-scattering effect plays a predominant role or there might be some new physics present in these collisions.

Keywords: Hadron-Hadron Collisions; Heavy-Ion Collisions; Charge Fluctuations

1. Introduction

Fluctuations measured in experiments, in general, depend on the property of the system under study and may contain important information about the system [1]. The interest in the studies involving event-by-event (ebe) fluctuations in hadronic (hh) and heavy ion *AA* collisions at relativistic energies is primarily connected to the idea that correlations and fluctuations of dynamical nature are believed to be associated with the critical phenomena of phase transition and leads to the local and global differences between the events produced under similar initial conditions [2]. Several different approaches [2-4] and ref. therein have been made to investigate the ebe fluctuations in *hh* and *AA* collisions at widely different energies, e.g., normalized factorial moments [5], multifractals [6,7], erracity [8,9], k-order rapidity spacing [10] and transverse momentum spectra [11]. Furthermore, fluctuations in the conserved quantities, like, electric charge, strangeness and baryonic numbers have emerged as new tools to estimate the degree of equilibration and criticality of the measured system [12]. The ebe electric charge fluctuations in *hh* and *AA* collisions have drawn considerable attention because;

1) *QGP signature*: A suppression in the net electric charge in the local phase has been theoretically predicted [13] as a signature of the plasma state. Such a decrease is expected as the charges are envisaged to be spread more evenly throughout the QGP volume than that in the hadronic gas.

2) *Thermodynamic signature*: An enhanced charge fluctuations has been observed [14] at RHIC and SPS energies, which might be due to anomalous proton number fluctuations at the critical point.

3) *ρ and ω mesons*: Charge fluctuations are influenced by the decay of hadronic resonances too. In the absence of QGP, the deviation of such fluctuations from the statistical behavior can be used to determine the abundance of *ρ* and *ω* mesons.

An attempt is, therefore made to carry out a systematic study of ebe charge fluctuations in *pp* and *AA* collisions by simulating the events using some of the popular Monte Carlo Generators (MCGs) which are based on the different physics pictures. Such studies may help establish the fluctuations as a robust variable and then to interpret the physical information contained in the measurement. Yet another advantage of the MCGs data is that the analysis could be carried out in both full and limited phase spaces, which in turn would lead to test the efficiencies of the detectors of limited acceptance [15].

2. Formalism

The charge fluctuations are generally studied in terms of two kind of measures [14,16]. The first one is the D-measure of the net charge fluctuations, the direct measure of which is the variance $V(Q) = \langle Q^2 \rangle - \langle Q \rangle^2$, where $Q = n^+ - n^-$; n^+ (n^-) being the multiplicities of positively (negatively) charged hadrons in a particular phase space of an event. The other measure is the charge ratio, $R^+ = n^+/n^-$ and(or) $R^- = n^-/n^+$. The D-measures of the charge ratio fluctuations are defined as $D(R^\pm) = \langle n_{ch} \rangle V(R^\pm)^2$, where $V(R) = \langle R^2 \rangle - \langle R \rangle^2$ is the variance. In the high multiplicity limit, the above measures are neatly equal and are expressed as;

$$D(R) = D(Q) = \frac{4V(Q)}{\langle n_{ch} \rangle} \qquad (1)$$

If each produced particle is assigned randomly a charge +1 or –1 with equal probability, then $V(Q) = \langle n_{ch} \rangle$ and $D(Q) = 4$. in order to account for a non-zero net charge due to baryon stopping and the charge conservation in the large pseudorapidity, η window, two corrections are applied to the D-measure and the redefined parameter is given by[14,16,17]

$$D_{corr}(Q) = \frac{D(Q)}{C_\mu C_y} \qquad (2)$$

where, $C_\mu = \dfrac{\langle n^+ \rangle_{\Delta\eta}}{\langle n^- \rangle_{\Delta\eta}}$ and $C_y = 1 - \dfrac{\langle n_{ch} \rangle_{\Delta\eta}}{\langle n_{ch} \rangle}$

It has been predicted that $D(Q) = 1$ for QGP, 2.9 for resonance gas 4 for uncorrelated pion gas [18-21].

3. Results and Discussion

3.1. Charge Fluctuations in *pp* Collisions

Values of $D(Q)$, $D_{corr}(Q)$ and $D(R^\pm)$ are calculated in *pp* data in the incident energy range $\sqrt{s} = 200$ GeV to 14 TeV using the MCGs, HIJING-1.35 [22,23], HIJING/$B\bar{B}$-1.10 [24] and URQMD-3.3p1 [25,26]. Each of these data samples consists of 10^5 events. Variations of $D(Q)$ with the width of the pseudorapidity window, $\Delta\eta$ for HIJING data samples at different incident energies are shown in **Figure 1**. Similar plots for the events generated by assigning the random charge is also shown in the same figure. For the purpose, all the charged particles of each event(taken from the 14 TeV sample) are randomly assigned a charge +1 or –1 with equal probabilities. It is noted from the figure that for this sample $D(Q)$ values are ~ 4 irrespective of the fact that how small (or large) is the multiplicity in a chosen narrow (or wide) η-window. This result indicates that the ebe analysis may be successfully applied to the narrow phase space bins having a very limited number of particles [14]. The values of $D(Q)$ for $\Delta\eta \leq 0.5$, for all the data sets are found to be ~4, *i.e.* in the region of hadronic gas. These values, with the widening of η-window, decrease to ~1 and even below this value, *i.e.* the one expected for QGP.

It is interesting to note that the data for different energies, *i.e.*, from SPS, RHIC to LHC, overlap, suggesting that there is no energy dependence at all. Almost similar trend of variation of $D(Q)$ with the size of the rapidity window has been observed [17] in the case of PYTHIA simulations of $\pi^+ p$ collisions at $\sqrt{s} = 22$ GeV. In order to compare the results from the three MCGs considered, $D(Q)$ vs. $\Delta\eta$ plots for the *pp* collisions at $\sqrt{s} = 200$ GeV and 14 TeV are shown in **Figure 2**. It may be noted from the figure that the URQMD data exhibit a slight energy dependence of $D(Q)$, while the HIJING and HIJING/$B\bar{B}$ data show no such trends. After applying the corrections to the $D(Q)$ values, as mentioned earlier, the values of D-measure, $D_{corr}(Q)$ are calculated and their dependence on the width of the η window are displayed in **Figure 3** for the HIJING data. Comparison of the findings for different MCGs are shown in **Figure 4**. Following observations may be made from these figures:

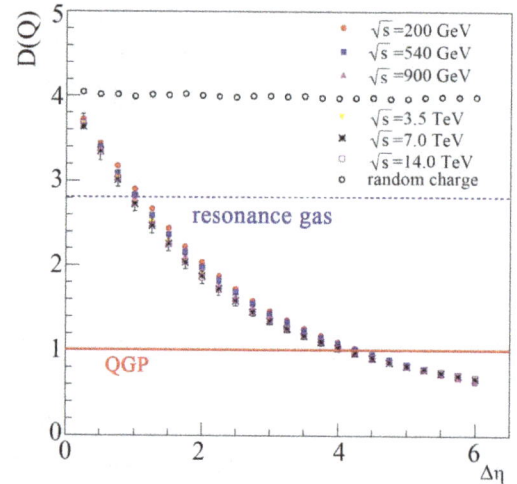

Figure 1. Variation of $D(Q)$ and with the size of central window for the HIJING events.

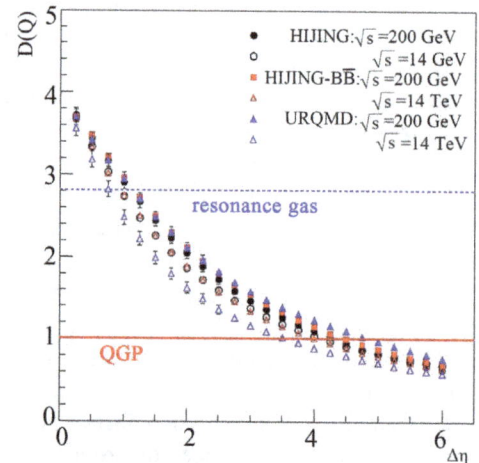

Figure 2. The same plot as in Figure 1 but for various MCGs.

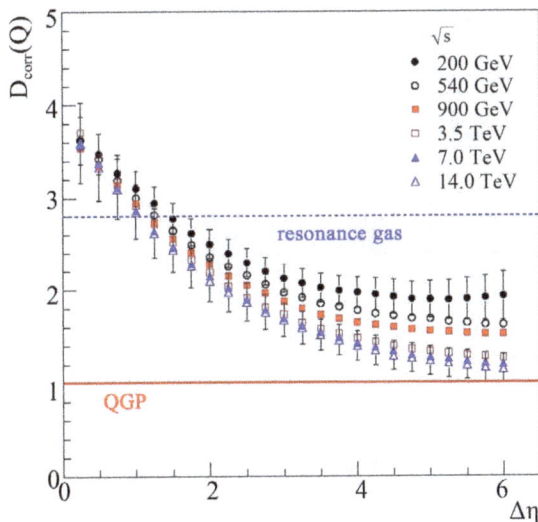

Figure 3. Dependence of $D_{corr}(Q)$ on $\Delta\eta$ for the HIJING pp data at different energies.

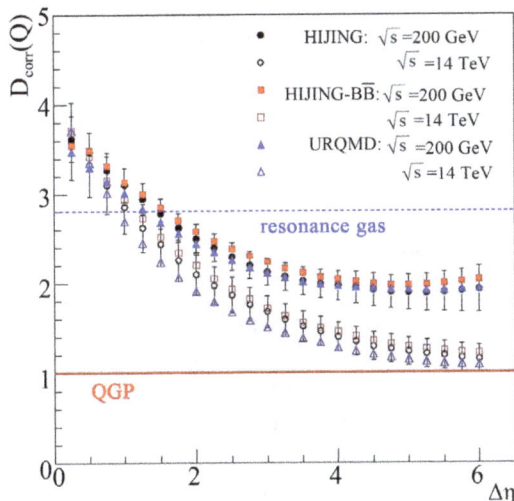

Figure 4. $D_{corr}(Q)$ vs $\Delta\eta$ for the HIJING , HIJING/$B\bar{B}$ and URQMD pp data at $\sqrt{s}$ = 200 GeV and 14 TeV.

- For a small η window, $D(Q) \sim D_{corr}(Q) \sim 4$. However, the $D_{corr}(Q)$ values decrease to a little above 1 and thereafter tend to acquire a saturation in the larger η windows. This indicates that the influence of global charge conservation and leading particle stopping is well taken into account by the corrected measure, $D_{corr}(Q)$.
- Dcorr(Q) values, for a given $\Delta\eta$ are found to decrease with energy and becomes more pronounced in the saturation region of $\Delta\eta$. Such dependence might be due to the increasing number of charged particles at higher energies rather larger charge asymmetry.
- Values of D-measures as obtained for the three MCGs are nearly the same.

Saturations in the $D_{corr}(Q)$ values are observed for the

pseudorapidity windows, $\Delta\eta \geq 4$. However, the correction factor,

$$1 - \frac{\langle n_{ch}\rangle_{\Delta\eta}}{\langle n_{ch}\rangle} \sim 0$$

for the entire kinematic phase space and hence can not be used for the larger η windows [15]. The results, therefore corresponding to $\Delta\eta \sim 4$ and beyond, presented in **Figures 3** and **4** should be overlooked. The values of $D_{corr}(Q)$ for a smaller η window are observed to be much larger than 1, which is expected because of the fact that if the η-bin width is small enough, it will not pick up all the particles decaying from a resonance [15]. Due to the presence of positively charged particles in the initial state, average number of positively charged particles in a particular η-window will be larger than that of negatively charged particles. This charge asymmetry becomes more pronounced with the widening of the η-windows. Leading to the fluctuations in the charge ratio, $R^+ = n^+/n^-$ and $R^- = n^-/n^+$ and that it is not a simple inverse relation. D-measures in terms of charge ratio, R^+ and R^- have been estimated and their variations with $\Delta\eta$ is presented in **Figure 5**; events with $n^\pm = 0$ have been excluded while evaluating $D(R^+)$ and $D(R^-)$ are found to depend on incident energy such that for a given $\Delta\eta$, values of both the measures are larger at higher energies. These values lie in the region of hadron gas and above for the $\Delta\eta$ range ~ 1 - 4. It has been reported [14,16] that for π^+-p and k^+-p collisions at 250 GeV, the values of $D(R^+)$ are larger than that of $D(R^-)$. However in the present study, values of both the parameters are found to have almost similar values except for the HIJING data sample for which $D(R^+)$ values are somewhat smaller than the corresponding $D(R^-)$ values.

It has been observed [14] that both $D(Q)$ and $D(R^\pm)$ exhibit almost similar multiplicity dependence in the region, $\Delta\eta \leq 2$ irrespective of the fact that whether the multiplicity in chosen η window is even or odd. However, for a wider η-window, $\Delta\eta \leq 3$, $D(Q)$ exhibits almost equal separation for the even and odd multiplicities. PHENIX collaboration [13,27], on the other hand has observed that $D(R)$ depends on the multiplicity while $D(Q)$ does not. Study of the multiplicity dependence of D-measures by analyzing the data sets considered in the present study may lead one to make the remark that whether both $D(Q)$ and $D(R)$ are equally good measures in recording the changes in charge fluctuations with multiplicity in different central rapidity windows.

3.2. Charge Fluctuations in AA Collisions

According to the participant model [28], average total charge Q of AA collisions is related to the charge produced in nucleon-nucleon (nn) collisions, Q_i as,

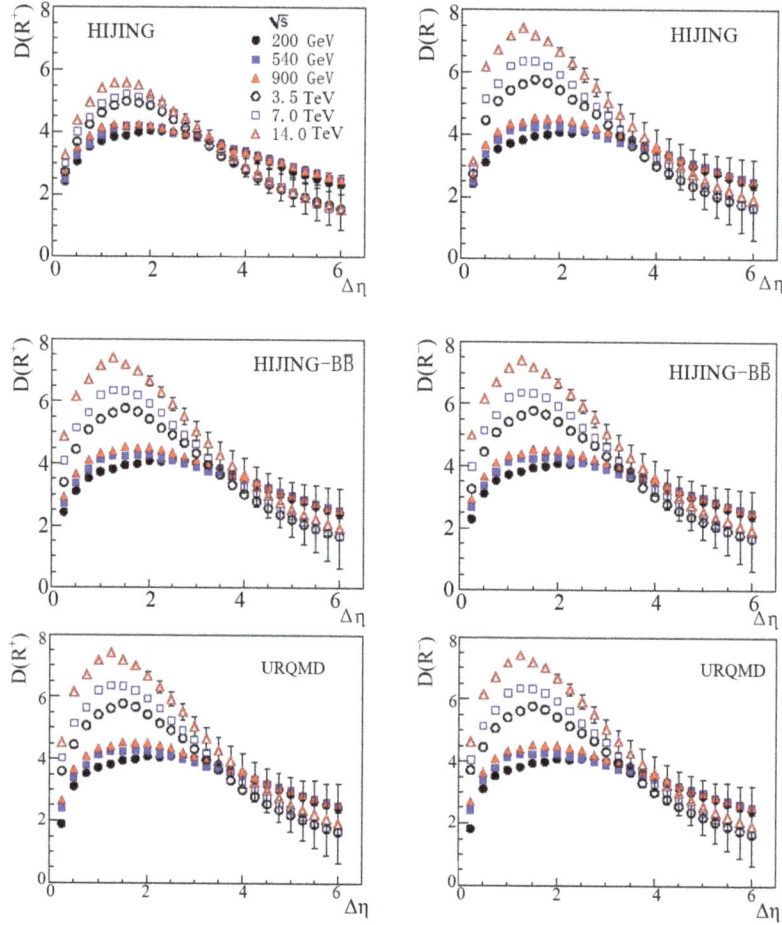

Figure 5. Variations of $D(R^+)$ and $D(R^+)$ with $\Delta\eta$ for the HIJING, HIJING/$B\bar{B}$ and URQMD pp data at different energies.

$\langle Q \rangle = <N_p><Q_i>$, where N_p is the total number of nn collisions while the quantities in angular brackets denote the averaged values over the event sample. The model predicts [15,28] that if re-scattering effects are neglected, D-measures for pp and AA collisions should acquire almost identical values. D-measures for pp and AA data simulated using different MCGs have been observed [15] to acquire nearly similar values with the re-scattering switched off, if the event generator takes it into account, e.g., VNIb and RQMD. In order to test the predictions with the participant model at LHC energies and compare the findings with the SPS and RHIC energy data, values of $D_{corr}(Q)$ are plotted against $\Delta\eta$ for the HIJING data sets corresponding to pp and AA collisions at RHIC and LHC energies. These variations are shown in **Figure 6**. It may be noted in the figure that D-measure for pp and Au-Au collisions at $\sqrt{s_{NN}} = 200$ GeV are nearly the same at least in the region of $\Delta\eta < 4$. This result incidentally, agrees fairly well with the one reported by PHENIX collaboration [13,27]. However, in the present study, the values of D-measure for the pp data are found to acquire relatively smaller values as compared to those obtained

for the AA data. Fluctuations in the $D_{corr}(Q)$ values corresponding to Pb-Pb data, as can be seen in the figure are due to the limited statistics. These findings, thus, reveal that the D-measures for both pp and AA exhibit significant energy dependence. These observations are, thus, in agreement with the idea [6,7] that at lower energies, a single nn collision is dominated by hadronic picture, whereas, at higher energies, such collision can experience the contents of nucleons, i.e., at higher energies gluons are expected to make larger contributions which would cause a decrease in the $D(Q)$ values with increasing incident energies.

Impact parameter, b dependence of D-measure is examined by plotting the variations of $D_{corr}(Q)$ with b for Pb-Pb collisions at 5.5A TeV. These results are shown in **Figure 7**. Data points for each of the two central η-windows are from the five sets of HIJING events, each generated for a different impact parameter range, e.g. b = 0 - 1 fm, 1 - 2 fm, etc. It may be noticed in the figure that the data points for both the η windows lie below the resonance gas and above that the one expected for a QGP. A slight centrality dependence may also be noticed such that the

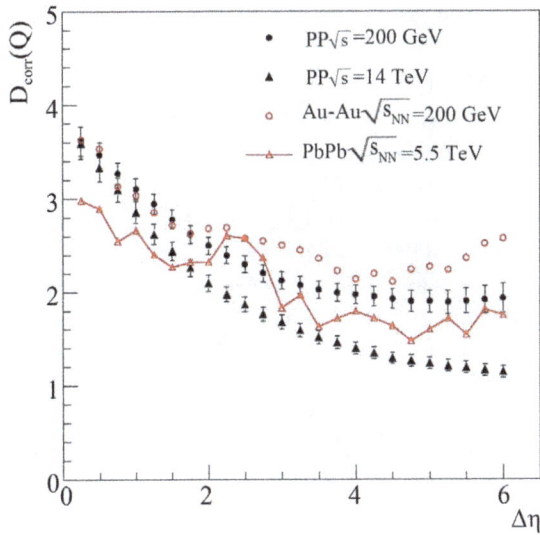

Figure 6. Dependence $D_{corr}(Q)$ on $\Delta\eta$ for the HIJING. Simulated events corresponding to pp and AA collisions at RHIC and LHC energies.

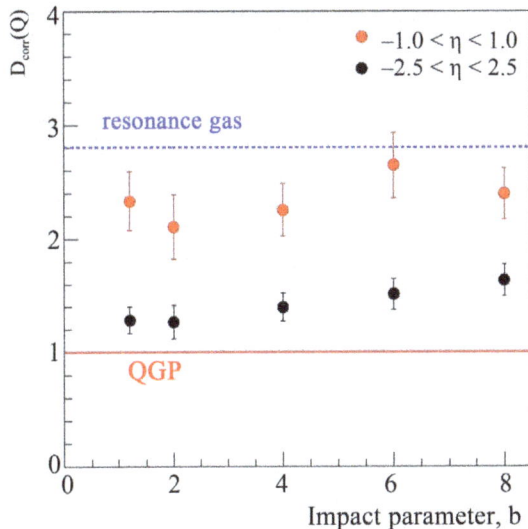

Figure 7. Dependence $D_{corr}(Q)$ on impact parameter for the HIJING simulated events corresponding to 5.5 A TeV. *Pb-Pb* collisions.

$D_{corr}(Q)$ increases with increasing centrality. It should be mentioned here that the D-measure observed in the present study, are relatively smaller as compared to those reported for *Au-Au* collisions at RHIC energies [13,14,27]. This difference in the D-measures might be due to the difference in gluon populations at RHIC and LHC energies.

4. Conclusions

On the basis of the findings of the present work the following conclusions may be drawn:

1) D-measures, after the correction, are found to ex-

hibit significant energy dependence for both pp and AA data for all the three MCGs.

2) D-measure in terms of the net charge fluctuations may be regarded as a better parameter for assessing the change in the charge fluctuations as compared to the D-measures of the charge ratio fluctuations.

3) There is no essential difference in the values of D-measures predicted by the different theoretical models considered in the present study, e.g., HIJING-1.35, HIJING/ $B\bar{B}$ -1.10 and URQMD-3.1p.

4) The observed energy dependence of the D-measure for pp collisions indicate that at higher energies gluons have higher contribution and reduced the values of the D-measures. D-measures for AA collisions also exhibit a similar trend of variation with the incident energy as the observed due to pp collisions.

It may be remarked that charge fluctuations are sensitive to the parton number embedded in different theoretical models, if the re-scattering effects are not essential and hence the D-measures may be regarded as a signature of QGP. Larger values of the D-measures for AA collisions as compared to those for pp collisions, if observed at LHC energies, would indicate that a stronger re-scattering effect is present in AA collisions. A smaller $D(Q)$ values for AA collisions in comparison to pp collisions, on the other hand, if observed, would lead to search of some new physics in future heavy-ion experiments.

REFERENCES

[1] M. Dooring and V. Koch, nucl-th/0204009.

[2] E. De, A. Wolf, *et al.*, *Physics Reports*, Vol. 27, 1996, p. 236.

[3] S. Ahmad, *Journal of Physics G*, Vol. 30, 2004, p. 1.

[4] M. M. Khan, *et al.*, Analysis beyond Intermittency in Multiparticle Production in Relativistic Nuclear Collisions," *International Journal of Modern Physics E*, Vol. 19, No. 11, 2010, p. 2219.

[5] A. Bialas and R. Peschanski, "Moments of Rapidity Distributions as a Measure of Short-Range Fluctuations in High-Energy Collisions," *Nuclear Physics B*, Vol. 273, No. 3-4, 1986, pp. 703-718.

[6] R. C. Hwa, "Fractal Measures in Multiparticle Production," *Physical Review D*, Vol. 41, No. 5, 1990, pp. 1456-1462.

[7] S. Ahmad, *et al.*, *Chaos, Solitons & Fractals*, Vol. 42, 2009, p. 538.

[8] R. C. Hwa, *Acta Physica Polonica B*, Vol. 27, 1996, p. 1789.

[9] F. Liu, *et al.*, *Physics Letters B*, Vol. 516, 2001, p. 293.

[10] M. L. Cherry, *et al.*, *Acta Physica Polonica B*, Vol. 29, 1998, p. 2129.

[11] J. C. Rider, *et al.*, *Nuclear Physics A*, Vol. 698, 2002, p. 611c.

[12] Q. J. Liu and T. A. Trainor, hep-ph/0301214.

[13] K. Adox, *et al.*, "PHENIX col.," *Physical Review Letters*, Vol. 89, 2001, Article ID: 082301.

[14] M. R. Atayan *et al.*, *Physical Review D*, Vol. 71, 2005, Article ID: 012002.

[15] Y. F. Wu and Z. M. Li, *Acta Physica Polonica B*, Vol. 35, 2004, p. 289.

[16] Q. H. Zhang, *et al.*, hep-ph/0202057.

[17] A. Feng, *et al.*, *International Journal of Modern Physics A*, Vol. 22, 2007, p. 2909.

[18] V. Koch, *et al.*, *Nuclear Physics A*, Vol. 698, 2002, p. 261.

[19] S. Jeon and V. Koch, *Physical Review Letters*, Vol. 85, 2000, p. 2076.

[20] M. Asakawa, *et al.*, *Physical Review*, Vol. 85, 2000, p. 2072.

[21] V. M. Martinis, nucl-th/0412007.

[22] X. N. Wang, *Physics Reports*, Vol. 280, 1997, p. 287.

[23] X. N. Wang and M. Gyulassy, *Computer Physics Communications*, Vol. 83, 1994, p. 307.

[24] S. E. Vance and M. Gyulassy, *Physical Review Letters*, Vol. 83, 1999, p. 1735.

[25] S. A. Bass, *et al.*, *Progress in Particle and Nuclear Physics*, Vol. 41, 1998, p. 225.

[26] M. Bleicher, *et al.*, *Journal of Physics G*, Vol. 25, 1999, p. 1859.

[27] J. Nystrand, nucl-ex/0209019.

[28] G. Baym and H. Heiselberg, *Physics Letters B*, Vol. 469, 1999, p. 7.

Measurement of Natural Radioactivity in Sand Samples Collected from Ad-Dahna Desert in Saudi Arabia

Abdulaziz S. Alaamer

Department of Physics, Al-Imam Mohammad Ibn Saud Islamic University (IMSIU), Riyadh, KSA

ABSTRACT

Natural radioactivity is a source of continuous exposure to human beings. The natural radioactivity due to the presence of ^{226}Ra, ^{232}Th and ^{40}K in sand samples collected from *Ad-Dahna* was measured by means of HPGe. The measured activity concentrations of radionuclides were compared with the worldwide reported data. Mean measured activity concentrations of ^{226}Ra, ^{232}Th and ^{40}K varied between 16.2 - 30.6, 15.8 - 36.7 and 285.3 - 533.2 Bq·kg^{-1} respectively with a mean value of 23.4 ± 4.3 Bq·kg^{-1}, 29.7 ± 5.9 Bq·kg^{-1} and 380 ± 65 Bq·kg^{-1} respectively. Mean values of radium equivalent activity, absorbed dose rate and external radiation hazard index were 106 ± 8 Bq·kg^{-1}, 51.4 nGy·h^{-1} and 0.29 respectively. The annual effective radiation dose was calculated to be 0.32 mSv·y^{-1}. The Ra$_{eq}$ values of sand samples are lower than the limit of 370 Bq·kg^{-1}, equivalent to a gamma dose of 1.5 mSv·yr^{-1}.This study shows that the measured sand samples do not pose any significant source of radiation hazard and are safe for use in building materials.

Keywords: Natural Radioactivity; Sand; Gamma-Ray Spectrometry; Dose Rate; Ad-Dahna

1. Introduction

Naturally occurring radioactive materials (NORM) is wide spread in the earth's environment. The presence of natural radioactivity in soil and other building materials results in internal and external exposure to the occupants. NORM existing in soil could pose potential health physics risk [1]. Terrestrial radioactivity, and the associated external exposure due to the gamma radiation depend primarily on the geological and geographical conditions and appear at different levels in the soils of each region [1,2]. The largest contribution to the radiation field is due to the cosmic rays, the natural radionuclides in soil, radioactivity of the ground and the radioactive decay products of radon in the air. Under normal conditions, artificial radioactivity emitted from the nuclear power plants, industrial plants and research facilities has smaller contribution to the overall radiation. Natural environmental radioactivity arises mainly from primordial radionuclides, such as^{226}Ra, ^{232}Th, ^{40}K and their decay products, which occur at trace levels in all ground formations [4]. Accumulation of these radionuclides in the environment raises many problems concerning safety of biotic life, food chain and ultimately humans. To address these problems, it is necessary to know the dose limits of public exposures and to measure the natural environmental radiation level for the estimation of the exposures to natural radiation sources [5]. Many studies have investigated the levels of natural background radiationby in situ measurements or by analysis of radionuclideconcentration in sand samples [6-15].

Desert is the most prominent feature of Arabian Peninsula of which Saudi Arabia is the largest country (**Figure 1**). About 35% of the land in Saudi Arabia is covered by sandy-deserts. The country has three major deserts. *Rub-al-Khali* extends over much of the southeast and beyond the southern frontier. *Rub-al-Khali* has an estimated area of about 650,000 km^2. *An-Nafud* is an upland desert of red sands, due to iron oxide coating, covering an area of 64,000 km^2. It lies at an elevation of 900 meters in the northern part of Saudi Arabia. *Ad-Dahna* is a narrow strip of sandy terrain. This reddish sandy desert is in the central Saudi Arabia, extending about 1300 km southward from the northeastern edge of *An-Nafud* to the northwestern borders of *Rub-al-Khali*. The sand available along *Ad-Dahna*is used as a construction material. Therefore, radiometric characterization provides a useful technique of acquiring better knowledge of the local environment and radiation doses to be received by the general public [16].

2. Materials and Methods

Sand samples were collected from twenty three sites of *Ad-Dahna* desert around Riyadh City. At every sampling site, samples were collected from about 30 cm deep of

Figure 1. Maps of deserts in Saudi Arabia.

four corners and the center of a square area corresponding to 1 m^2. The five samples were mixed together in situ; and this sand mixture, weighing approximately 1.25 kg, was considered representative of the sampling site [13]. They were placed in plastic bags, labeled and carried to the laboratory. They were oven dried at a temperature of 110°C for 12 hours, and sieved through a 1 mm mesh. A 200 g of the homogenous samples were then packed in standard Marinelli beakers, weighed and carefully sealedto prevent the escape of gaseous ^{222}Rn and ^{220}Rn from the samples. They were stored for at least 4 weeks before counting to allow time for ^{238}U and ^{232}Th to reach equilibrium with their respective radionuclide daughters [17]. The measurement of activity concentrations of naturally occurring radionuclides of ^{238}U ^{226}Ra, ^{232}Th and ^{40}K in the samples were carried out using a high purity germanium (HPGe) detector coupled with a multi-channel analyzer (MCA). The measurement procedures and activity calculations performed, were as described by [14]. Similarly, the assessment of radiation hazards: the radium equivalent activity (Ra$_{eq}$), the absorbed gamma radiation dose rate in air (D), the annual effective dose (E), the external radiation hazard index (H$_{ex}$).

3. Assessment of Radiation Hazards

3.1. Radium Equivalent Activity

The radiation hazards associated with the radionuclides are estimated by calculating the radium equivalent activity (Ra$_{eq}$). It is a weighted sum of activities of ^{226}Ra, ^{232}Th and ^{40}K; and it is based on the assumption that 370 Bq·kg^{-1} of ^{226}Ra, 259 Bq·kg^{-1} of ^{232}Th and 4810 Bq·kg^{-1} of ^{40}K produce the same gamma radiation dose rate [18]. To avoid radiation hazards, materials whose Ra$_{eq}$ is greater than 370 Bq·kg^{-1} should not be used. Ra$_{eq}$ is defined by the following formula:

$$Ra_{eq} = A_{Ra} + 1.43 A_{Th} + 0.077 A_K \qquad (1)$$

where A_{Ra}, A_{Th} and A_K are the activity concentrations of ^{226}Ra, ^{232}Th and ^{40}K, respectively.

3.2. Air Absorbed Gamma Radiation Dose Rate

Effects of gamma radiation are normally expressed in terms of the absorbed dose rate in air, which originate from radioactive sources in the soil. The activity concentrations in soil correspond to total absorbed dose rate in air at 1 m above the ground level. The absorbed dose rate in air (D) for the population living in the studied area is calculated using the following equation [18]:

$$D = \left(F_{Ra} \cdot A_{Ra} + F_{Th} \cdot A_{Th} + F_K \cdot A_K \right) \times 10^{-6} \qquad (2)$$

where D is the absorbed dose rate in air (nGy·h^{-1}) at 1 m height above the ground level. F_{Ra}, F_{Th} and F_K are the dose conversion factors for ^{226}Ra, ^{232}Th and ^{40}K respectively. They are taken as 4.27, 6.62 and 0.43 for ^{226}Ra, ^{232}Th and ^{40}K respectively as assessed by UNSCEAR [3].

3.3. Annual Effective Dose

The annual effective dose received by the population is calculated using the following formula:

$$E = T \cdot Q \cdot D \times 10^{-6} \qquad (3)$$

where D is the absorbed dose rate in air, Q is the conversion factor of 0.7 Sv·Gy^{-1}, which converts the absorbed dose rate in air to human effective dose received, and T is the time for 1 year, i.e. 8760 hrs.

3.4. Internal and External Radiation Hazard Index

Radiation hazards due to natural radionuclides of ^{40}K, ^{232}Th and ^{226}Ra may be internal or external depending upon the location of a receptor indoor (inside a dwelling) or outdoor (outside a dwelling) on the ground. These hazards are defined in terms of internal or indoor and external or outdoor radiation hazard index and are denoted by H_{in} and H_{ex}, respectively. These are computed by using the following expressions:

$$H_{in} = \frac{A_{Ra}}{185} + \frac{A_{Th}}{259} + \frac{A_K}{4810} \qquad (4)$$

$$H_{ex} = \frac{A_{Ra}}{370} + \frac{A_{Th}}{259} + \frac{A_K}{4810} \qquad (5)$$

where A_K, A_{Th} and A_{Ra} are the activity concentrations of ^{40}K, ^{232}Th and ^{226}Ra respectively. The indoor hazard index is calculated to determine the radiation hazards from ^{40}K, ^{232}Th and ^{226}Ra. There are no wooden houses in Riyadh. All houses are built with soil and concrete. All floors are lined with soil beneath tiles. So, internal radiation hazard index has been calculated.

4. Results and Discussion

The mean values of measured activity concentrations of selected radionuclides of ^{226}Ra, ^{232}Th and ^{40}K in sand samples from all twenty three sites in *Ad-Dahna* are shown in **Table 1**. The activity concentrations of ^{226}Ra, ^{232}Th and ^{40}K are in the range from 32.55 - 16.20 Bq·kg^{-1}, 28.30 - 39.95 Bq·kg^{-1}, 333 – 533 Bq·kg^{-1}, with a mean value of 23.4 ± 4.3, 29.7 ± 5.9 and 380 ± 65 Bq·kg^{-1}, respectively. The measured activity concentrations of ^{226}Ra, ^{232}Th and ^{40}K were compared with world-wide reported values as shown in **Table 2**. It is found that the measured activity concentrations of the three naturally occurring radionuclides in this study are lower than most of the reported values from other countries as well as the world's average values. The results shown in **Table 1** indicate that mean value of ^{226}Ra (23.4 ± 4.3 Bq·kg^{-1}) < ^{232}Th (29.7 ± 5.9 Bq·kg^{-1}) < ^{40}K (380 ± 65 Bq·kg^{-1}).

Radium equivalent activity (Ra$_{eq}$) owing to activity concentration of the three natural radionuclides from all sites varies from 90.5 to 119.5 Bq·kg^{-1}. The mean value of Ra$_{eq}$ is 106 ± 8 Bq·kg^{-1}, which is much less than the threshold value of 370 Bq·kg^{-1}. The mean values of air absorbed gamma radiation dose rate (*D*), annual effective dose (*E*$_{air}$), and external radiation hazard index (*H*$_{ex}$) calculated in this work are shown in **Table 2**. It is shown that mean value of *D*, *E*$_{air}$ and *H*$_{ex}$ are 51.4 nGy·h^{-1}, 0.32 mSv·y^{-1} and 0.13 respectively. Mean annual effective radiation dose of 0.32 mSv·y^{-1} computed in this work is much less than the dose rate reported world-wide.

Table 1. Activity concentrations (Bq·kg^{-1}) of ^{226}Ra, ^{232}Th and ^{40}K in sand samples; and their corresponding radium equivalent activity (Ra$_{eq}$), Internal radiation hazard index (*H*$_{in}$), external radiation hazard index (*H*$_{ex}$), absorbed dose rate (*D*) and annual effective dose (*E*$_{air}$).

Site	Activity concentrations (Bq·kg^{-1})			Ra$_{eq}$ (Bq·kg^{-1})	H_{in}	H_{ex}	D (nGy·h^{-1})	E_{air} (mSv·y^{-1})
	^{226}Ra	^{232}Th	^{40}K					
S 1	22.0	23.9	285	78.1	0.27	0.21	37.48	0.23
S 2	20.8	34.8	347	97.3	0.32	0.26	46.83	0.29
S 3	23.1	31.4	368	96.4	0.32	0.26	46.48	0.29
S 4	21.6	32.0	479	104.2	0.34	0.28	50.96	0.31
S 5	30.2	29.9	533	114.0	0.39	0.31	55.61	0.34
S 6	20.5	32.2	333	92.1	0.30	0.25	44.35	0.27
S 7	20.7	35.3	402	102.1	0.33	0.28	49.49	0.30
S 8	25.0	28.0	375	93.8	0.32	0.25	45.30	0.28
S 9	18.6	16.1	295	64.4	0.22	0.17	31.30	0.19
S 10	23.4	32.4	478	106.6	0.35	0.29	52.01	0.32
S 11	16.2	31.9	297	84.6	0.27	0.23	40.77	0.25
S 12	16.9	29.5	456	94.1	0.30	0.25	46.32	0.28
S 13	29.7	34.4	429	111.8	0.38	0.30	53.86	0.33
S 14	28.5	36.1	393	110.2	0.37	0.30	52.90	0.32
S 15	25.9	31.5	316	95.3	0.33	0.26	45.50	0.28
S 16	24.3	36.8	390	106.9	0.35	0.29	51.49	0.32
S 17	17.2	20.0	413	77.5	0.26	0.21	38.28	0.23
S 18	30.6	31.6	307	99.3	0.35	0.27	47.15	0.29
S 19	27.4	35.4	378	107.1	0.36	0.29	51.39	0.32
S 20	26.1	31.6	330	96.6	0.33	0.26	46.21	0.28
S 21	19.1	15.8	408	73.0	0.25	0.20	36.12	0.22
S 22	23.9	28.3	340	90.5	0.31	0.24	43.56	0.27
S 23	27.7	24.6	385	92.5	0.32	0.25	44.65	0.27
Mean	23.4	29.7	380	95	0.32	0.26	46.00	0.28
Std	4.3	5.9	65	13	0.04	0.03	6.11	0.04
UNSCEAR	35	30	400	108.7	0.39	0.29	52.01	0.32
World-wide	30 ± 14	37 ± 20	397 ± 220	113.5	0.39	0.31	54.38	0.33

Table 2. Activity concentrations (Bq· kg^{-1}) of ^{226}Ra, ^{232}Th and ^{40}K measured worldwide.

Ref.	Activity concentration (Bq·kg^{-1})			Region
	^{40}K	232Th	226Ra	
[19]	398.3	-	26.3	Taiwan
[6]	425.5	33.3	70.3	Malaysia
[7]	528	27	24	Italy
[20]	200	10.6	8.1	The Netherlands
[21]	-	22.8	22.8	Mexico
[8]	842	27	24	Hong Kong
[9]	807	18	14.3	Brazil
[22]	714	26	24	Zambia
[10]	188	8	25	Jordan
[23]	188.1	14.6	25.1	Jordan
[11]	158	25	14	Bangladesh
[24]	456	64	44	India
[25]	367	17	18	Greece
[12]	618	21.4	25.3	Egypt
[26]	508.8	43.2	24.5	Pakistan
[13]	859	39	22.1	China
[27]	188	16	17	Cuba
[28]	586	31	14	Cameroon
[15]	441	26	44	Turkey

5. Conclusion

The present study has been carried out to establish a base line data regarding concentration levels of naturally occurring radionuclides of ^{226}Ra, ^{232}Th and ^{40}K in soils and the corresponding radiation doses in Riyadh, Saudi Arabia. Measured mean activity concentrations of the three radionuclides are found less than the world's average values. Calculated values of external radiation doses are also lower than the world average of about 0.5 mSv per year. It is concluded that there is no potential radiological health risk associated with the soils of area investigated during this study. The data generated here may be useful for the introduction of radiation safety standards by the State Authorities for the protection of general population from radiation hazards owing to terrestrial sources.

REFERENCES

[1] M. J. Willson, "Anthropogenic and Naturally Occurring Radioactive Materials Detected on Radiological Survey of Properties in Monticello, Utah. Environmental Health Physics," 26th Midyear Topical Meeting, 24-28 January 1993, p. 564.

[2] United Nations Scientific Committee on the Effects of Atomic Radiation, "Sources and Effects of Ionising Radiation," Unscear Report, New York, 1993.

[3] United Nations Scientific Committee on the Effects of Atomic Radiation, "Exposures from Natural Radiation Sources," UNSCEAR Report, New York, 2000.

[4] M. Tzortzis, E. Svoukis and H. Tsertos, "A Comprehensive Study of Natural Gamma Radioactivity Levels and Associated Dose Rates from Surface Soils in Cyprus," Radiation Protection Dosimetry, Vol. 109, No. 3, 2004, pp. 217-224.

[5] P. McDonald, G. T. Cook and M. S. Baxter, "Natural and Anthropogenic Radioactivity in Coastal Regions of the UK," Radiation Protection Dosimetry, Vol. 45, No. 1-4, 1992, pp. 707-710.

[6] S. Chong and G. U. Ahmad, "Gamma Activity of Some Building Materials in West Malaysia," Health Physics, Vol. 43, No. 2, 1982, pp. 272-273.

[7] G. Sciocchetti, F. Scacco, P. G. Baldassini, L. Monte and R. Sarao, "Indoor Measurements of Airborne Natural Radioactivity in Italy," Radiation Protection Dosimetry, Vol. 7, No. 1-4, 1984, pp. 347-351.

[8] M. Chung-Keung, L. Shun-Yin, A. Shui-Chun and N. Wai-Kwok, "Radionuclide Contents in Building Materials in Hong Kong," Health Physics, Vol. 57, No. 3, 1989, pp. 397-401.

[9] A. Malanca, V. Pessina and G. Dallara, "Radionuclide Content of Building Materials and Gamma Ray Dose Rates in Dwellings of Rio Grande Do Norte, Brazil," Radiation Protection Dosimetry, Vol. 48, No. 2, 1993, pp. 199-203.

[10] A. J. A. H. Khatibeh, A. Maly, N. Ahmad and Matiullah, "Natural Radioactivity in Jordanian Construction Materials," Radiation Protection Dosimetry, Vol. 69, No. 2, 1997, pp. 143-147.

[11] M. I. Chowdury, M. N. Alam and A. K. S. Ahmed, "Concentration of Radionuclides in Building and Ceramic Materials of Bangladesh and Evaluation of Radiation Hazard," Journal of Radioanalytical and Nuclear Chemistry, Vol. 231, No. 1-2, 1998, pp. 117-123.

[12] A. M. El-Arabi, "Natural Radioactivity in Sand Used in Thermal Therapy at the Red Sea Coast," Journal of Environmental Radioactivity, Vol. 81, No. 1, 2005, pp. 11-19.

[13] X. W. Lu and X. L. Zhang, "Measurement of Natural Radioactivity in Sand Samples Collected from the Baoji Weihe Sands Park, China," Environmental Geology, Vol. 50, No. 7, 2006, pp. 977-982.

[14] A. S. Alaamer, "Assessment of Radiological Hazards Owing to Natural Radioactivity Measured in Soil of Riyadh, Saudi Arabia," Turkish Journal of Engineering & Environmental Sciences, Vol. 32, 2008, pp. 229-234.

[15] U. Cevik, N. Damla, A. I. Koby, N. Celik, A. Celik and A. A. Van, "Assessment of Natural Radioactivity of Sand Used in Turkey," Journal of Radiological Protection, Vol. 29, No. 1, 2009, p. 61.

[16] R. Trevisi, M. Bruno, C. Orlando, R. Ocone, C. Paolelli, M. Amici, A. Altieri and B. Antonelli, "Radiometric Characterization of More Representative Natural Building Materials in the Province of Rome," Radiation Protection

Dosimetry, Vol. 113, 2005, pp. 168-172.

[17]　G. Gonzalez-Chornet and J. Gonzalez-Labajo, "Natural Radioactivity in Beach Sands from Donana National Park and Mazagon (SPAIN)," *Radiation Protection Dosimetry*, Vol. 112, No. 2, 2004, pp. 307-310.

[18]　Matiullah, A. Ahad, S. ur Rehman, S. ur Rehman and M. Fahee, "Measurement of Radioactivity in the Soil of Bahawalpur Division, Pakistan," *Radiation Protection Dosimetry*, Vol. 112, No. 3, 2004, pp. 443-447.

[19]　T. Y. Chang, W. L. Cheng and P. S. Weng, "Potassium Uranium, and Thorium Contents in Building Material of Taiwan," *Health Physics*, Vol. 27, No. 4, 1974, pp. 385-387.

[20]　J. G. Ackers, J. F. Den-Boer, P. De-Jong and R. A. Wolschrijn, "Radioactivity and Radon Exhalation Rates of Building Materials in the Netherlands," *Science of the Total Environment*, Vol. 45, 1985, pp. 151-156.

[21]　G. Espinosa, J. I. Golzarri, I. Gamboa and I. Jacobson, "Natural Radioactivity in Mexican Building Material by SSNTD," *Nuclear Tracks and Radiation Measurements*, Vol. 12, No. 1-6, 1986, pp. 767-770.

[22]　P. Hayumbu, M. B. Zaman, N. C. H. Lubaba, S. S. Munsanje and D. Nuleya, "Natural Radioactivity in Zambian Building Materials Collected from Lusaka," *Journal of Radioanalytical and Nuclear Chemistry*, Vol. 199, No. 3, 1995, pp. 229-238.

[23]　A. N. Matiullah and A. J. A. J. Hussain, "Natural Radioactivity in Jordanian Soil and Building Materials and the Associated Radiation Hazards," *Journal of Environmental Radioactivity*, Vol. 39, No. 1, 1998, pp. 9-22.

[24]　V. Kumar, T. V. Ramachandran and R. Prasad, "Natural Radioactivity of Indian Building Materials and By-Products," *Applied Radiation and Isotopes*, Vol. 51, No. 1, 1999, pp. 93-96.

[25]　S. Stoulos, M. Manolopoulou and C. Papastefanou, "Assessment of Natural Radiation Exposure and Radon Exhalation from Building Materials in Greece," *Journal of Environmental Radioactivity*, Vol. 69, No. 3, 2003, pp. 225-240.

[26]　K. Khalid, P. Akhter and S. D. Orfi, "Estimation of Radiation Doses Associated with Natural Radioactivity in Sand Samples of the North Western Areas of Pakistan Using Monte Carlo Simulation," *Journal of Radioanalytical and Nuclear Chemistry*, Vol. 265, No. 3, 2005, pp. 371-375.

[27]　F. O. Brigido, E. N. Montalvan and Z. J. Tomas, "Natural Radioactivity in Some Building Materials in Cuba and Their Contribution to the Indoor Gamma Dose Rate," *Radiation Protection Dosimetry*, Vol. 113, No. 2, 2005, pp. 218-222.

[28]　M. Ngachin, M. Garavaglia, C. Giovani, M. KwatoNjock G. and A. Nourreddine, "Assessment of Natural Radioactivity and Associated Radiation Hazards in Some Cameroonian Building Materials," *Radiation Measurements*, Vol. 42, No. 1, 2007, pp. 61-67.

28

Reconstruction of the Neutron Flux in a Slab Reactor

Adilson Costa da Silva[*], Aquilino Senra Martinez, Alessandro da Cruz Gonçalves
Nuclear Engineering Department, Federal University of Rio de Janeiro, Rio de Janeiro, Brazil

ABSTRACT

In this electronic article we use the one-dimensional multigroup neutron diffusion equation to reconstruct the neutron flux in a slab reactor from the nuclear parameters of the reactor, boundary and symmetry condition, initial flux and k_{eff}.

The diffusion equation was solved analytically for one single homogeneous fuel region and for two regions considering fuel and reflector. To validate the method proposed, the results obtained in this article were compared using reference methods found in the literature.

Keywords: Neutron Diffusion Equation; Neutron Flux; Slab Reactor

1. Introduction

In the analysis of the neutronic behaviour of a nuclear reactor, one of the most relevant parameters is the determining of the neutron flux in any region of the reactor core, as a precise assessment of this neutron flux will allow determining the spatial distribution of the reactor's power, as well as other parameters of equal relevance for safe reactor operation such as reactor switching off margin and the value for the control rods [1].

Due to the need of precisely determining the neutron flux, some methods were created with this purpose, which take into consideration the geometry and composition of the reactor core. Amongst the many methods, we can point the calculation of the neutron flux from the multigroup diffusion equation through the finite difference method. This method is simple and of easy implementation, although it requires great computing effort for cases of practical interest, given that there is a need to use an extremely fine mesh.

With the aim of avoiding the large computing effort inherent to the finite difference method, several methods have come to the fore, called nodal methods, which allow the use of a coarse mesh (node); however, the use of these methods provides only average flux values for a given region. This way, one needs to use reconstruction methods [2,3] to obtain the neutron flux at any point of the reactor core.

There are many pin power reconstruction methods for few-group, some methods use polynomial expansions for representation of the intranodal flux distribution [3] and others new methods employs the analytical solution of

Helmholtz equation satisfying a given set of boundary conditions. According [3] the analytical solution renders superior accuracy compared to polynomial based method.

Aimed at the need for new reconstruction methods that are fast and accurate [4,5], this work presents a method based on the analytical solution for the neutron flux from the solution of the neutron diffusion equation for two energy groups in one dimension, using boundary and symmetry conditions.

2. Multigroup Neutron Diffusion Equation

We have seen advances in recent decades in the development of coarse-mesh nodal methods aimed at numerically solving the one-dimensional multigroup neutron diffusion equation. These methods calculate with great precision the eigenvalue and neutron flux in each node as seen in the comparison with some reference methods (usually fine-mesh calculations). Amongst them, we point the nodal expansion method (NEM) [6-8] that uses the continuity equation and Fick's Law [9]. The neutron continuity equation and Fick's Law in one dimension, two groups of energy and stationary state are expressed by the following equations, respectively:

$$\frac{d}{dx}J_g(x) + \Sigma_{rg}(x)\phi_g(x)$$
$$= \sum_{\substack{g'=1 \\ g'\neq g}}^{2} \Sigma_{gg'}(x)\phi_{g'}(x) + \frac{\chi_g}{k_{eff}}\sum_{g'=1}^{2}\nu\Sigma_{fg'}(x)\phi_{g'}(x); \quad (1)$$
$$g = 1,2$$

and

---
[*]Corresponding author.

$$J_g(x) = -D_g(x)\frac{d}{dx}\phi_g(x) \tag{2}$$

By replacing Equation (2) in Equation (1) we have the neutron diffusion equation [1] expressed in terms of the neutron flux, such that,

$$\frac{d}{dx}\left(-D_g(x)\frac{d}{dx}\phi_g(x)\right) + \Sigma_{rg}(x)\phi_g(x)$$

$$= \sum_{\substack{g'=1 \\ g' \neq g}}^{2} \Sigma_{gg'}(x)\phi_{g'}(x) + \frac{\chi_g}{k_{eff}}\sum_{g'=1}^{2} \nu\Sigma_{fg'}(x)\phi_{g'}(x) \tag{3}$$

where $D_g(x)$ is the diffusion coefficient for group g, $\Sigma_{rg}(x)$ is the macroscopic removal cross section, $\Sigma_{gg'}(x)$ is the macroscopic scattering cross section from group g' to group g, $\Sigma_{fg'}(x)$ is the product of the average number of neutrons emitted by fission by the macroscopic fission cross section, χ_g is the fission spectrum, $\phi_g(x)$ is the neutron flux and k_{eff} is the eigenvalue of the problem.

Due to the fact of the reactor to be homogeneous in the case of one single region (fuel) or homogeneous per part in the case of more than one region (fuel and reflector), these nuclear parameters are constant in each region, i.e.:

$$D_g(x) \Box D_g, \Sigma_{rg}(x) \Box \Sigma_{rg}, \Sigma_{gg'}(x) \Box \Sigma_{gg'}$$

and $\nu\Sigma_{fg'}(x) \Box \nu\Sigma_{fg'}$. With this, Equation (3) it becomes:

$$-D_g\frac{d^2}{dx^2}\phi_g(x) + \Sigma_{rg}\phi_g(x)$$

$$= \sum_{\substack{g'=1 \\ g' \neq g}}^{2} \Sigma_{gg'}\phi_{g'}(x) + \frac{\chi_g}{k_{eff}}\sum_{g'=1}^{2} \nu\Sigma_{fg'}\phi_{g'}(x) \tag{4}$$

We shall seek an analytical solution from Equation (4) for the following cases: Homogeneous slab reactor consisting only of fuel and a heterogeneous slab reactor consisting of fuel and reflector as we shall describe in the next sections.

3. Analytical Solution for the Neutron Diffusion Equation for a Homogeneous Slab Reactor

The multigroup neutron diffusion equation defined as a eigenvalue problem can be written in the matrix form as follows:

$$\begin{bmatrix} -D_1\nabla^2 + \Sigma_{r1} - \dfrac{1}{k_{eff}}\nu\Sigma_{f1} & -\dfrac{1}{k_{eff}}\nu\Sigma_{f2} \\ -\Sigma_{21} & -D_2\nabla^2 + \Sigma_{r2} \end{bmatrix} \cdot \begin{bmatrix} \phi_1(x) \\ \phi_2(x) \end{bmatrix} = \begin{bmatrix} 0 \\ 0 \end{bmatrix}$$

$$\tag{5}$$

where $\phi_1(x)$ and $\phi_2(x)$ represent respectively the fast and thermal neutron flux. If we study the equation above it is possible to see that the equations both for the fast flux and for the thermal flux satisfy Helmholtz's equation, given by:

$$\nabla^2\phi_g(x) + B^2\phi_g(x) = 0; \quad g = 1,2 \tag{6}$$

where B^2 denotes any one of the two roots of the equation characteristic of the second-order equation in B^2. By replacing Equation (6) in Equation (5), it results that:

$$\begin{bmatrix} D_1B^2 + \Sigma_{r1} - \dfrac{1}{k_{eff}}\nu\Sigma_{f1} & -\dfrac{1}{k_{eff}}\nu\Sigma_{f2} \\ -\Sigma_{21} & D_2B^2 + \Sigma_{r2} \end{bmatrix} \cdot \begin{bmatrix} \phi_1(x) \\ \phi_2(x) \end{bmatrix} = \begin{bmatrix} 0 \\ 0 \end{bmatrix} \tag{7}$$

For Equation (7) to be solved, it is enough that the determinant of the matrix is null, i.e.,

$$\left(D_1B^2 + \Sigma_{r1} - \frac{1}{k_{eff}}\nu\Sigma_{f1}\right)\left(D_2B^2 + \Sigma_{r2}\right) - \frac{1}{k_{eff}}\nu\Sigma_{f2}\Sigma_{21} = 0$$

from which it produces

$$\left(B^2\right)^2 + \left(\frac{\Sigma_{r1}}{D_1} + \frac{\Sigma_{r2}}{D_2} - \frac{1}{k_{eff}}\frac{\nu\Sigma_{f1}}{D_1}\right)B^2$$

$$+ \frac{\Sigma_{r1}\Sigma_{r2}}{D_1D_2} - \frac{1}{k_{eff}}\left(\frac{\nu\Sigma_{f1}\Sigma_{r2} + \nu\Sigma_{f2}\Sigma_{21}}{D_1D_2}\right) = 0 \tag{8}$$

Note that we have a second-degree equation for B^2 whose solution is given by:

$$B_1 = \pm\sqrt{\frac{b}{2}\left(-1 + \sqrt{1 - \frac{4c}{b^2}}\right)}$$

and

$$B_2 = \pm\sqrt{\frac{b}{2}\left(-1 - \sqrt{1 - \frac{4c}{b^2}}\right)}$$

with

$$b = \frac{\Sigma_{r1}}{D_1} + \frac{\Sigma_{r2}}{D_2} - \frac{1}{k_{eff}}\frac{\nu\Sigma_{f1}}{D_1}$$

and

$$c = \frac{\Sigma_{r1}\Sigma_{r2}}{D_1D_2} - \frac{1}{k_{eff}}\left(\frac{\nu\Sigma_{f1}\Sigma_{r2} + \nu\Sigma_{f2}\Sigma_{21}}{D_1D_2}\right).$$

Note that until now we only seek to find the roots of the characteristic equation. As we have not defined the type of solution and we know that the diffusion equation is a second-order differential equation, the types of solution that satisfy this equation are many, although we will consider that this solution has an exponential behaviour for both fast and thermal flux. Thus, the general solution for the thermal flux is given by:

$$\phi_2(x) = C_1e^{-B_1x} + C_2e^{B_1x} + C_3e^{-B_2x} + C_4e^{B_2x} \tag{9}$$

Due to the fact that the system provided by Equation (5) is coupled, the solution for the fast flux can be obtained by replacing Equation (9) in (4) to $g = 2$, of which we have,

$$\phi_1(x) = \left(\frac{\Sigma_{r2} - D_2 B_1^2}{\Sigma_{21}}\right)\left(C_1 e^{-B_1 x} + C_2 e^{B_1 x}\right)$$
$$+ \left(\frac{\Sigma_{r2} - D_2 B_2^2}{\Sigma_{21}}\right)\left(C_3 e^{-B_2 x} + C_4 e^{B_2 x}\right) \tag{10}$$

We will now seek some conditions that have to be applied to Equation (9) or (10) in order to determine the coefficients of the equation. Note that the coefficients are the same for the fast and thermal flow, so one only needs to impose contour and symmetry conditions for only one of the equations. In the case of a slab reactor [9] (plain slab in direction x) of dimension a as shown in **Figure 1**, we can impose the following conditions:

1) Null flux in the boundary the left, such that,
   $\phi_g(-a/2) = 0$;

2) Null flux in the boundary the right, such that,
   $\phi_g(a/2) = 0$;

3) Maximum flux in the origin, *i.e.*,
   $\phi_g(0) = \phi_{go}$;

4) Null net current in the origin, $J_g(0) = 0$.

From these conditions, we can build a system with four conditions and four unknowns such as, for example, for the thermal group the neutron flux coefficients can be determined, such that,

$$C_1 = C_2 = -\frac{1}{2}\frac{\phi_{2o}\cosh\left(\frac{1}{2}B_2 a\right)}{\cosh\left(\frac{1}{2}B_1 a\right) - \cosh\left(\frac{1}{2}B_2 a\right)} \tag{11}$$

and

$$C_3 = C_4 = \frac{1}{2}\frac{\phi_{2o}\cosh\left(\frac{1}{2}B_1 a\right)}{\cosh\left(\frac{1}{2}B_1 a\right) - \cosh\left(\frac{1}{2}B_2 a\right)} \tag{12}$$

We will now analyze the case where the roots for B_1 or B_2 have negative values. We will first verify if

$$\sqrt{1 - \frac{4c}{b^2}} > 0$$

For that, it is enough to analyze if $b^2 - 4c > 0$, *i.e.*:

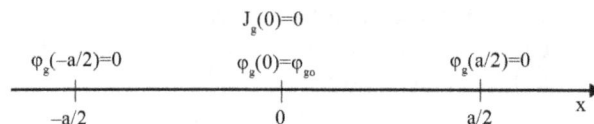

Figure 1. Slab reactor with an a dimension.

$$b^2 - 4c = \left(\frac{\Sigma_{r1}}{D_1} - \frac{\Sigma_{r2}}{D_2} - \frac{\nu\Sigma_{f1}}{k_{eff}D_1}\right)^2 + 4\frac{\nu\Sigma_{f2}\Sigma_{21}}{k_{eff}D_1 D_2}$$

As we could see in the previous equation, the first term is always positive, as it is squared whereas the second term, as it only depends of nuclear parameters and of the eigenvalue, thus it is also positive. With this, we ensure that possible values for B_1 and B_2 are actual positive or negative values.

In a more general way we can simplify these notations using hyperbolic or trigonometric functions with the possible values for B_n, with $n = 1, 2$, such that the neutron flux is:

$$\phi_2(x) = c_1 sn(B_1 x) + c_2 cn(B_1 x) + c_3 sn(B_2 x) + c_4 cn(B_2 x) \tag{13}$$

and

$$\phi_1(x) = \left(\frac{\Sigma_{r2} \pm D_2 B_1^2}{\Sigma_{21}}\right)\left(c_1 sn(B_1 x) + c_2 cn(B_1 x)\right)$$
$$+ \left(\frac{\Sigma_{r2} \pm D_2 B_2^2}{\Sigma_{21}}\right)\left(c_3 sn(B_2 x) + c_4 cn(B_2 x)\right) \tag{14}$$

where these functions are given by:

$$sn(B_n x) \equiv \begin{cases} \sin(B_n x) & \text{if } B_n^2 > 0, \text{ with } \pm = - \\ \sinh(B_n x) & \text{if } B_n^2 < 0, \text{ with } \pm = - \end{cases}$$

and

$$cn(B_n x) \equiv \begin{cases} \cos(B_n x) & \text{if } B_n^2 > 0, \text{ with } \pm = - \\ \cosh(B_n x) & \text{if } B_n^2 < 0, \text{ with } \pm = - \end{cases}$$

Note that for each case mentioned above, the equations for the fast and thermal flux will be different, that is, they may be expressed in terms of hyperbolic or trigonometric functions, or combinations of hyperbolic or trigonometric functions. This will depend of the nuclear parameters used in the calculations and of the eigenvalue. It is worth remembering that the coefficients depend of the values for the roots of B_n. However, the conditions of the problem do not change, *i.e.*, a null flux in the boundary, null net current in the symmetry axis and constant initial flux.

4. Analytical Solution of the Neutron Diffusion Equation for a Heterogeneous Slab Reactor

The one-dimensional multigroup neutron diffusion equation for the fuel region was presented in the previous section, whose solution was given by Equations (9) and (10). For the reflector region this equation undergoes some modifications both in the nuclear parameters that will be different in relation to the nuclear parameters for the fuel, as well as for the form of the diffusion equation, given that for this region there is no neutron fission. Therefore, this equation can be written as follows:

$$-D_1 \frac{d^2}{dx^2}\psi_1(x) + \Sigma_{r1}\psi_1(x) = 0 \qquad (15)$$

and

$$-D_2 \frac{d^2}{dx^2}\psi_2(x) + \Sigma_{r2}\psi_2(x) = \Sigma_{21}\psi_1(x) \qquad (16)$$

The solutions of the diffusion equation for the fast and thermal group for the reflector region are, respectively,

$$\psi_1(x) = R_1 e^{L_1 x} + R_2 e^{-L_1 x} \qquad (17)$$

and

$$\psi_2(x) = \left(\frac{D_1 \Sigma_{21}}{\Sigma_{r2}D_1 - \Sigma_{r1}D_2}\right)\left(R_1 e^{L_1 x} + R_2 e^{-L_1 x}\right) \\ + R_3 e^{-L_2 x} + R_4 e^{L_2 x} \qquad (18)$$

where $L_1 = \sqrt{\dfrac{\Sigma_{r1}}{D_1}}$ and $L_2 = \sqrt{\dfrac{\Sigma_{r2}}{D_2}}$ are the respective diffusion lengths for the fast and thermal groups.

Equations (17) and (18) represent the solutions for the neutron flux in the reflector region, while Equations (9) and (10) represent the solutions for the neutron flux in the fuel region. These regions are shown in **Figure 2**.

As we already have the solutions for the flux in the two regions of the reactor, we will now seek to find which condition we should impose to the flux in order to determine the coefficients for Equations (9), (10), (17) and (18). As the coefficients for Equation (17) are present in Equation (18) and the coefficients for Equations (9) and (10) are identical, we will apply the conditions only for thermal flux in the two regions, fuel and reflector, such that

1) Continuity of flux in the interface between regions to the left, such that,

$$\phi_{2c}(-a/2) = \psi_{2r}(-a/2);$$

2) Continuity of flux in the interface between regions to the right, such that,

$$\phi_{2c}(a/2) = \psi_{2r}(a/2);$$

3) Continuity of current in the interface between regions to the left, such that,

$$J_{2c}(-a/2) = J_{2r}(-a/2);$$

4) Continuity of current in the interface between regions to the right, such that,

$$J_{2c}(a/2) = J_{2r}(a/2);$$

5) Null flux in the contour to the left, such that,

$$\psi_{2r}(-a/2 - b) = 0;$$

6) Null flux in the contour to the right, such that,

$$\psi_{2r}(a/2 + b) = 0;$$

7) Maximum flux in the origin, i.e.,

$$\phi_{2c}(0) = \phi_{2oc};$$

8) Null net current in the origin

$$J_{2c}(0) = 0.$$

Subscripts *c* and *r* represent respectively, the fuel and reflector regions.

Note that, due to the fact that the equations have eight coefficients to be determined, it became necessary to impose eight conditions in order to determine these coefficients as mentioned earlier.

5. Results

This section presents the results obtained in the analytical solution of the neutron diffusion equation and compares the results obtained with the nodal expansion method (NEM) [1] and with the finite differences method (FDM) [10]. **Table 1** shows the nuclear parameters used in the calculation of a homogeneous slab reactor of a = 100 cm dimension and eigenvalue $k_{eff} = 0.7586362$ obtained by the finite difference method.

Figures 3 and **4** show the results obtained for fast and thermal flux, respectively, comparing the analytical solution (Analytic) with the nodal expansion method (NEM) and finite difference method (FDM).

We can see that the results are quite satisfactory for both fast and thermal fluxes. The only input data used in the analytical solution from the numerical results was: eigenvalue k_{eff} and initial flux ϕ_{go}. With this data we were able to reconstruct the entire neutron flux point-by-point in the homogeneous reactor.

We will now present the results obtained by the analytical solution for a slab reactor with two distinct regions, i.e., fuel and reflector. The nuclear parameters used in this calculation are shown in **Table 2**.

$J_{2c}(0)=0$ $J_{2c}(a/2)=J_{2r}(a/2)$

$\varphi_{2c}(0)=\varphi_{2o}$ $\varphi_{2c}(a/2)=\psi_{2r}(a/2)$ $\psi_{2r}(a/2+b)=0$

Fuel   Reflector

0   a/2   b   x

Figure 2. Slab reactor with two regions of dimension a + 2b.

Table 1. Nuclear parameters for a homogeneous slab reactor.

g	Σ_{rg}	$\nu\Sigma_{fg}$	D_g	$\Sigma_{gg'}$
1	0.02935	0.000242	1.4380	0.00000
2	0.10490	0.155618	0.3976	0.01563

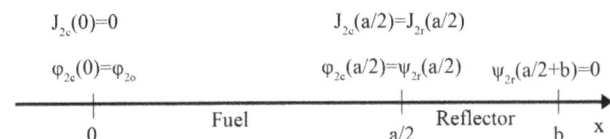

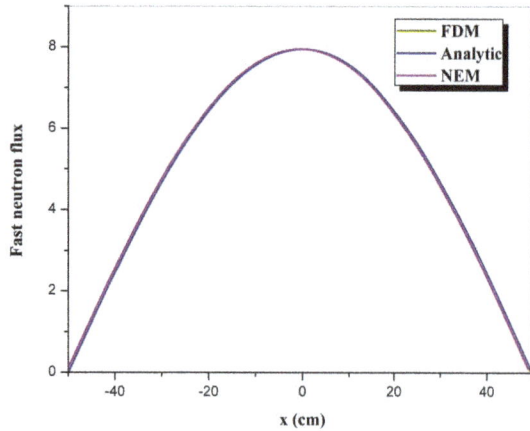

Figure 3. Comparison of fast neutron flux.

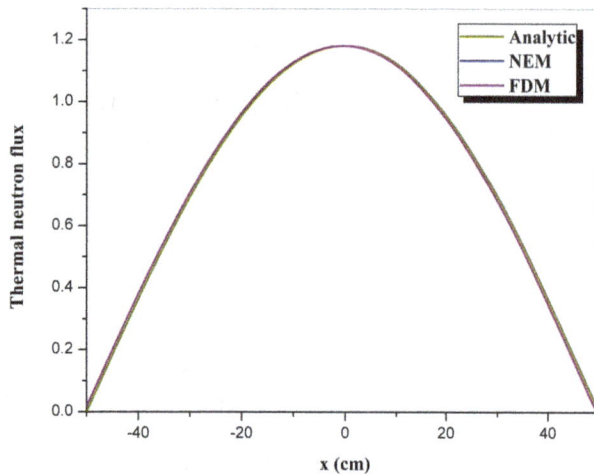

Figure 4. Comparison of thermal neutron flux.

Table 2. Nuclear parameters for heterogeneous slab reactor.

Type	g	Σ_{rg}	$\nu\Sigma_{fg}$	D_g	$\Sigma_{gg'}$
1a	1	0.02935	0.000242	1.4380	0.00000
	2	0.10490	0.155618	0.3976	0.01563
2b	1	0.035411	0.000000	1.871420	0.00000
	2	0.031579	0.000000	0.283409	0.034340

a—Fuel, b—Reflector.

We used the following dimensions in this calculation: $a = 30$ cm (fuel region) and $b = 20$ cm (reflector region). The eigenvalue obtained by the finite difference method is $k_{eff} = 0.7346988$. **Figures 5** and **6** show the results obtained for fast and thermal flux, respectively, comparing the analytical solution (Analytic) with the numerical methods of finite difference (FDM) and the nodal expansion method (NEM).

We can see that the results remained good for the two cases. This shows that when it is possible to obtain an analytical solution for a simplified geometry, the results obtained by the analytical solution are quite close to the

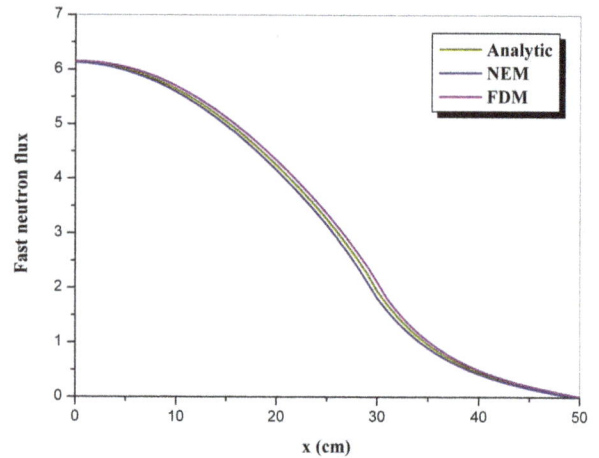

Figure 5. Comparison of fast neutron flux for two regions.

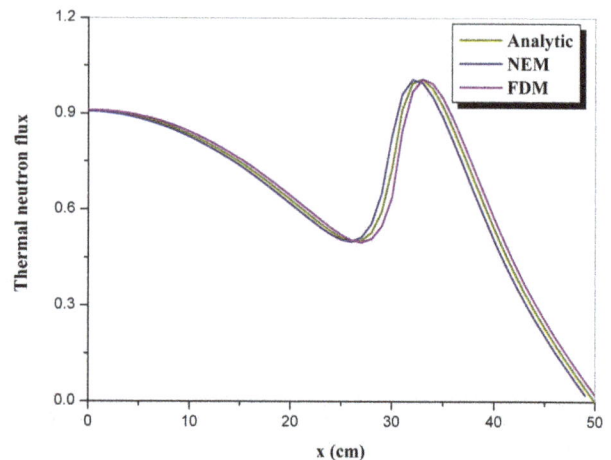

Figure 6. Comparison of thermal neutron flux for two regions.

numerical results. The information that the analytical solution needs to reconstruct the flux point-by-point that comes from the numerical method are: the eigenvalue k_{eff} that is an information intrinsic of the reactor due to its geometry and nuclear parameters and the initial flux ϕ_{g0} that is obtained by the numerical solution.

6. Conclusions

The objective of this work was to verify, when an analytical solution can be obtained, this solution can reproduce point-by-point the results of the numerical method, in this case, finite difference method (reference method). With this, some simplified cases were presented to compare analytical results with numerical ones. External data from the numerical method were incorporated as input data for the analytical solution, that were k_{eff} and the initial neutron flux. With such information, with the conditions of boundary and symmetry of the problem and with the nuclear parameters, it was possible to reconstruct the multigroup neutron flux in a slab reactor for the case of a single region (fuel) and two regions (fuel and reflector) where in both cases the analytical solution produced very satisfactory results.

Based on the conclusions presented, it is hoped that the methodology implemented in this article can be extended to bi-dimensional cases of greater practical interest, as well as its use in the axial reconstruction [11] of the neutron flux in PWR reactors.

REFERENCES

[1]   W. M. Stacey, "Nuclear Reactor Physics," Wiley, New York, 2001.

[2]   R. Boer and H. Finnemann, "Fast Analytical Flux Reconstruction Method for Nodal Space-Time Nuclear Reactor Analysis," *Annals of Nuclear Energy*, Vol. 19, No. 10-12, 1992, pp. 617-628.

[3]   H. G. Joo, J. I. Yoon and S. G. Baek, "Multigroup Pin Power Reconstruction with Two-Dimensional Source Expansion and Corner Flux Discontinuity," *Annals of Nuclear Energy*, Vol. 36, No. 1, 2009, pp. 85-97.

[4]   J. M. Aragones and N. Garcia-Herranz, "The Analytic Coarse-Mesh Finite Difference Method for Multi-Group and Multidimensional Diffusion Calculations," *Nuclear Science and Engineering*, Vol. 157, No. 1, 2007, pp. 1-15.

[5]   S. G. Baek, H. G. Joo and U. C. Lee, "Two-Dimensional Semi-Analytic Nodal Method for Multigroup Pin Power Reconstruction," *International Congress on Advances in Nuclear Power Plants* (*ICAPP* 2007), Vol. 1, Nice, 13-18 May 2007, p. 1323.

[6]   H. Finnemann, F. Bennewitz, and M. R. Wagner, "Interface Current Techniques for Multidimensional Reactor Calculations," *Atomkernenergie*, Vol. 30, No. 2, 1977, pp. 123-128.

[7]   R. D. Lawrence, "Progress in Nodal Methods for the Solution of the Neutron Diffusion and Transport Equations," *Progress in Nuclear Energy*, Vol. 17, No. 3, 1986, pp. 271-301.

[8]   A. C. Silva, F. C. Silva and A. S. Martinez, "Prediction of the neutrons subcritical multiplication using the diffusion hybrid equation with external neutron sources," *Annals of Nuclear Energy*, Vol. 38, No. 7, 2011, pp. 1667-1675.

[9]   J. J. Duderstadt and L. J. Hamilton, "Nuclear Reactor Analysis," John Wiley and Sons, New York, 1976.

[10] O. Rübenkönig, "The Finite Difference Method an Introduction," Albert Ludwigs University of Freiburg, Freiburg, 2006.

[11] L. Yu, D. Lu, S. Zhang and Y. Chao, "Group Decoupled Multi-Group Pin Power Reconstruction Utilizing Nodal Solution 1D flux Profiles," Physor-2010, Pittsburg, 2010.

29

Development of New Methodology for Distinguishing Local Pipe Wall Thinning in Nuclear Power Plants

Kyeong Mo Hwang, Hun Yun, Chan Kyoo Lee
Power Engineering Research Institute, Yongin, Korea

ABSTRACT

To manage the wall thinning of carbon steel piping in nuclear power plants, the utility of Korea has performed thickness inspection for some quantity of pipe components during every refueling outage and determined whether repair or replacement after evaluating UT data. Generally used UT thickness data evaluation methods are Band, Blanket, and PTP (Point to Point) methods. Those may not desirable to identify wall thinning on local area caused by erosion. This is because the space between inspecting points of those methods are wide for covering full surface being inspected components. When the evaluation methods are applied to a certain pipe component, unnecessary re-inspection may also be generated even though wall thinning of components does not progress. In those cases, economical loss caused by repeated inspection and problems of maintaining the pipe integrity followed by decreasing the number of newly inspected components may be generated. EPRI (Electric Power Research Institute in USA) has suggested several statistical methods such as FRIEDMAN test method, ANOVA (Analysis of Variance) method, Monte Carlo method, and TPM (Total Point Method) to distinguish whether multiple inspecting components have been thinned or not. This paper presents the NAM (Near Area of Minimum) method developed by KEPCO-E & C for distinguishing whether multiple inspecting components have been thinned or not. In addition, this paper presents the analysis results for multiple inspecting ones over three times based on the NAM method compared with the other methods suggested by EPRI.

Keywords: Pipe Wall Thinning; Component; Multiple Inspection; ANOVA-1 Method; TPM (Total Point Method); NAM (Near Area of Minimum) Method

1. Introduction

Pipe components made by carbon or low alloy steel placed on secondary system in nuclear power plants have experienced wall thinning caused by FAC (Flow Accelerated Corrosion) or erosion such as cavitation, flashing, and LDIE (Liquid Droplet Impingement Erosion) [1]. FAC is the corrosion phenomenon accelerated by high temperature and high velocity water and wet steam circumstance. LDIE is caused by the impact of high velocity droplets or liquid jets [2]. To manage the wall thinning in nuclear power plants, plant utilities have performed thickness inspection for some quantity of components during every refueling outage (RFO) and determined whether to perform repair or replacement after evaluating UT data.

Generally used UT data evaluation methods are Band, Blanket, and PTP (Point to Point) methods [3]. However, those are not desirable to identify wall thinning on local area caused by erosion. This is because the space between inspecting points are wide for covering full surface being inspected components. In addition, unnecessary re-inspection may be generated due to manufacturing or measurement errors for components in spite of no wear-

ing. In such cases, economical loss caused by repeated inspection and problems of maintaining the pipe integrity followed by decreasing the number of newly inspected components may be generated. EPRI (Electric Power Research Institute in USA) has suggested several statistical methods [4] such as FRIEDMAN test method, ANOVA (Analysis of Variance) method, Monte Carlo method, and Total Point method to distinguish whether multiple inspecting components have been thinned or not. Although the suggested ones are applicable for distinguishing large area of wall thinning such as damages by FAC, those may not be applicable for distinguishing local area of wall thinning such as by erosion.

This paper presents the NAM (Near Area of Minimum) method developed for applicable both wide range of wall thinning caused by FAC and small range of wall thinning caused by erosion.

2. Features of Wall Thinning Area and UT Data Evaluation Methods

2.1. Features of Wall Thinning Area

For carbon steel pipe components installed secondary

system in nuclear power plants, once FAC has been progressed, relatively large area of components has been affected. Meanwhile, once erosion has been occurred, relatively small area of components has been affected. **Figure 1** shows wall thinned surface of elbow caused by FAC. Wide range of extrados surface was damaged. **Figure 2** shows erosion surface caused by liquid droplet impingement. The wet steam, coming from the feedwater heater vent pipeline, collided with the other side surface of the vent line (middle of lower one in **Figure 2**) and generated local erosion.

2.2. Evaluation Methods for UT Data

The utility of Korea perform thickness inspection with a full-coverage grid layout on components being inspected. **Figure 3** shows the grid layout for an elbow. With the full-coverage grid layout, local progressed erosion damage could not be found. When the existing UT data evaluation methods, such as Band, Blanket, PTP (Point to Point) Methods, etc. [3] are applied to a certain pipe component, unnecessary re-inspecting may be generated due to manufacturing or measurement errors for components even though wall thinning does not progress.

Therefore, for components being inspected several times, it is very important to distinguish whether the components have been thinned or not in order to decide whether future inspection should be done or not in light of cost effectiveness and maintaining pipe integrity.

Figure 1. Damaged surface by FAC.

Figure 2. Damaged surface by LDIE.

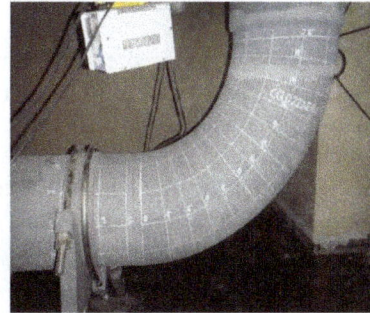

Figure 3. Full-coverage grid layout.

3. Review of Wall Thinning Distinction Methods

In all utilities through the world, unnecessary re-inspection for components even though wall thinning does not progress is an unsolved problem. EPRI suggested several methods for distinguishing whether multiple inspection components have been thinned or not. The suggested ones are FRIEDMAN test method, ANOVA-1 (One-way Analysis of Variance), ANOVA-2 (Two-way Analysis of Variance), Monte Carlo Method [4], and Total Point Method (TPM) [5].

FRIEDMAN test method uses a ranking of the measurements rather than the measurements themselves to determine whether components have been thinned or not. It is concluded that the Freidman test method is inappropriate for the determination desired purpose. This is because the method predicts many cases of wear while evidence indicates that the components are not wearing.

ANOVA-2 method uses the measurements themselves to determine whether components have been thinned or not by comparing two independent data sets of one case. It is concluded that the ANOVA-2 method is inappropriate for the determination desired purpose. This is because only fluctuation of thickness data by inspection terms can be considered as an independent variable.

Monte Carlo Method uses fictious data to calculate average wear rate in order to determinate the components will be thinned or not for next inspection period. It is concluded that the method is inappropriate for the determination whether the inspected components will experience wall thinning or not.

LSS (Least Square Slope) method is not explained in this paper. The results of LSS method are used to TPM for the desired determination.

It remains to be seen what is applicable for determination whether pipe component has been wall thinned or not. It is concluded that the ones capable for the purpose are ANOVA-1, TPM, and NAM methods. The NAM method is the new approach developed by KEPCO-E & C. Section 3.1 through 3.3 give explanations of those methods. Section 4 presents overview of analysis results for multiple inspection data.

3.1. TPM

In order to apply TPM for the desired determination, UT data matrix at least inspected two different terms should be presented. And, the number of rows and columns of data matrix should be same.

LSS method [6] is a way to obtain wear rate from the magnitude of least square slope in the same point on a component. The histogram is acquired by the magnitude of the wear rates obtained from the least squares slopes in the same points for all points on a component. **Figure 4** shows the acquired histogram.

Components are "Wear" when the number of positive values is more than the number of negative values on inspected points. Components are "No Wear" when the number of negative values is more than the number of positive values on inspected all points. The component on **Figure 4** is "Wear" which is indicated by most wear rate magnitude is positive, placed on 4.4 mils/year.

3.2. ANOVA-1 Method

This section presents the ANOVA-1 [4] approach for determination wether components has been thinned or not. The associations between mean values of data set and inspect outages are taken into account when the one way ANOVA performs analysis data sets.

Table 1 shows inspected measurement obtained at the four different outages on 38 components. The calculated results by ANOVA-1 are presented on **Table 2**. The null hypothesis for the desired determination is that there is no association between measurements and inspection outages. In other words, it is expected that the mean value of measurements at every inspection outages are identical. As shown in **Table 2**, the null hypothesis is not rejected because F ratio (0.026) is less than the critical F (2.67). The same conclusions can also be reached by examining P value. The P value (0.99) is greater than 0.05 with a 95% confidence level. It means that wall thinning is progressed.

Table 1. UT data for each inspection outage.

Inspection Points	R-15	R-16	R-17	R-18
a01	0.512	0.513	0.514	0.513
a02	0.508	0.507	0.508	0.510
a03	0.512	0.513	0.513	0.515
a04	0.513	0.513	0.514	0.515
a05	0.515	0.516	0.515	0.516
a06	0.514	0.514	0.514	0.516
a07	0.507	0.506	0.507	0.507
a08	0.506	0.507	0.508	0.508
a09	0.508	0.508	0.509	0.511
a10	0.510	0.511	0.511	0.512
a11	0.506	0.508	0.505	0.507
a12	0.502	0.503	0.503	0.504
a13	0.494	0.496	0.495	0.497
a14	0.501	0.500	0.502	0.504
a15	0.504	0.505	0.505	0.506
a16	0.508	0.508	0.509	0.509
a17	0.508	0.508	0.509	0.509
a18	0.506	0.506	0.506	0.506
a19	0.506	0.507	0.507	0.507
b01	0.505	0.506	0.507	0.508
b02	0.509	0.508	0.508	0.510
b03	0.507	0.506	0.507	0.507
b04	0.500	0.499	0.500	0.501
b05	0.502	0.502	0.503	0.505
b06	0.503	0.502	0.503	0.505
b07	0.401	0.401	0.401	0.400
b08	0.398	0.385	0.398	0.397
b09	0.475	0.473	0.475	0.475
b10	0.501	0.502	0.501	0.502
b11	0.506	0.503	0.504	0.502
b12	0.502	0.501	0.502	0.504
b13	0.497	0.497	0.497	0.498
b14	0.503	0.502	0.503	0.503
b15	0.502	0.504	0.502	0.502
b16	0.508	0.507	0.508	0.510
b17	0.508	0.508	0.508	0.510
b18	0.503	0.501	0.504	0.503
b19	0.504	0.505	0.504	0.505
Sum	18.974	18.961	18.989	19.019
Average	0.499	0.499	0.500	0.501
SD	0.02476	0.02633	0.02488	0.02533
Variance	0.00061	0.00069	0.00062	0.00064

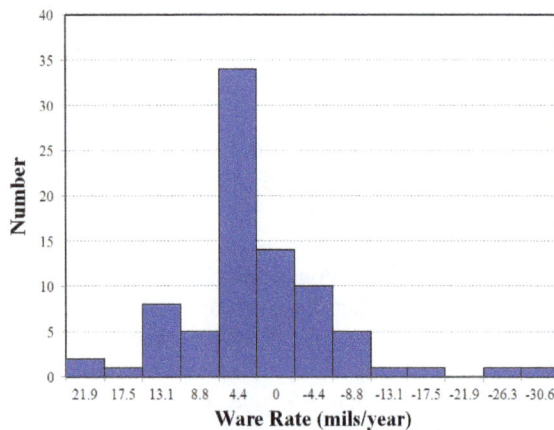

Figure 4. Sample histogram of TPM.

Table 2. Calculation results for ANOVA-1.

Source of Variation	Degrees of Freedom	Sum of Squares	Mean Square	F Ratio	P-Value	F Crit
Between Groups	3	4.91E−05	1.64E−05			
Within Groups	148	0.095	6.42E−04	0.026	0.99	2.67
Total	151	0.095	-			

3.3. NAM Method

NAM method is the new approach developed by KEP-CO-E & C for determination whether components have been thinned or not. The calculation for NAM method is following. For last inspection period, select the minimum point and neighboring 8 points of it and calculate average slop of those 9 points by inspections outages. It is determined that thinning is progressed when the mean value of slops is negative. The evaluation work by NAM meth- od is based on minimum point and neighboring 8 points of it. Therefore, the method is applicable to determine not only wide range of wall thinning caused by FAC but also small range of wall thinning caused by erosion. Followings are estimated results by NAM method. **Tables 3-5** present measurement data obtained from a 90° elbow at RFO 13, RFO 16, and RFO 18. The minimum thickness data during RFO 18 is 0.434 inch on the point (a, 6). The extracted slops of the point and neighboring 8 points of it are presented on **Figure 5**. The average value of slops is 0.0003 that is, wear is not occurring.

Another 90° elbow (B) was examined at RFO 13, RFO 15 and RFO 18. Data set at each RFO is presented on **Tables 6-8**. The minimum thickness data during RFO 18 is 0.453 inch on the point (h, 7). The extracted slops by RFO of the point and neighboring 8 points of it are presented on **Figure 6**. Wear is occurring as the average value of slops is –0.0169.

Table 3. UT data of a 90° elbow (A) at RFO 13.

	a	b	c	d	e	f	g	h	i	j	k	l
1	0.464	0.493	0.516	0.507	0.501	0.539	0.511	0.510	0.496	0.511	0.466	0.488
2	0.458	0.492	0.494	0.501	0.499	0.522	0.526	0.497	0.508	0.518	0.490	0.488
3	0.449	0.480	0.493	0.488	0.499	0.534	0.53	0.501	0.506	0.511	0.482	0.475
4	0.441	0.470	0.478	0.489	0.507	0.543	0.579	0.516	0.515	0.516	0.488	0.483
5	0.431	0.474	0.483	0.496	0.521	0.556	0.551	0.532	0.521	0.516	0.493	0.480
6	0.431	0.486	0.500	0.512	0.523	0.561	0.554	0.539	0.523	0.512	0.491	0.476
7	0.438	0.491	0.508	0.514	0.528	0.560	0.559	0.532	0.508	0.509	0.484	0.490
8	0.449	0.498	0.499	0.503	0.524	0.551	0.560	0.530	0.509	0.509	0.494	0.504
9	0.467	0.509	0.500	0.505	0.521	0.553	0.556	0.526	0.516	0.513	0.508	0.507
10	0.473	0.501	0.502	0.489	0.510	0.546	0.548	0.529	0.516	0.528	0.522	0.505
11	0.471	0.494	0.498	0.484	0.500	0.541	0.543	0.523	0.516	0.537	0.530	0.513
12	0.479	0.506	0.513	0.49	0.508	0.534	0.564	0.520	0.511	0.540	0.529	0.506

Table 4. UT data of a 90° elbow (A) at RFO 16.

	a	b	c	d	e	f	g	h	i	j	k	l
1	0.467	0.494	0.504	0.503	0.500	0.525	0.513	0.489	0.487	0.511	0.469	0.491
2	0.461	0.492	0.494	0.502	0.500	0.525	0.523	0.496	0.512	0.514	0.492	0.493
3	0.449	0.479	0.489	0.489	0.498	0.527	0.530	0.498	0.508	0.511	0.485	0.489
4	0.439	0.471	0.480	0.490	0.504	0.541	0.543	0.519	0.512	0.514	0.489	0.476
5	0.436	0.477	0.484	0.500	0.516	0.555	0.559	0.533	0.523	0.516	0.494	0.486
6	0.436	0.487	0.500	0.519	0.530	0.561	0.562	0.538	0.521	0.515	0.490	0.481
7	0.440	0.49	0.504	0.515	0.525	0.563	0.556	0.533	0.510	0.517	0.485	0.478
8	0.453	0.498	0.503	0.510	0.521	0.558	0.553	0.529	0.510	0.506	0.496	0.494
9	0.469	0.505	0.503	0.506	0.514	0.554	0.552	0.527	0.510	0.518	0.514	0.508
10	0.470	0.505	0.503	0.489	0.509	0.547	0.546	0.530	0.521	0.530	0.524	0.510
11	0.468	0.498	0.491	0.483	0.501	0.546	0.546	0.523	0.515	0.539	0.531	0.509
12	0.480	0.507	0.515	0.493	0.506	0.533	0.543	0.519	0.512	0.545	0.529	0.511

Table 5. UT data of a 90° elbow (A) at RFO 18.

	a	b	c	d	e	f	g	h	i	j	k	l
1	0.464	0.494	0.511	0.505	0.503	0.518	0.52	0.499	0.493	0.508	0.476	0.486
2	0.459	0.493	0.493	0.499	0.497	0.527	0.526	0.500	0.516	0.518	0.489	0.490
3	0.447	0.481	0.489	0.487	0.494	0.537	0.548	0.497	0.509	0.509	0.480	0.488
4	0.437	0.471	0.479	0.486	0.499	0.542	0.558	0.514	0.513	0.512	0.485	0.476
5	0.438	0.477	0.486	0.498	0.523	0.558	0.564	0.531	0.515	0.515	0.493	0.483
6	0.434	0.487	0.499	0.517	0.529	0.560	0.551	0.538	0.506	0.508	0.487	0.478
7	0.440	0.491	0.501	0.511	0.527	0.562	0.556	0.530	0.505	0.506	0.476	0.474
8	0.451	0.499	0.498	0.507	0.518	0.558	0.549	0.530	0.503	0.503	0.491	0.494
9	0.469	0.503	0.500	0.505	0.517	0.553	0.546	0.526	0.517	0.512	0.510	0.512
10	0.469	0.498	0.502	0.493	0.509	0.544	0.544	0.529	0.527	0.527	0.521	0.511
11	0.465	0.493	0.490	0.481	0.501	0.545	0.542	0.525	0.533	0.534	0.528	0.509
12	0.476	0.503	0.511	0.496	0.506	0.534	0.545	0.518	0.538	0.539	0.528	0.508

Table 6. UT data of a 90° elbow (B) at RFO 13.

	a	b	c	d	e	f	g	h	i	j	k	l
1	0.611	0.619	0.640	0.698	0.661	0.651	0.653	0.657	0.652	0.668	0.661	0.594
2	0.612	0.552	0.605	0.707	0.653	0.573	0.568	0.632	0.641	0.506	0.558	0.594
3	0.542	0.516	0.497	0.491	0.464	0.464	0.473	0.475	0.473	0.511	0.500	0.526
4	0.549	0.534	0.525	0.499	0.468	0.462	0.468	0.486	0.500	0.537	0.519	0.545
5	0.562	0.536	0.502	0.489	0.459	0.457	0.463	0.473	0.497	0.537	0.524	0.544
6	0.548	0.520	0.495	0.494	0.468	0.470	0.472	0.493	0.508	0.544	0.522	0.541
7	0.533	0.510	0.497	0.499	0.482	0.477	0.482	0.503	0.519	0.554	0.526	0.537
8	0.535	0.515	0.509	0.506	0.489	0.484	0.487	0.504	0.522	0.561	0.523	0.533
9	0.544	0.519	0.518	0.507	0.487	0.484	0.496	0.509	0.526	0.566	0.520	0.535
10	0.547	0.528	0.521	0.507	0.479	0.484	0.497	0.504	0.523	0.562	0.514	0.535
11	0.547	0.526	0.506	0.494	0.485	0.483	0.494	0.506	0.516	0.553	0.535	0.534

Table 7. UT data of a 90° elbow (B) at RFO 15.

	a	b	c	d	e	f	g	h	i	j	k	l
1	0.531	0.516	0.505	0.488	0.484	0.482	0.492	0.504	0.511	0.556	0.532	0.530
2	0.537	0.518	0.510	0.496	0.481	0.482	0.493	0.506	0.517	0.555	0.518	0.529
3	0.529	0.523	0.514	0.506	0.483	0.483	0.489	0.507	0.514	0.561	0.521	0.534
4	0.532	0.520	0.507	0.501	0.486	0.481	0.488	0.503	0.513	0.557	0.523	0.528
5	0.528	0.511	0.497	0.492	0.48	0.479	0.482	0.497	0.508	0.549	0.522	0.529
6	0.544	0.523	0.495	0.487	0.468	0.463	0.468	0.496	0.503	0.532	0.516	0.537
7	0.556	0.535	0.506	0.489	0.461	0.452	0.463	0.477	0.495	0.533	0.515	0.563
8	0.549	0.533	0.506	0.495	0.466	0.452	0.462	0.477	0.500	0.530	0.513	0.539
9	0.525	0.521	0.497	0.500	0.463	0.460	0.473	0.474	0.475	0.496	0.499	0.524
10	0.609	0.561	0.598	0.617	0.638	0.575	0.600	0.579	0.631	0.627	0.610	0.612
11	0.605	0.623	0.642	0.703	0.663	0.635	0.667	0.66	0.664	0.657	0.610	0.612

Table 8. UT data of a 90° elbow (B) at RFO 18.

	a	b	c	d	e	f	g	h	i	j	k	l
1	0.543	0.532	0.542	0.565	0.516	0.503	0.493	0.478	0.485	0.492	0.504	0.519
2	0.528	0.533	0.519	0.562	0.527	0.504	0.493	0.483	0.481	0.497	0.515	0.524
3	0.529	0.538	0.522	0.555	0.521	0.503	0.498	0.484	0.49	0.502	0.517	0.522
4	0.529	0.531	0.522	0.562	0.514	0.500	0.49	0.482	0.492	0.507	0.509	0.513
5	0.526	0.532	0.524	0.568	0.529	0.500	0.486	0.483	0.482	0.503	0.497	0.511
6	0.546	0.536	0.521	0.529	0.508	0.491	0.474	0.469	0.475	0.490	0.492	0.518
7	0.557	0.537	0.518	0.538	0.499	0.481	0.462	0.453	0.465	0.490	0.505	0.535
8	0.547	0.54	0.514	0.531	0.506	0.488	0.464	0.456	0.467	0.497	0.507	0.535
9	0.586	0.529	0.517	0.528	0.507	0.498	0.487	0.478	0.484	0.505	0.506	0.531
10	0.531	0.516	0.477	0.494	0.467	0.476	0.481	0.453	0.459	0.488	0.496	0.521
11	0.53	0.532	0.535	0.532	0.538	0.547	0.534	0.543	0.530	0.539	0.535	0.537

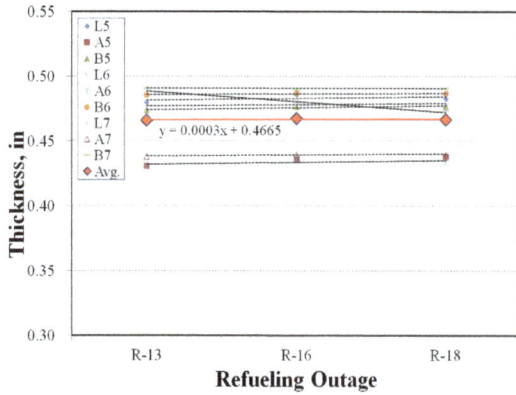

Figure 5. Application of NAM method for a 90° elbow (A).

$y = 0.0003x + 0.4665$

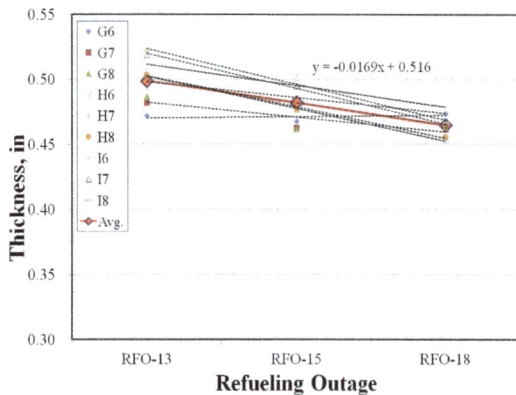

Figure 6. Application of NAM method for a 90° elbow (B).

$y = -0.0169x + 0.516$

When components are indicated wear is occurring by NAM method, further inspection is performed. As shown in **Figure 7**, to point out the presence of wear, the space between inspecting points is reduced to a fifth on the thinned area. By conducting evaluation process by NAM method based on the re-inspected data, it is determined whether the component has experienced wear or not.

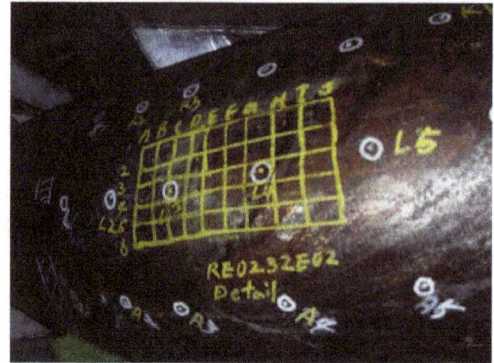

Figure 7. Grid layout for a detailed UT examination.

4. Discussion of the Various Methods

This section presents availability of ANOVA-1, TPM, and NAM method for answering wall thinning determination on multiple inspection components. The availability above three methods explored based on at least three different multiple inspection data of 41 components in NPP (nuclear power plant) A and 40 components in NPP B. Then a close reading based on visual identification of the data made it clear that whether the actual wall thinned or not.

Table 9 shows the thickness data obtained from TE0-17AXA (Expander) at each refueling outage. The evaluation results are "No Wear" by ANOVA-1, "Wear" by TPM, and "No Wear" by NAM. A close reading of the data table makes the component "No wear". Average thickness of the minimum measurement area (inside small rectangular in the data matrix) at three different RFO is 1.060, 1.047, and 1.060. Also, the minimum thickness data is 0.966, 0.971, and 0.973. This indicates that there are no specific parts which have been gradually thinned. Therefore, the results by ANOVA-1 and NAM are appropriately evaluated.

Table 9. UT data on TE017AXA at each RFO.

RFO	UT Data												
R-15	1.003	0.998	0.996	0.981	0.989	0.993	1.003	0.991	0.966	0.980	0.987	0.974	0.989
	1.143	1.112	1.121	1.128	1.114	1.119	1.147	1.133	1.157	1.134	1.136	1.121	1.133
	1.351	1.324	1.302	1.321	1.358	1.372	1.361	1.373	1.380	1.351	1.349	1.347	1.350
	1.538	1.570	1.605	1.691	1.714	1.714	1.682	1.696	1.611	1.605	1.543	1.523	1.526
	1.986	2.025	1.956	1.932	1.919	1.916	1.917	1.927	1.946	1.961	1.963	1.971	1.997
	1.725	1.611	1.664	1.643	1.601	1.679	1.667	1.656	1.691	1.681	1.724	1.691	1.719
R-17	1.001	0.998	0.993	0.987	0.992	0.989	0.971	0.988	0.979	0.98	0.988	0.985	0.988
	1.182	1.098	1.105	1.087	1.113	1.114	1.079	1.127	1.131	1.074	1.127	1.129	1.131
	1.336	1.324	1.301	1.321	1.364	1.363	1.360	1.365	1.372	1.350	1.342	1.335	1.332
	1.524	1.548	1.611	1.696	1.739	1.619	1.643	1.678	1.585	1.560	1.542	1.518	1.518
	1.987	2.023	1.957	1.931	1.916	1.911	1.904	1.923	1.944	1.960	1.965	1.970	1.993
	1.686	1.612	1.638	1.591	1.683	1.666	1.687	1.645	1.676	1.691	1.706	1.655	1.678
R-19	1.015	1.002	1.001	0.984	0.998	0.993	1.008	0.982	0.973	0.993	0.980	0.976	0.984
	1.157	1.087	1.080	1.136	1.041	1.076	1.141	1.134	1.156	1.122	1.074	1.093	1.091
	1.352	1.323	1.303	1.324	1.357	1.375	1.355	1.373	1.379	1.356	1.347	1.344	1.339
	1.558	1.571	1.611	1.694	1.696	1.705	1.668	1.702	1.592	1.561	1.526	1.514	1.529
	1.990	2.034	1.962	1.934	1.928	1.919	1.926	1.930	1.956	1.964	1.970	1.976	1.999
	1.731	1.686	1.639	1.608	1.636	1.644	1.667	1.631	1.672	1.749	1.725	1.681	1.665

Table 10 shows the thickness data obtained from TF124AXA (Expander) at each refueling outage. The evaluation results are "No Wear" by ANOVA-1, "No Wear" by TPM, and "Wear" by NAM method. A close reading of the table makes the component "Wear". Average thickness of the minimum measurement area (inside small rectangular in the data matrix) at three different RFO is 0.350, 0.336, and 0.331. Also, the minimum thickness data is 0.231, 0.217, and 0.190. This indicates that there are specific parts which have been gradually thinned. Therefore, the results by NAM are only appropriately evaluated. TE017AXA, TF124AXA were given as component names.

The results of evaluation by the different methods based on 41 components in NPP A are tabulated Table 11. The results of evaluation by the different methods based on 40 components in NPP B are tabulated Table 12. The following two acronyms are tabulated in the column of TPM, ANOVA-1, and NAM. "W" represents wall thinning is occurring, "NW" represents wall thinning is not occurring. The last column on those tables presents the judgments by a close reading of the thickness data, visual identification. The following three features are tabulated in the column of visual identification. ○ represents wall thinning is occurring, × represents wall thinning is not occurring, and △ represents ambiguous to determine whether the wall thinning is occurring or not.

Assuming that the visual identification results for 41 components placed on NPP A is correct, results from ANOVA-1, TPM, and NAM methods shows 39%, 71%, and 85% accuracy, respectively. If ambiguous ones are being considered as wearing ones, the accuracy is 39%, 80%, and 100%. Assuming that the visual identification results for 40 components placed on NPP B is correct, results from ANOVA-1, TPM, and NAM methods shows 43%, 58%, and 68% accuracy. If ambiguous ones are being considered as a wearing, the accuracy is 43%, 70%, and 93%. Considering ambiguous ones as wearing ones, and put evaluation results from NPP A and NPP B together, the accuracy of ANOVA-1, TPM, NAM method is 42%, 76.5% and 97.5%. It clearly shows evaluation results from NAM are more accurate than results from the other two methods.

Table 10. UT data on TF124AX at each RFO.

RFO	UT Data											
R-14	0.347	0.345	0.347	0.349	0.346	0.346	0.346	0.346	0.346	0.346	0.346	0.345
	0.760	0.764	0.765	0.762	0.764	0.767	0.764	0.766	0.765	0.765	0.766	0.767
	0.769	0.768	0.767	0.766	0.767	0.770	0.770	0.769	0.769	0.768	0.770	0.769
	0.762	0.761	0.760	0.764	0.757	0.757	0.760	0.770	0.761	0.761	0.761	0.758
	0.536	0.534	0.534	0.535	0.535	0.533	0.537	0.533	0.535	0.555	0.494	0.492
	0.239	0.231	0.278	0.302	0.299	0.323	0.319	0.372	0.322	0.280	0.282	0.265
	0.281	0.262	0.258	0.290	0.286	0.322	0.327	0.337	0.322	0.292	0.273	0.283
	0.293	0.274	0.274	0.278	0.271	0.306	0.296	0.313	0.298	0.281	0.286	0.293
R-15	0.346	0.345	0.344	0.341	0.344	0.345	0.348	0.343	0.344	0.345	0.344	0.345
	0.758	0.762	0.764	0.764	0.766	0.763	0.764	0.767	0.765	0.768	0.742	0.762
	0.767	0.768	0.767	0.767	0.766	0.768	0.765	0.769	0.770	0.768	0.770	0.766
	0.758	0.759	0.760	0.757	0.754	0.756	0.754	0.759	0.758	0.761	0.759	0.745
	0.509	0.530	0.510	0.495	0.534	0.531	0.530	0.495	0.551	0.481	0.465	0.508
	0.226	0.217	0.259	0.295	0.286	0.310	0.309	0.321	0.319	0.271	0.278	0.261
	0.271	0.252	0.246	0.277	0.273	0.310	0.314	0.333	0.319	0.279	0.263	0.263
	0.281	0.269	0.264	0.272	0.264	0.295	0.286	0.292	0.281	0.272	0.270	0.283
R-17	0.347	0.344	0.344	0.347	0.343	0.346	0.351	0.346	0.345	0.346	0.346	0.346
	0.787	0.787	0.762	0.789	0.791	0.786	0.786	0.789	0.790	0.764	0.765	0.767
	0.769	0.791	0.794	0.795	0.788	0.792	0.794	0.792	0.792	0.768	0.796	0.797
	0.795	0.792	0.792	0.787	0.787	0.764	0.763	0.794	0.789	0.794	0.769	0.790
	0.535	0.561	0.534	0.531	0.534	0.538	0.536	0.536	0.533	0.535	0.512	0.534
	0.211	0.190	0.256	0.283	0.274	0.303	0.299	0.317	0.312	0.268	0.274	0.255
	0.247	0.221	0.226	0.250	0.253	0.289	0.298	0.328	0.323	0.257	0.249	0.251
	0.255	0.242	0.234	0.238	0.235	0.259	0.258	0.266	0.258	0.237	0.249	0.264

Table 11. Comparison of accuracy among wall thinning distinction methods for NPP A.

No	Component Name	TPM	ANOVA-1	NAM	Visual Identification
1	TD005AEC	W	NW	W	△
2	TD011AEC	W	NW	NW	×
3	TE003AEA	W	NW	NW	×
4	TE020FEE	NW	NW	NW	×
5	TE018AEA	NW	NW	W	△
6	TE018AEC	NW	NW	NW	×
7	TE017AXA	W	NW	NW	×
8	TE016AXA	NW	NW	NW	×
9	TE020DEA	NW	NW	W	△
10	TE027AEG	NW	NW	NW	×
11	TE054AEA	W	NW	W	○
12	TE037CEA	NW	W	W	○
13	TE019APH	W	NW	W	○
14	TE023AED	W	NW	W	○
15	TE024APH	W	W	W	○
16	TE037ERA	W	NW	W	○
17	TE037FRA	W	NW	W	○
18	TE024BXA	W	NW	W	○
19	PL025APO	NW	NW	NW	×
20	TF176AEI	W	NW	W	△
21	TF435AEA	W	NW	W	△
22	TF190AXB	NW	NW	NW	×
23	TF214AXA	W	NW	W	○
24	TF455ATB	W	NW	W	○
25	TF470ATA	W	NW	W	○
26	TF079BTA	W	NW	W	○
27	TF091CTA	W	NW	W	○
28	TF530AEJ	W	NW	W	○
29	TF530AEI	W	NW	W	○
30	TF531BEK	W	NW	W	○
31	TF610ATB	W	NW	W	○
32	TF522AEJ	W	W	W	○
33	TF540AXA	W	W	W	○
34	TF604CTC	W	NW	W	△
35	TF559AED	W	NW	NW	×
36	TF597AXA	NW	NW	NW	×
37	TF098AXA	W	NW	W	○
38	TF167AXA	W	NW	W	○
39	TF163AEC	NW	NW	NW	×
40	TF124AXA	NW	NW	W	○
41	TF122AXA	W	NW	W	○

Table 12. Comparison of accuracy among wall thinning distinction methods for NPP B.

No	Component Name	TPM	ANOVA-1	NAM	Visual Identification
1	TF604CTC	W	NW	W	○
2	AM010AXB	W	NW	W	○
3	AM011AXB	W	NW	NW	×
4	AM012AXB	NW	NW	W	△
5	CB003AEE	NW	NW	NW	×
6	CE021AEB	NW	NW	NW	×
7	PL025APO	NW	NW	NW	×
8	TB004AEA	NW	NW	W	○
9	TB113ETB	NW	NW	W	○
10	TC077APH	W	NW	W	△
11	TC097AP1	NW	NW	W	△
12	TD023CEA	W	NW	NW	×
13	TD071AEE	W	W	W	○
14	TD076ARA	NW	NW	NW	○
15	TE018AEC	NW	NW	W	△
16	TE018AXA	NW	NW	NW	×
17	TE020DEA	NW	NW	NW	×
18	TE024AEB	W	W	W	○
19	TE024AED	W	W	W	○
20	TE024AXA	W	W	W	○
21	TE027AEE	W	NW	W	△
22	TE037BEC	W	NW	W	×
23	TF079BTA	NW	NW	W	×
24	TF086BXA	NW	NW	NW	×
25	TF097ATA	W	NW	W	○
26	TF164AEF	W	NW	W	○
27	TF165AEA	W	W	W	○
28	TF177AEI	W	NW	W	△
29	TF190AXB	W	NW	W	○
30	TF212AEA	W	NW	W	○
31	TF237ATA	W	NW	W	○
32	TF269AEC	W	NW	W	○
33	TF423AEF	NW	NW	W	△
34	TF426AEA	W	NW	W	○
35	TF453ATA	W	NW	W	△
36	TF453ATB	NW	NW	W	△
37	TF534AXA	W	W	W	○
38	TF542AEA	W	NW	NW	×
39	TF542AEB	W	NW	W	○
40	TF604CTA	W	NW	W	△

5. Conclusions

To manage the wall thinning of carbon steel piping, thickness inspection for a certain amount of components has performed with the full-grid layout during every re-fueling outage. Then continue to use or replacement are determined by evaluating the UT inspection data. Yet, occasionally local wall thinning such as erosion damage cannot be examined with the full-grid layout due to the distance between inspection points is too large. In addition, unnecessary re-inspection may be generated due to manufacturing or measurement errors for components in spite of no wearing. Problems may appear economical loss due to unnecessary repeated inspection on components, wall thinning does not evidently progress, and unstable pipe integrity due to reduced new inspection components by increasing components needed to be checked in limited inspection period.

The NAM method, developed by KEPCO-E&C, is for distinguishing whether at least three different inspecting components have been thinned or not. The evaluation process of NAM method is first selecting the minimum thickness point and neighboring 8 points of it from the last order of inspection thickness data and then calculating average value of slopes for the 9 points by inspection period. Wear is indicated when the average value of slops is negative. When components are indicated wear is occurring by NAM method, detailed UT inspection is performed with reduced space between inspecting points.

A total of 81 components, multiple inspection ones, placed on different two nuclear power plants were evaluated by ANOVA-1, TPM, and NAM method. The analysis results from NAM method show more high accuracy than results from the other two methods

REFERENCES

[1] J. Horowitz, "Recommendations for Controlling Cavitation, Flashing, Liquid Droplet Impingement Erosion, and Solid Particle Erosion in Nuclear Power Plant Piping Systems," Electric Power Research Institute 1011231, Final Report, 2004.

[2] N. A. Barton, "Erosion in elbows in hydrocarbon production systems: Review document," Health & Safety Executive, TUV NEL Limited, Glasgow, 2003.

[3] A. Machiels and D. Munson, "CHECWORKS^{TM} Steam/Feedwater Application, Guidelines for Plant Modeling and Evaluation of Component Inspection Data," Electric Power Research Institute 1009599, Final Report, 2004.

[4] J. Horowitz, "Statistical Methods for the Analysis of Multiple-Inspection Flow-Accelerated Corrosion Data," Electric Power Research Institute 1019175, Final Report, 2009.

[5] D. G. Gusso, "Implementation Test Project on the Total Points Method," Electric Power Research Institute 1022575, Technical Update, 2011.

[6] J. Horowitz, "Least Squares Methods for Evaluation Inspection Data," Electric Power Research Institute 1018456, Technical Update, 2008.

30

Correlations in Nuclear Interactions between E_{CM}/u and Unexplained Experimental Observables

Wolfram Westmeier[1,2*], Reinhard Brandt[2], Reza Hashemi-Nezhad[3], Reinhard Odoj[4], Wolfgang Ensinger[5], Maria Zamani-Valasiadou[6], Andrei Sosnin[7]

[1]Gesellschaft für Kernspektrometrie mbH, Ebsdorfergrund, Germany
[2]Fachbereich Chemie, Philipps-Universität Marburg, Marburg, Germany
[3]Institute of Nuclear Science, School of Physics, University of Sydney, Sydney, Australia
[4]Forschungszentrum Jülich GmbH, Jülich, Germany
[5]Fachbereich Materialwissenschaften, Technische Universität, Darmstadt, Germany
[6]Physics Department, Aristotle University of Thessaloniki, Thessaloniki, Hellas
[7]Joint Institute for Nuclear Research, Dubna, Russia

ABSTRACT

A new concept is introduced for the classification of "unresolved problems" in the understanding of interactions in thick targets irradiated with relativistic ions: The centre-of-mass energy per nucleon of a hypothetical compound nucleus from a primary interaction, E_{CM}/u, is calculated and correlated with experimental observations in thick target irradiations. One observes in various reactions of relativistic primary ions with thick targets that there appears to be a threshold energy for reactions leading to "unresolved problems" which lies around $E_{CM}/u \sim 150$ MeV. All "unresolved problems" are exclusively observed above this threshold, whereas below this threshold no "unresolved problems" are found. A similar threshold at 158 ± 3 MeV exists for massive pion production in nuclear interactions. Hagedorn had proposed this threshold decades ago and it is known as the Hagedorn limit. In this paper we will only mention, but not elaborate on Hagedorn's theoretical concept any further. Some considerations will be presented and further studies in this field are suggested.

Keywords: Thick Target; High Energy Projectile; Neutron Multiplicity; Mass Distribution

1. Introduction

Spallation mass-yield curves in nuclear interactions with thin targets were systematically studied in many nuclear chemistry laboratories for decades around the world. These observed spallation mass-yield curves strictly obey well-known concepts of "limiting fragmentation" and "factorisation" (see Section 2.2) and are thus well understood within current theoretical models. This applies for nuclear reaction studies induced by ions from $E_{kinetic} < 1$ GeV and is extending up to 80 GeV ^{40}Ar irradiations. Limited studies extend up to proton induced reactions with $E_{kinetic} = 300$ GeV (see [1] for details).

Several articles have recently appeared describing "unresolved problems" in the study of nuclear interactions in thick targets induced by relativistic ions and their secondary reaction products [1,2]. Product yield distributions in thick copper targets from irradiations with 72 GeV ^{40}Ar (at the LBNL, Berkeley), 44 GeV ^{12}C (at the JINR, Dubna), and 48 GeV ^{4}He (at CERN, Geneva) [3]

cannot be understood with well-established theoretical concepts, thus constituting "unresolved problems". Moreover, exceedingly large neutron emission during the irradiation of thick copper, lead and uranium targets with high energy heavy ion beams having $E_{kinetic} > 30$ GeV have been observed in several laboratories; where an exceedingly large neutron multiplicity is also considered to be an "unresolved problem".

Several authors [4,5] confirm the existence of experimentally observed unresolved problems, however, they reject in very clear and strong terms any attempt to interpret these unresolved problems, even with unconventional approaches. Hartmann and Brandt have recently published one such unconventional approach [6].

All attempts to characterise unresolved problems in thick-target nuclear reactions since about 1954 [7] have borne no fruit; the problem being that there are no defined combinations of ion energy, projectile mass, and target mass where these unresolved problems systematically occur.

In this paper the following approach will be intro-

---

*Corresponding author.

duced:

One calculates on a purely hypothetical basis the centre-of-mass energy E_{CM} per nucleon in the entrance channel of the nuclear interaction. This entrance channel is defined by the kinetic energy E_P of the primary ion (projectile) with mass A$_P$ and the target mass A_T. The value of E_{CM}/u in units of MeV is calculated as:

$$E_{CM}/u = E_p \frac{A_T}{\left(A_p + A_T\right)^2} \qquad (1)$$

In thick targets experimentally observed phenomena are produced both by primary ions (primaries) up to the end of their range and in addition by secondary fragments (secondaries) making nuclear interactions in the thick target. The relative importance of nuclear reactions in thick targets due to secondaries compared to primaries increases with the thickness of the target [8]. One may correlate the value E_{CM}/u—which might be taken as the hypothetical average excitation energy of each nucleon in the entrance channel of the reaction—with experimentally observed phenomena.

Some correlations are presented in Section 2 for increasing E_{CM}/u. Obviously any observed correlation between E_{CM}/u in the entrance channel and interactions of secondary fragments in thick targets will not explain the reason for unresolved problems. However, one does find a systematic dependence which allows a priori classification and selection of experiments where unresolved results are to be expected. In Section 3 we will present some considerations which may be helpful to understanding the observed order as presented in Section 2. Section 4 contains our conclusions on the subject and new experiments are suggested which may help to shed light onto this rather old and complex set of unresolved problems. In the Appendix few known and published experiments on thick targets irradiated with very high-energy ions having E$_{kinetic}$ > 100 GeV will be reported.

2. Correlations between E_{CM}/u and Unresolved Problems

Unresolved problems as discussed in detail in [1,2] are observed only in high energy nuclear interactions with thick targets. Three types of experiments which reveal unresolved problems are described in more detail below.

2.1. Production of ^{24}Na in Two Copper Discs in Contact

The quantification of the isotope ^{24}Na (T$_{1/2}$ = 15 h) produced in a thick copper target consisting of two Cu-disks of 8 cm diameter and 1 cm thickness each in irradiations with relativistic ions requires just conventional gamma-ray spectrometry. Irradiations of two copper disks at various accelerators lasted only a few hours. After the irra-

diation, radioactive decay of ^{24}Na was measured in order to calculate with an accuracy of about ±1% the activity ratio:

$$R_0\left(^{24}\text{Na}\right) = \frac{\left(^{24}\text{Na } in \ downstream \ \text{Cu}\right)}{\left(^{24}\text{Na } in \ upstream \ \text{Cu}\right)} \qquad (2)$$

where "upstream Cu" denotes the Cu-disk which is first hit by the beam and "downstream Cu" is the following Cu disk. There may be several downstream disks in a very thick target stack.

Correlations of E_{CM}/u with experimental R$_0$(^{24}Na)-values are presented in **Table 1** (column 3). In [1] it was shown that R$_0$(^{24}Na) > 1.0 constitutes an unresolved problem, i.e. one would normally expect more production of the very far spallation product ^{24}Na in the first disk than in the second one. This case of an unresolved problem is systematically observed for E_{CM}/u > 192 MeV as seen from the data presented in the third column of **Table 1**. The third column is divided into two sub-columns in which consistent and inconsistent (= unresolved) data are listed separately. R$_0$(^{24}Na) was not determined in reactions of 44 GeV ^{12}C on Pb and U.

2.2. Maximum of Spallation Product Yields in Two Copper Discs in Contact

Similar to the activity ratio for the very distant spallation product ^{24}Na, one can determine the yield ratio for any spallation product having nuclear charge Z and mass A as:

$$R_0\left(A\right) = \frac{\left(^A\text{Z } in \ downstream \ \text{Cu}\right)}{\left(^A\text{Z } in \ upstream \ \text{Cu}\right)} \qquad (3)$$

Most product cross-sections that were measured some time after the end-of-bombardment are actually cumulative. One expects from the standard-model of nuclear reactions, based on the concept of "limiting fragmentation" and "factorisation", that R$_0$(A) distributions have a maximum value close to the mass of the target nucleus, i.e. close to A = 63 in a Cu target. The R$_0$(A) distribution should then decrease continuously with decreasing product mass A, i.e. with increasing mass difference ΔA from the target mass. Detailed supporting arguments for this statement are given in [1,2], and in particular in [6].

An example of thin-target reactions induced by relativistic particles is shown in **Figure 1** where mass distributions measured in thin copper targets that were irradiated with various projectiles are shown. The distributions of reaction products are characterised by the principles of "limited fragmentation" and "factorisation". The term "limiting fragmentation" means that the shapes of spallation product mass distributions do not change between reaction systems and the term "factorisation" means that

Table 1. Correlations in nuclear interactions between E_{CM}/u and several observables.

Reaction	E_{CM}/u MeV	$R_0(^{24}Na)$ in Cu(*)		$R_0(A)$ in Cu (**)	Neutron emission in GAMMA-2 (***)
		≤ 1.00	> 1.00		
48 GeV ^{4}He + Cu	664		**1.21 ± 0.02**	**Problem**	
44 GeV ^{12}C + Cu	488		**1.24 ± 0.02**	**Problem**	**Problem**
72 GeV ^{40}Ar + Cu	426		**1.50 ± 0.02**	**Problem**	**Problem**
24 GeV ^{1}H + Cu	333		**1.10 ± 0.02**	**Problem**	
25.2 GeV ^{12}C + Cu	279		**1.13 ± 0.03**	**Problem**	
36 GeV ^{40}Ar + Cu	213		**1.17 ± 0.02**	**Problem**	
18 GeV ^{12}C + Cu	194		**1.08 ± 0.10**	**Problem**	
22.4 GeV ^{22}Ne + Cu	192		**1.08 ± 0.02**		
44 GeV ^{12}C + Pb	189				**Problem**
44 GeV ^{12}C + U	168				**Problem**
7.3 GeV ^{2}H + Cu	107	0.90 ± 0.05		o.k.	o.k.
4.5 GeV ^{1}H + Cu	69	0.98 ± 0.05			
14.7 GeV ^{4}He + Pb	68				o.k.
4 GeV ^{4}He + Cu	55	0.92 ± 0.01			
3 GeV ^{2}H + Cu	43	0.90 ± 0.05		o.k.	o.k.
2.6 GeV ^{1}H + Cu	39	0.96 ± 0.02			
7.4 GeV ^{2}H + Pb	35				o.k.
6 GeV ^{4}He + Pb	27				o.k.
1.3 GeV ^{1}H + Cu	19	0.99 ± 0.03			
3 GeV ^{2}H + Pb	14				o.k.
2.0 GeV ^{1}H + Pb	10				o.k.
1.0 GeV ^{1}H + Pb	5				o.k.

Notes for **Table 1**: Data in bold font stand for "unresolved problems". Only $R_0(^{24}Na) ≤ 1.00$ is compatible with model calculations (see [1] which contains data from the PhD Thesis of Lerman [9]). The value for 22.4 GeV ^{22}Ne on Cu is from the PhD Thesis of Haase as also quoted in [1], the value for 18 GeV ^{12}C on Cu is from the PhD Thesis of Ochs [10]. $R_0(A)$ in Cu is compatible with the experimental concept of "limiting fragmentation" and can be understood with model calculations, such as MCNPX 2.7a when the maximum in $R_0(A)$ is close to $A_{target} = 63$. When A_{max} is significantly below target mass the system exhibits an unresolved problem. For details see text and **Figures 2** & **3**, and [2]. The experimental neutron emission from 1 GeV and 2 GeV protons on lead is perfectly compatible with results from MCNPX 2.7a calculations [11]. Further details are given in **Table 2**.

Table 2. Correlations in nuclear interactions between E_{CM}/u and the observed neutron multiplicities on GAMMA-2.

Reaction	E_{CM}/u MeV	Neutron emission in GAMMA-2	Arguments
72 GeV ^{12}C + Cu	426	**Problem**	Note 1 (see below)
44 GeV ^{12}C + Cu	488	**Problem**	Note 2: V = (108/115) = 0.94 ± 0.09
44 GeV ^{12}C + Pb	189	**Problem**	Note 2: V = (266/275) = 0.97 ± 0.07
44 GeV ^{12}C + U	168	**Problem**	Note 2: V = (500/726) = 0.69
7.4 GeV ^{2}H + Cu	107	o.k.	
14.7 GeV ^{4}He + Pb	68	o.k.	
3 GeV ^{2}H + Cu	43	o.k.	
7.4 GeV ^{2}H + Pb	35	o.k.	For detailed description of experiments and data, see [2]
6 GeV ^{4}He + Pb	27	o.k.	
3 GeV ^{2}H + Pb	14	o.k.	
2.0 GeV ^{2}H + Pb	10	o.k.	Note 2: V = (8.86/26.4) = 0.34 ± 0.03
1.0 GeV ^{2}H + Pb	5	o.k.	Note 2: V = (4.69/15.0) = 0.31 ± 0.02

Note 1: During the irradiation of a 20 cm thick Cu target with 72 GeV ^{40}Ar at the LBNL in Berkeley on March 10, 1987 a massive neutron emission was observed all around the laboratory. This highly excessive neutron fluence was definitely beyond expectation; however, no quantitative results on this event have ever been published, as described in [1]. **Note 2**: The term V is defined as $V = B(^{140}La)/(n/ion) \times [10^{-5} \cdot g^{-1}$ neutron$^{-1}]$ where B is the experimentally determined amount of product atoms per beam particle per gram of target, and the amount of neutrons per beam particle (n/ion) comes from a model calculation. The term V is based on definitions and considerations introduced in [2]. The finding $V ≤ 0.35 ± 0.03$ on the GAMMA-2 target indicates good agreement between experimental and theoretical neutron emission, measured via the $B(^{140}La)$ value and the theoretically calculated value using MCNPX 2.7a [14]. Values of $V > 0.35$ on GAMMA-2 indicate that neutron production exceeds calculation, which is an unresolved problem.

the height (cross-section) of the distributions is simply dependent on the projectile mass.

For thick targets two figures shall clarify the situation of resolved vs. unresolved results: In **Figure 2** the $R_0(A)$ distribution is shown for the reaction of 7.3 GeV ^{2}H on a thick Cu target [10]. The maximum value of spallation product ratios is found around mass A = 57 from where on the distribution goes slowly down with rising ΔA, which is perfectly consistent with standard theoretical model results. On the other hand, the $R_0(A)$ distribution measured from interactions of 72 GeV ^{40}Ar on a thick Cu target [12] shown in **Figure 3** has its maximum around mass A = 51 which is far below the target mass and which cannot be reproduced by current models. In the former experiment (**Figure 2**) the cross-section ratio for the very distant spallation product ^{24}Na is below unity (0.90 ± 0.05), whereas it exceeds unity (1.51 ± 0.02) in the latter experiment (**Figure 3**). The reaction system of **Figure 2** is well resolved and in agreement with calculations, whereas the reaction system of **Figure 3** is an unresolved problem.

Figure 4 gives slightly modified $R(^AZ)$ distributions from various projectiles hitting a 20 cm thick copper target, consisting of 20 disks of 1 cm thickness each. This Figure is taken from [1] and it is discussed in detail therein. Suffice it to say that the observed $R(^AZ)$ distributions for 44 GeV ^{12}C and 18 GeV ^{12}C constitute "unresolved problems" whereas the distributions for 7.3 GeV ^{2}H and 3 GeV ^{2}H indicate no problem, as A_{max} is close to the target mass. Identical conclusions derived from the distribution of ^{24}Na throughout thick targets in the same reaction systems as shown in **Figure 4** are drawn in [13].

2.3. Neutron Emission from "GAMMA-2" Target

The "GAMMA-2" target as described in several publications [e.g. 1, 3, 6, and 11] is a 20 cm long (thick) copper or lead spallation target, consisting of 20 metallic disks of 1 cm thickness and 8 cm diameter each. The metallic core is surrounded by a 6 cm thick paraffin moderator on all sides—with the exception of the front side, where the ion beam enters directly into the metallic target. The moderator surface contains small holes for plastic vials containing low-energy neutron (n, γ) sensors, for example

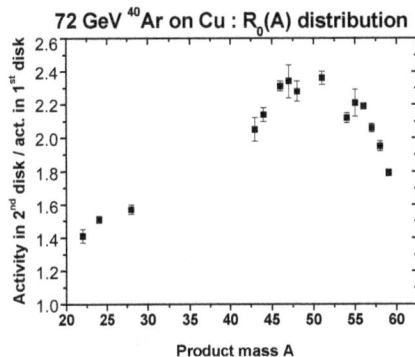

Figure 1. Spallation mass distributions from interactions of various relativistic projectiles with a thin copper target. For details, see text.

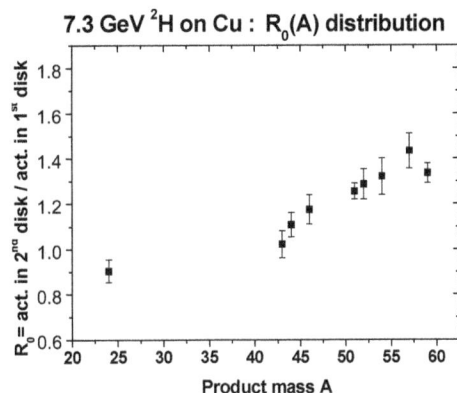

Figure 3. $R_0(A)$ distribution for the reaction of 72 GeV^{40}Ar on two 1-cm thick Cu-disks. The maximum is around mass 51 which is NOT consistent with model calculations and this distribution is an unresolved problem.

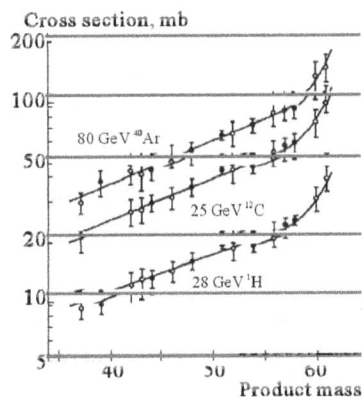

Figure 2. $R_0(A)$ distribution for the reaction of 7.3 GeV^{2}H on a thick two-disks Cu target. The maximum is around mass 57 which is consistent with model calculations and it constitutes NO unresolved problem.

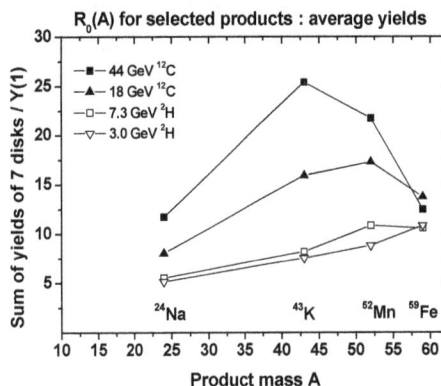

Figure 4. Yield ratio $R_0(A)$ for various masses A in a 20 cm thick Cu-target. Note the definition of $R_0(A)$ on the ordinate of this picture.

stable lanthanum salt. La-sensors measure directly the low-energy neutron fluence via the reaction ^{139}La (n,γ) ^{140}La. The radioactive decay of ^{140}La is measured, thus allowing determination of the neutron production in the "GAMMA-2" target during irradiation. This target system and the integral data measured are an IAEA benchmark for transmutation since 2007. The target allows determination of the neutron production in the system during high energy irradiations, and simultaneous measurement of the spallation product yield distribution inside the metallic core. All experimental results are compared with modern computer simulations, for example using the MCNPX 2.7a code [14].

Results of comparisons between experiments on the GAMMA-2 target and model calculations are listed in **Table 2**. The agreement between experimental and calculated neutron yields is excellent for systems such as (2 GeV ^{2}H + Pb) and (1 GeV ^{2}H + Pb) [11]. No unresolved problems are encountered. However, one observes about a factor of 3 times more neutrons experimentally than calculated in reactions of 44 GeV ^{12}C onto Cu, Pb, and U-targets [2]. This again constitutes another unresolved problem. These results are also included in the last column of **Table 1**. Again, it seems to be the same threshold E_{CM}/u of primary interactions that separates resolved from unresolved thick target results.

In summary one can see in **Tables 1** and **2**:

All nuclear reactions—without any exception—having

$$E_{CM}/u > 150\,\text{MeV}$$

are associated with "unresolved problems", when their interactions are investigated in thick targets.

All nuclear reactions—without any exception—having

$$E_{CM}/u < 110\,\text{MeV}$$

exhibit NO "unresolved problems", when their interactions are investigated in thick targets. There is no problem with the latter class of nuclear interactions as theoretical model calculations agree with experimental findings. This is an important statement from a practical point-of-view. All low-energy nuclear interactions having practical relevance, such as in nuclear reactors or most accelerator applications are understood and modelled sufficiently well. They are not associated with any unresolved problem, especially with respect to excessive neutron production.

3. Tentative Interpretation of Observed Correlations

The essential result presented so far is the clear separation of unresolved experimental thick target results using the newly introduced E_{CM}/u limit value. This limit separates results into two classes where those results from experiments below ~150 MeV agree with model calcula-

tions and those above this value are not reproduced by models. At present it is not clear to us why this is so, but it is interesting to note agreement with statements from Hagedorn who also considered an energy (or temperature T_0) of 158 ± 3 MeV ("Hagedorn limit") in a paper dating back to 1965 [15]. He pointed out that above this limit one should expect to produce pions easily (their rest mass is just below 150 MeV). Therefore there should be some kind of upper limit in nuclear temperatures that are allowed to exist in nucleonic matter. We will not pursue this line-of-thinking any further at this point.

A review of physics experiments studying nuclear interactions in thin targets and the emission of secondary elementary particles and projectile fragments was presented by Friedlander and Heckmann [16]. They describe a broad range of experimental techniques used and the data measured in these investigations. They report of no experimental result that lies outside the understanding in the frame of standard models—however, with the exception of certain nuclear emulsion studies, which are in fact thick target studies. In these studies some evidence for a "short mean-free-path" of projectile fragments was observed, called "anomalons" in those days. This evidence was observed for interactions of secondary fragments and they also have recently been considered [1] as unresolved problems.

We continue our considerations with the empirical approach to separate the FIRST nuclear interaction induced by the primary ion from SECOND nuclear interactions induced by secondary fragments in a thick target. Furthermore, we focus on the interactions of primary ions in nuclear interactions in thick targets above the critical limit of $E_{CM}/u \sim 150\,\text{MeV}$ which systematically show evidence of unresolved problems. Our interpretation of these phenomena suggests three hypothetical postulations, similar to the approach in [6]. However, our concept of interpretation is different.

1st postulation:

Small secondary fragments (probably mainly p, n, $\pi^{0,+,-}$) may have properties that lead to reactions creating unresolved problems. In thin targets there is no possibility for secondaries to interact, whereas further reactions in thick targets expose the unresolved problems.

2nd postulation:

Secondary fragments created in a FIRST interaction with $E_{CM}/u > 150$ MeV may have properties that change along their flight path.

How this change occurs is unclear at this stage, however, the fact that this change happens seems to be experimental fact, as will be shown below. We do not even know what the change is, but it could for example be a decay of the secondary fragment.

So far, there are few experiments known to the authors that show such a time-dependent change in nuclear properties without any external material interaction or particle emission within a time interval of 10^{-10} to 10^{-9} seconds. Evidence has been observed at the JINR in Dubna (Russia) in the irradiation of 20 Cu foils of 1 mm thickness each with 44 GeV ^{12}C ions. The resulting $R_i(^{24}Na)$ yield distribution ratio is shown in **Figure 5**, taken from [17]. One observes almost constant $R_i(^{24}Na)$ values in the first foils which start to increase after about 5 mm target thickness and seem to reach equilibrium after about 15 mm. This may indicate that something "NEW" is entering into the experimentally observable physical reality.

The extension of results from **Figure 5** is shown in **Figure 6** where the production of ^{24}Na by 44 GeV ^{12}C in a stack of 20 Cu-disks of 1 cm thickness each is shown [17]. The absolute value of the activity ratio R_i measured for the 2 cm target slab in **Figure 6** matches nicely the average of the 10 mm to 20 mm values from **Figure 5**, and then the ratio continues to increase up to about 10 cm target thickness. From there on the amount of product goes down as would be expected in a thick target where projectile density decreases through interactions.

3rd postulation:

The last postulation is a radical hypothesis.

When secondary fragments from an interaction in the $E_{CM}/u > 150$ MeV regime interact with another target nucleus, the interaction cross-section and energy transfer is higher than that normally expected. This is the reason for the observation of shorter than expected mean free path and an enhanced neutron production as well as the concurrent enhanced production of very far spallation products.

Future experiments will determine whether this third postulation proves to be an acceptable or fruitful suggestion. More experiments and theoretical considerations are needed.

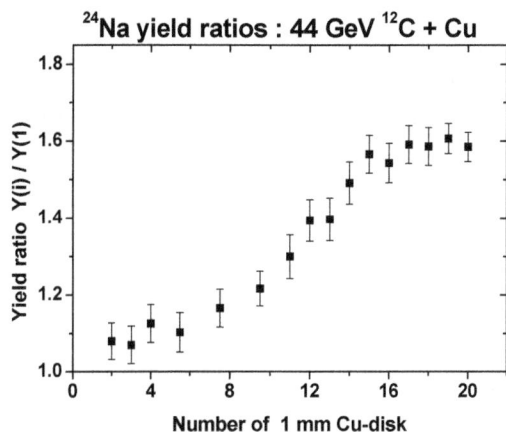

Figure 5. Yield ratio of ^{24}Na in a stack of 20 foils (each 1 mm thick) of Cu irradiated with 44 GeV ^{12}C at the Synchrophasotron in Dubna (Russia). The ^{24}Na contents in each foil is measured (Y_i) and compared to the 1st Cu foil (Y_1). Some data points are averages of 2 foils.

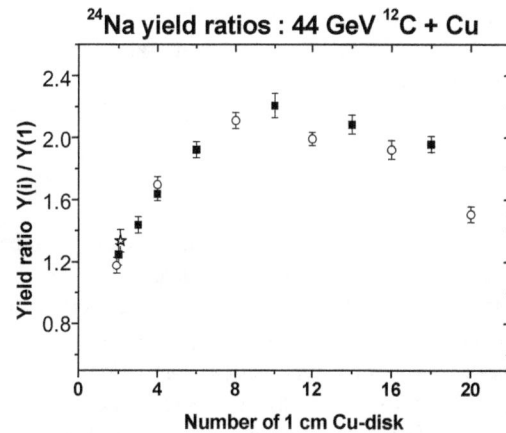

Figure 6. Yield ratio of ^{24}Na in a stack of 20 Cu-disks of 1 cm thickness each irradiated with 44 GeV ^{12}C at the Synchrophasotron in Dubna (Russia). Different symbols denote different experiments. The data point for the second disk plotted as a star is calculated from average values from Figure 5. Two data points for the second disk are displaced by 1 mm.

Some experiments that may provide additional information to this problem have already been published, and will be presented in the Appendix.

4. Conclusions

Various phenomena have been found in thick-target experiments induced by relativistic particles which are not in agreement with known reaction product systematics and cross-sections. These phenomena are found when projectiles of very high energies interact with the thick target or a stack of thin targets. The experimental results are divided into two groups according to their agreement or not with theoretical model calculations as well as conformity with results from thin-target experiments.

Phenomena are denoted as "resolved" when:

—the mass distribution of spallation products follows the principles of limiting fragmentation and factorisation as exemplified in **Figure 1** and the cross-section of very far spallation products, such as ^{24}Na, is largest in the beginning of the target;

—the measured neutron multiplicity is in agreement with model calculations;

—the mean free path of secondary particles follows known systematics.

On the other hand, phenomena are "unresolved" when

—the mass distribution of spallation products in a second (or following) interaction is skewed towards large ΔA and the cross-section of very far spallation products, such as ^{24}Na, is significantly enhanced;

—the neutron multiplicity exceeds expectation values from model calculations, which seems to indicate a very high energy transfer in the interaction of a secondary fragment;

—the mean free path of secondary fragments is significantly shorter than known from range-energy systematics (see Appendix).

These experimental phenomena are produced in thick targets by primary ions up to the end of their path and by secondary fragments making additional nuclear interactions in the thick target.

A hypothetical description of the average centre-of-mass excitation energy of each nucleon (E_{CM}/u) in the entrance channel of the reaction is correlated with experimentally observed phenomena. When we accept this concept of E_{CM}/u of a hypothetical compound system as a means of classification, there seems to be a threshold value of E_{CM}/u below which no unresolved phenomena at all are found in thick targets and all reaction product cross-sections are in agreement with model calculations, whereas reactions with energies above the threshold value, without any exception, lead to unresolved phenomena. However, due to scarcity of experimental results, the threshold energy is not very well defined. Results from all reactions having $E_{CM}/u \leq 107$ MeV show no unexpected results whereas the outcome of all reactions having $E_{CM}/u \geq 168$ MeV is clearly "unresolved". We adopt the energy of $E_{CM}/u \sim 150$ MeV as the threshold separating resolved from unresolved reaction systems. This value is close to another limiting energy (or limiting temperature T_0) of 158 ± 3 MeV which was introduced through considerations of statistical thermodynamics by Hagedorn long ago and which marks the transition energy beyond which real hadrons are produced and "the number and longitudinal momentum of the secondaries produced would increase" [15].

In order to define the threshold energy more precisely, we wish to propose an experiment which scans over the limiting energy and will hopefully allow us to define it well. A simple and straightforward experiment is a thick-target experiment employing three copper disks of 1 cm thickness each in contact, which are irradiated with a beam of ^{12}C. Beam energies of 0.6 GeV/u (7.2 GeV kinetic energy), 0.9 GeV/u, 1.1 GeV/u, 1.2 GeV/u, 1.3 GeV/u, 1.5 GeV/u and 1.8 GeV/u are recommended. The experiments scan the kinetic energy range around 1.2 GeV/u which corresponds to $E_{CM}/u = 158$ MeV for the ^{12}C + ^{65}Cu reaction. If the hypothetical average excitation energy around 150 MeV is actually a relevant limit, then $R_0(A)$ should indicate significant enhancement for masses A far away from the target mass, the ^{24}Na cross Section ratio should go from below to above unity and the neutron dose-rate should rise over-proportionally (it is 3 times more than expected for 44 GeV ^{12}C projectiles) when the threshold energy is exceeded. As interactions between nucleons inside a nucleus can be described by statistical thermodynamics, the limiting energy (or temperature) is not expected to be a sharp boundary but one rather expects to find a smooth onset of "unresolved" data around the limit value.

REFERENCES

[1]   R. Brandt, V. A. Ditlov, K. K. Dwivedi, W. Ensinger, E. Ganssauge, S.-L. Guo, M. Haiduc, S. R. Hashemi-Nezhad, H. A. Khan, M. I. Krivopustov, R. Odoj, E. A. Pozharova, V. A. Smirnitzki, A. N. Sosnin, W. Westmeier and M. Zamani-Valasiadou, "Interactions of Relativistic Heavy Ions in Thick Heavy Element Targets—and Some Unresolved Problems," *Physics of Elementary Particles and Nuclei*, Vol. 39, No. 2, 2008, pp. 507-526 (original), reprinted in *Physics of Particles and Nuclei*, Vol. 39, No. 2, pp. 259-285.

[2]   S. R. Hashemi-Nezhad, M. Zamani-Valasiadou, M. I. Krivopustov, R. Brandt, W. Ensinger, R. Odoj and W. Westmeier, "Neutron Production in Thick Targets Irradiated with High Energy Ions," *Physics Research International*, Vol. 2011, 12 Pages, Article ID: 128429.

[3]   G. Dersch, "Kernchemische Studien zum Anomalonenphänomen bei Kernreaktionen mit Relativistischen Schwerionen," PhD Dissertation, Philipps Universität, Marburg, 1986.

[4]   P. B. Price, "Overview of a Conference (23rd International Conference on SSNTDs, Beijing, September 2006)," *Radiation Measurements*, Vol. 43, 2008, pp. 662-664.

[5]   R. C. Barber, H. Gäggeler, P. J. Karol, H. Nakahara, E. Vardaci and E. Vogt, "Discovery of the Element with Atomic Number 112 (IUPAP Technical Report)," *Pure and Applied Chemistry*, Vol. 81, No. 7, 2009, pp. 1331-1343.

[6]   I. Hartmann and R. Brandt, "Noise between ASAT and SAT," In: Dr. P. Gschwind and B. Hugiweg, Eds., *A Model for Nuclear Phenomena, Called Anomalons, Mathematisch-Physikalische Korrespondenz*, Dornach, 2009, pp. 3-35.

[7]   E. Rössle and E. Schopper, "Absorptionsanalyse der Sternauslösenden Komponenten der Ultrastrahlung," *Zeitschrift für Naturforschung Teil A*, Vol. 9, 1954, pp. 837-851.

[8]   S. R. Hashemi-Nezhad, R. Brandt, W. Westmeier, V. P. Bamblevski, M. I. Krivopustov, B. A. Kulakov, A. N. Sosnin, J.-S. Wan and R. Odoj, "Monte Carlo Analysis of Accelerator Driven Systems: Studies on Spallation Neutron Yield and Energy Gain," *Kerntechnik*, Vol. 66, No. 1-2, 2001, pp. 47-53.

[9]   L. Lerman, "On the Symmetry of Nuclear Identity between Relativistic Primary and Secondary Nuclei," PhD Dissertation, Philipps-Universität, Marburg, 2002.

[10]   M. Ochs, "Experimente zur Neutronenfreisetzung durch Spallationsreaktionen in Massiven Kupferund Bleitargets," PhD Dissertation, Philipps-Universität, Marburg, 1997.

[11]   W. Westmeier, *et al.*, (in Preparation).

[12]   K. Aleklett, R. Brandt, G. Dersch, G. Feige, E. M. Friedlander, E. Ganssauge, G. Haase, D. C. Hoffmann, J. Herrmann, B. Judek, W. Loveland, P. L. McGaughey, N.

T. Porile, W. Schulz and G. T. Seaborg, "Unusual Behaviour of Projectile Fragments Produced by the Interactions of Relativistic Ar Ions with Copper," *Physical Review C*, Vol. 38, 1988, pp. 1658-1673 and *Erratum*: *Physical Review C*, Vol. 44, No. 1, 1991, p. 566.

[13] M. Ochs, I. G. Abdullaev, J. Adam, J. C. Adloff, I. G. Bersina, V. Bradnova, R. Brandt, M. Bognitzki, V. S. Butsev, M. Debeauvais, K. K. Dwivedi, F. Fernandez, S.-L. Guo, M. I. Krivopustov, B. A. Kulakov, E.-J. Langrock, G. Modolo, R. Odoj, V. P. Perelygin, A. N. Premyshev, V. S. Pronskikh, Th. Schmidt, A. N. Sosnin, V. I. Stegailov, R. Sudowe, P. Vater, J.-S. Wan, M. Zamani and V. M. Tsoupko-Sitnikov, "SSNTD and Radiochemical Studies on the Transmutation of Nuclei Using Relativistic Ions," *Radiation Measurements*, Vol. 28, No. 1-6, 1997, pp. 255-268.

[14] D. B. Pelowitz, J. S. Hendricks, J. W. Durkee, M. R. James, M. L. Fensin, G. W. McKinney, S. G. Mashnik and L. S. Waters, "MXNPX 2.7A Extensions," Report LA-UR-08-07182, Los Alamos National Laboratory, 2008.

[15] R. Hagedorn, "Statistical Thermodynamics of Strong Interactions at High Energies," *Supplemento Al Nuovo Cimento*, Vol. III, 1965, pp. 147-186.

[16] E. M. Friedlander and H. H. Heckmann, "Relativistic Heavy Ion Collisions," In: D. A Bromley, Ed., *Heavy Ion Science*, Plenum Press, New York, 1983.

[17] R. Brandt, G. Dersch, E. M. Friedlander, G. Haase, W. S. Butsev, M. I. Krivopustov, B. A. Kulakov, E. Ganssauge, W. Schulz, E. J. Langrock and F. Pille, "Untersuchung des Anormalen Verhaltens Relativistischer Fragmente aus Schwerionenreaktionen (Russian)," *Isotopenpraxis*, Vol. 25, 1989, pp. 433-438.

[18] R. Brandt, V. S. Butsev, H. H. Cui, G. Dersch, E. M. Friedlander, E. Ganssauge, G. Haase, M. Heck, M. I. Krivopustov, B. A. Kulakov, E.-J. Langrock and F. Pille, "Enhanced Production of ^{24}Na by Wide-Angle Secondaries Produced in the Interaction of Relativistic Carbon Ions with Copper," *Physical Review C*, Vol. 45, No. 3, 1992, pp. 1194-1208.

[19] G. Alexander, R. H. W. Johnston and C. O'Ceallaigh, "The Relative Frequencies of the Decay Modes of Positive K-Mesons and the Decay Spectra of Modes $K_{\mu 3}$, τ' and K_β," *Physics and Astronomy*, Vol. 6, No. 3, 1957, pp. 478-500.

[20] E. Ganssauge, "Der Experimentelle Stand der Anomalonen-Forschung," *Annalen der Physik*, 7, Folge, 11, 1987, pp. 202-246. and "The Experimental Status of the Anomalon Research," *Nuclear Tracks and Radiation Measurement*, Vol. 15, No. 1-4, 1988, pp. 371-379.

Appendix

Experiments Using Thick Targets Irradiated with Ions Having Very Large Energy of $E_{kinetic} > 100$ GeV

Experimental results published from the study of thick targets irradiated with very high-energy ions above 100 GeV are scarce. We report only of a few experiments which appear to be significant and where unresolved problems may also play a role.

Evidence Based on the Radiochemical Cu-Disks Technique [18]

The $R_i(^{24}Na)$ ratios for several Cu disks in contact were determined for ions having up to 7000 GeV kinetic beam energy. No drastic or even surprising results have been reported. $R_0(^{24}Na)$ for two Cu disks in contact in 7 TeV ^{32}S irradiations was measured as $R_0(^{24}Na) = (1.8 \pm 0.1)$, not much larger than for 72 GeV ^{40}Ar with $R_0(^{24}Na) = (1.50 \pm 0.02)$. It should be noted that both reactions fall into the same regime of $E_{CM}/u > 150$ MeV and similar results may be expected.

Evidence Based on Reduced Mean Free Path

A strong difference in the mean-free-path of an elementary particle was observed in the very early days of investigations of high-energy nuclear reactions. Alexander *et al.* observed in 1957 [19] that π^+-pions created in the decay of kaons [$K^+ \rightarrow \pi^+ + \pi^0$] (called "young" pions) have a mean free path of [Λ (young) = (11.0 ± 2.4) cm] in nuclear emulsion. However, conventional pions (called old pions) of the same momentum are known to have a mean free path of [Λ (old) = (31 ± 2) cm] in the same nuclear emulsion. Unfortunately, this crucial experiment has never been reproduced.

Similar evidence of reduced mean free path measured in nuclear emulsion was published by Friedlander and others, as reviewed by Ganssauge [20], for reactions induced by projectiles with kinetic energies up to 100 GeV.

In summary, one needs further experiments and new concepts in this field in order to have sufficient understanding of the underlying physics of reactions induced by secondary fragments.

31

Reduction of the Beam Hardening Artifacts in the X-Ray Computer Tomography: Energy Discrimination with a Photon-Counting Detector

Yukino Imura[1,2]*, Takuto Yanagida[1], Hisashi Morii[1], Hidenori Mimura[1,2], Toru Aoki[1,2]
[1]Research institute of Electronics, Shizuoka University, Hamamatsu, Japan
[2]Graduate School of Science and Technology, Shizuoka University, Hamamatsu, Japan

ABSTRACT

The material identification is a pressing requirement for the sensitive security applications. Dual-energy X-ray computer tomography (DXCT) has been investigated for material identification in the medical and security fields. It requires two tomographic images at sufficiently different energies. To discriminate dangerous materials of light elements such as plastic bombs in luggage, it is needed to measure accurately with several tens of kilo electron volts where such materials exhibit significant spectral differences. However, CT images in that energy region often include artifacts from beam hardening. To reduce these artifacts, a novel reconstruction method has been investigated. It is an extension of the Algebraic Reconstruction Technique and Total Variation (ART-TV) method that reduces the artifacts in a lower-energy CT image by referencing it to an image obtained at higher energy. The CT image of a titanium sample was reconstructed using this method in order to demonstrate the artifact reduction capability.

Keywords: X-Ray Computer Tomography; Artifacts; Photon Counting; Iterative Reconstruction; ART-TV; Titanium Sample; Security Applications

1. Introduction

Dual-energy X-ray computer tomography (DXCT) is a powerful material identification technique in the field of medical engineering [1-6]. In DXCT, the effective atomic number and electron density of a target object are derived from the attenuation coefficient or CT value at two different energies [7]. CT systems of this field have dual sources or the ability to rapidly switch the tube current and voltage.

DXCT is also expected to be applied to the security field. For medical use the target object is body tissue, consisting of light elements. However, these medical systems cannot be directly applied to security inspection, because the target object is composed of various elements. In addition, for space and cost reasons, medical equipment is unsuitable. The applications of DXCT technology have been applied to material discrimination in the security field [8]. An X-ray CT system using a conventional X-ray tube and a photon-counting line sensor has been developed and shown to be feasible for material discrimination [9], but one problem was the presence of artifacts due to beam hardening.

In DXCT, the effective atomic number and electron density are derived from the difference between the attenuation coefficients at different energies. Therefore, two CT images need to be obtained at a significant energy difference. In addition, the target object in the security field can include light elements such as plastic bombs. Thus, the lower X-ray energy needs to be several tens of kilo-electron volts. Depending on the thickness of the target, such low-energy X-rays may be completely absorbed and not reach the detector. In this case the attenuation is not a linear function of absorber thickness [10,11]. This nonlinear relationship induces artifacts as a beam-hardening effect, evident as variations in pixel values even within the same material. An example of an image showing this artifact [12] is the CT image of a copper sample at 40 keV, for which a circular hole in it is blurred in comparison to its image at 90 keV. To identify the material accurately, it is necessary to reduce such artifacts and derive the correct attenuation coefficient from the low-energy CT images.

To reduce such artifacts, an iterative reconstruction method that uses CT images obtained at different energies was investigated. Higher energy images have fewer artifacts than those of lower energy [12]. This iterative

Reduction of the Beam Hardening Artifacts in the X-Ray Computer Tomography: Energy Discrimination
with a Photon-Counting Detector

199

reconstruction method uses a higher-energy CT image obtained simultaneously using polychromatic X-rays as the reference. As the basis for this method, Algebraic Reconstruction Technique and Total Variation (ART-TV) was adopted because it has been verified to be capable of overcoming data insufficiencies [13,14]. As an extension of ART-TV, we developed an improved method referred to here as ART-FG-TV, where FG stands for *flatten gradient*. This method utilizes a CT system that can simultaneously image an object at different X-ray energies.

Experimental projection data from a titanium sample were reconstructed with this method and evaluated in terms of artifact reduction. Although titanium is a heavier element than the atoms contained in plastics, it is selected to evaluate the effect of the ART-FG-TV because it clearly reveals artifacts from beam hardening. Also, as titanium is often included in electronic products and cameras, it is a key target material in the security field.

2. Reconstruction Method and Experiment

2.1. Reconstruction Method

The FG step is an iterative optimization process added to the ART-TV method. It flattens the gradient of a CT image by referencing the gradient of a higher-energy CT image. Following the notation of the original algorithm [14], this method can be summarized by the following four steps:

1) ART reconstruction

$$f_{n+1} = f_n + \frac{g_i - H_i f_n}{\|H_i\|^2} H_i, i = 1, \cdots, P \quad (1)$$

2) FG step

$$f_{n+1,m+1} = f_{n+1,m} + \mathrm{FG}\left(f_{n+1,m}, h, \theta, \alpha, R\right), m = 1, \cdots, S \quad (2)$$

3) Positive constraint

$$f_{n+1} = \begin{cases} f_{n+1} & f_{n+1} \geq 0 \\ 0 & f_{n+1} < 0 \end{cases} \quad (3)$$

4) TV minimization

$$f_{n+1,l+1} = f_{n+1,l} + \mathrm{d}\frac{\partial TV\left(f_{n+1,l}\right)}{\partial f_{n+1,l}}, l = 1, \cdots, Q \quad (4)$$

where f and g respectively denote a reconstructed image and the projection data, where H is the projection matrix. The number of projections is P, the TV iteration number is Q, and the FG iteration number is S.

In FG procedure, argument f represents a processed image vector, whereas h is the gradient vector of the higher-energy CT image, calculated by applying a Laplacian filter to the image. θ and α are, respectively, a threshold and a constant that represents the strength of this procedure, which affect the resulting image and should be

chosen appropriately. R is a set of vectors of the indices of f that the X-ray path crosses, as in **Figure 1**.

The FG procedure is described below: the image f is traced, on a pixel by pixel basis, along the X-ray path R, and focused pixel is increased as to the strength parameter α and the difference s between that pixel and its adjacent pixels, if the gradient of the reference image at the same position is less than the threshold value θ. This iterative step modifies the values of the pixels on the image f corresponding to the positions of the flat area, or the same material area, of the reference image; the decline of CT value, or the artifact mentioned in Section 1 is thereby reduced.

2.2. Experiment

The CT images were reconstructed from the data measured using the X-ray CT system shown in **Figure 2**. It consists of a continuous X-ray tube (running at 150 kV and 10 μA) and a cadmium telluride (CdTe) X-ray/gamma-ray detector (Amptek X-123CdTe). The sample is a titanium cylinder 15 mm in diameter. To avoid pile-up at the detector, a 30-mm-thick aluminum plate is placed between the X-ray tube and the sample. The CdTe detector is wrapped with lead shielding to avoid scattered rays. Since the X-rays are collimated at the tube window and by the hole in the lead shield, they are assumed to constitute a pencil-shaped beam. Assigning the line connecting the X-ray tube and the detector to be 0 mm, data are measured by one second for each 1 mm from −16 to +16 mm. As the sample shape is cylinder, the measurement is done once and the data were repeated for 36 times to obtain a complete rotation. Examples of spectral data are shown in **Figure 3**.

The measured spectrum is divided into energy bands of 10 keV and the midpoint of each energy band is taken to define the effective energy [9]. Dual-source CT systems in the medical field often approximate the mean photon energy of the spectrum in terms of the total number of counts in the whole spectrum [1,2]. In contrast, our method counts the number of photons at the actual energy.

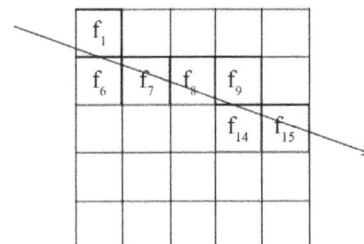

Figure 1. Example of an element *r* of *R* for the case of a 5 × 5 pixel image *f*. Element *r* is a vector that consists of pixel indices intersected by the ray passing through the image. In this example, *r* = {1, 6, 7, 8, 9, 14, 15}.

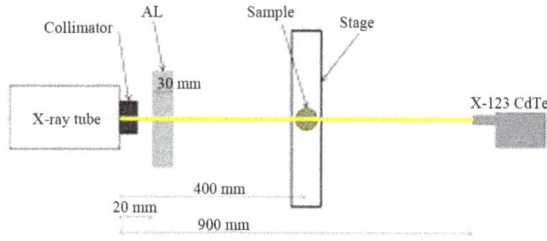

Figure 2. Schematic diagram of the adopted X-ray CT system. To avoid the pile-up phenomenon, a 30-mm-thick aluminum plate was placed between the X-ray tube and the sample.

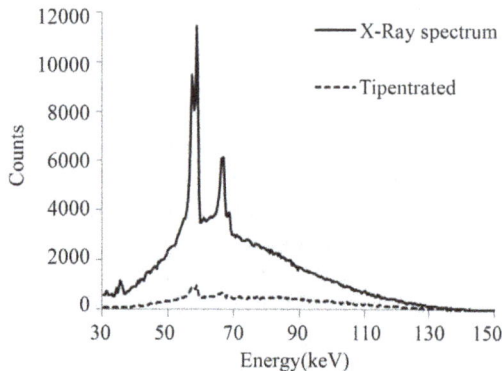

Figure 3. Spectra output by the CT system. The solid line represents the spectrum measured without the sample and the dashed line represents that with the Ti sample.

3. Results and Discussion

Measured CT data of the sample were reconstructed using ART-TV and ART-FG-TV at each energy band. The higher-energy image referenced in the FG procedure was also reconstructed using ART-TV. In **Figure 4**, the CT images of 50 keV and the reference image of 110 keV are shown; (a) is ART-TV, (b) is ART-FG-TV and (c) is the referenced image. Profiles along the centers of the images (dashed line in **Figure 4**) are plotted in **Figure 5**. Since the sample is a uniform titanium cylinder, the CT value should be uniform. Whereas the CT value obtained using ART-TV was non-uniform, it became more nearly uniform with ART-FG-TV. This improvement verifies that the method reduces artifacts.

In **Figure 5**, the lowest value in the ART-TV image is 0.098, and the difference between the highest and lowest values is 0.025, which is 20% of the highest value in the image. On the other hand, the lowest value in the ART-FG-TV image is 0.106, and the difference between the highest and lowest values is 0.011, which is only 9% of the largest value. Furthermore, the depression in the CT image in **Figure 4**, which is displayed in 256 shades of gray, is barely visible.

The ART-FG-TV image was also compared to one measured by conventional filtered back projection (FBP).

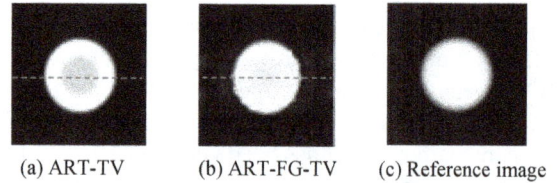

(a) ART-TV   (b) ART-FG-TV   (c) Reference image

Figure 4. CT images of a 15-mm-diameter titanium cylinder using (a) ART-TV at 50 keV; (b) ART-FG-TV at 50 keV; and (c) ART-TV at 110 keV as the reference. The parameters used for ART-TV were $P = 80$ and $Q = 100$. The parameters used for the FG step were $\theta = 0.001$, $\alpha = 0.05$, and $S = 50$. A gray circle is evident at the middle of the image shown in (a), despite the sample being uniform titanium. By using ART-FG-TV, this artifact was reduced.

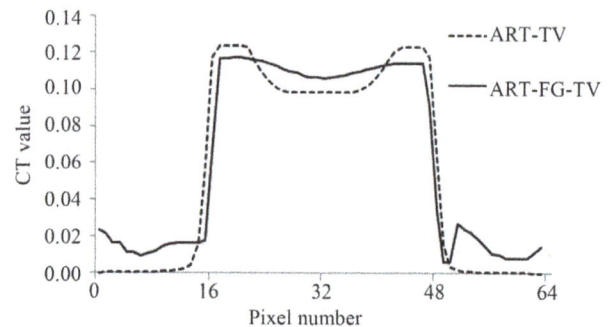

Figure 5. Profiles along the center line (dashed line in Figure 4) of CT images by ART-TV (dashed lines) and ART-FG-TV (solid line) methods.

The image in **Figure 6(a)** is identical to that in **Figure 4(b)**, whereas the image in **Figure 6(b)** is reconstructed from the same data by FBP. The image profile along the center of each image (dashed line) is shown in **Figure 6(c)**. The CT value by FBP is large at the edges of the cylinder but declines sharply toward the center. Meanwhile, the CT value by ART-FG-TV is fairly flat across the cylinder, demonstrating how ART-FG-TV can flatten an artifact.

The influence of α and θ on the image was investigated. The value of α was varied from 0.01 to 0.10 in steps of 0.01 and images were reconstructed. The artifacts are reduced more by using increasingly larger values of the parameter α (**Figure 7**). However, when α is larger than 0.05, the image does not converge. Changes in the value of the threshold θ did not noticeably affect the image of this sample.

In this method, the artifact reduction is done by referencing the higher-energy CT image. Measuring the X-ray with the photon counting detector, the data are obtained as the spectrums. As previously described, the measured spectrum is split into the energy bands of 10 keV. This is enabled by measuring the energy of each photon accurately by photon counting detector. With this accuracy of measuring the energy of photon, it is possible to evaluate the reduction of artifacts numerically.

Reduction of the Beam Hardening Artifacts in the X-Ray Computer Tomography: Energy Discrimination
with a Photon-Counting Detector

201

(a) ART-FG-TV　　　(b) FBP

(c)

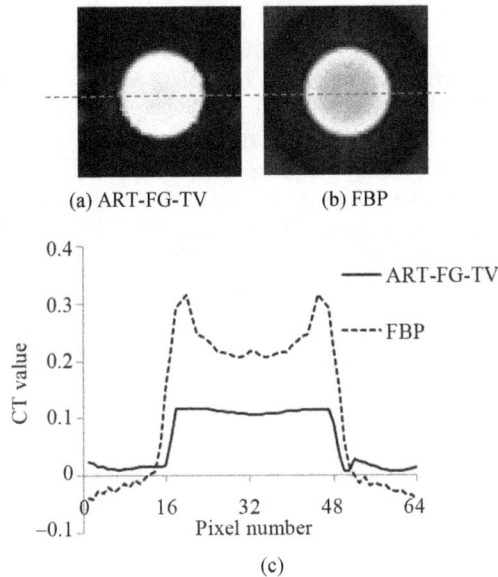

Figure 6. Comparison of CT image of Ti (diameter: 15 mm) by (a) ART-FG-TV and (b) FBP. (c) Is the image profile along the centers of each image in the horizontal directions. The image of (a) is same with Figure 4(b).

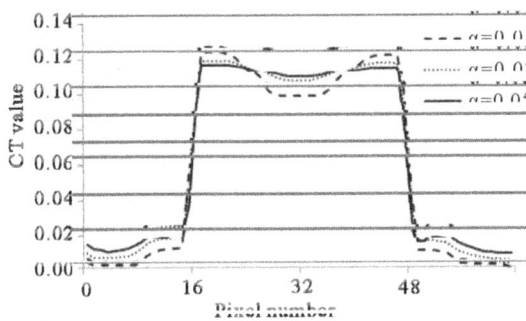

Figure 7. Image profile along the center of the images in the horizontal directions with ART-FG-TV, with varying the strength α from 0.01 to 0.05.

4. Conclusions

An ART-FG-TV algorithm is proposed as a method of reducing artifacts in CT images due to the nonlinear relationship between the attenuation and the material thickness. The method uses a high-energy CT image as a reference, and this CT image is available from arbitrary energy band of X-ray spectra. The technique relies on accurate energy measurements using a photon-counting detector.

The artifact in a CT image of a titanium sample at low energy was suppressed in CT value from 0.025 to 0.008 (20% to 7%). As a first step, it has been demonstrated that this method flattens the artifact. In the future, we will evaluate this method with samples of lighter elements. We will further evaluate the method using complex samples, and investigate the effects of the chosen parameters on the attenuation coefficient.

REFERENCES

[1] T. R. C. Johnson, B. Krauß, M. Sedlmair, M. Grasruck, H. Bruder, D. Morhard, C. Fink, S. Weckbach, M. Lenhard, B. Schmidt, T. Flohr, M. F. Reiser and C. R. Becker, "Material Differentiation by Dual Energy CT: Initial Experience," *European Radiology*, Vol. 17, No. 6, 2007, pp. 1510-1517.

[2] R. T. Gupta, L. M. Ho, D. Marin, D. T. Boll, H. X. Barnhart and R. C. Nelson, "Dual-Energy CT for Characterization of Adrenal Nodules: Initial Experience," *American Journal of Roentgenology*, Vol. 194, No. 6, 2010, pp. 1479-1483.

[3] A. Graser, T. R. C. Johnson, H. Chandarana and M. Macari, "Dual Energy CT: Preliminary Observations and Potential Clinical Applications in the Abdomen," *European radiology*, Vol. 19, No. 1, 2009, pp. 13-23.

[4] K. Engelke, C. Libanati, Y. Liu, H. Wang, M. Austin, T. Fuerst, B. Stampa, W. Timm and H. K. Genant, "Quantitative Computed Tomography (QCT) of the Forearm Using General Purpose Spiral Whole-Body CT Scanners: Accuracy, Precision and Comparison with Dual-Energy X-Ray Absorptiometry (DXA)," *Bone*, Vol. 45, No. 1, 2009, pp. 110-118.

[5] G. Hides, R. Eliahou, M. Dovdevani, P. Coulon, L. Lemaitre, O. N. Gofrit, D. Pode and J. Sosna, "Determination of Renal Stone Composition with Dual-Energy CT: *In Vivo* Analysis and Comparison with X-Ray Diffraction," *Radiology*, Vol. 257, No. 2, 2010, pp. 394-401.

[6] B. Ruzsics, H. Lee, P. L. Zwerner, M. Gebregziabher, P. Costello and U. J. Schoepf, "Dual-Energy CT of the Heart for Diagnosing Coronary Artery Stenosis and Myocardial Ischemia-Initial Experience," *European Radiology*, Vol. 18, No. 11, 2008, pp. 2414-2424.

[7] M. Torikoshi, T. Tsunoo, M. Sasaki, M. Endo, Y. Noda, Y. Ohno, T. Kohno, K. Hyodo, K. Uesugi and N. Yagi, "Electron Density Measurement with Dual-Energy X-Ray CT Using Synchrotron Radiation," *Physics in Medicine and Biology*, Vol. 48, No. 5, 2003, pp. 673-685.

[8] G. Zentai, "X-Ray Imaging for Homeland Security," *International Journal of Signal and Imaging Systems Engineering*, Vol. 3, No. 1, 2010, pp. 13-20.

[9] W. Zou, T. Nakashima, Y. Onishi, A. Koike, B. Shinomiya, H. Morii, Y. Neo, H. Mimura and T. Aoki, "Atomic Number and Electron Density Measurement Using a Conventional X-Ray Tube and a CdTe Detector," *Japanese Journal of Applied Physics*, Vol. 47, No. 9, 2008, pp. 7317-7323.

[10] R. A. Brooks and G. D. Chiro, "Beam Hardening in X-Ray Reconstructive Tomography," *Physics in Medicine and Biology*, Vol. 21, No. 3, 1976, pp. 390-398.

[11] Van de Casteele, E. E. V. Casteele, D. V. Dyck, J. Sijbers and E. Raman, "An Energy-Based Beam Hardening

Model in Tomography," *Physics in Medicine and Biology*, Vol. 47, No. 23, 2002, pp. 4181-4190.

[12] G. Matsumoto, Y. Imura, H. Morii, A. Miyake and T. Aoki, "Analysis of Artifact With X-Ray CT Using Energy Band by Photon Counting CdTe Detector," *Nuclear Instruments and Methods in Physics Research A*, Vol. 621, No. 1-3, 2010, pp. 292-294.

[13] E. Y. Sidky, C. Kao and X. Pan, "Accurate Image Reconstruction from Few-Views and Limited-Angle Data in Divergent-Beam CT," *Journal of X-Ray Science and Technology*, Vol. 14, No. 2, 2006, pp. 119-139.

[14] X. Duan, J. Cheng, L. Zhang, Y. Xing, Z. Chen and Z. Zhao, "X-Ray Cargo Container Inspection System with Few-View Projection Imaging," *Nuclear Instruments and Methods in Physics Research A*, Vol. 598, No. 2, 2009, pp. 439-444.

Permissions

The contributors of this book come from diverse backgrounds, making this book a truly international effort. This book will bring forth new frontiers with its revolutionizing research information and detailed analysis of the nascent developments around the world.

We would like to thank all the contributing authors for lending their expertise to make the book truly unique. They have played a crucial role in the development of this book. Without their invaluable contributions this book wouldn't have been possible. They have made vital efforts to compile up to date information on the varied aspects of this subject to make this book a valuable addition to the collection of many professionals and students.

This book was conceptualized with the vision of imparting up-to-date information and advanced data in this field. To ensure the same, a matchless editorial board was set up. Every individual on the board went through rigorous rounds of assessment to prove their worth. After which they invested a large part of their time researching and compiling the most relevant data for our readers. Conferences and sessions were held from time to time between the editorial board and the contributing authors to present the data in the most comprehensible form. The editorial team has worked tirelessly to provide valuable and valid information to help people across the globe.

Every chapter published in this book has been scrutinized by our experts. Their significance has been extensively debated. The topics covered herein carry significant findings which will fuel the growth of the discipline. They may even be implemented as practical applications or may be referred to as a beginning point for another development. Chapters in this book were first published by Scientific Research Publishing Inc.; hereby published with permission under the Creative Commons Attribution License or equivalent.

The editorial board has been involved in producing this book since its inception. They have spent rigorous hours researching and exploring the diverse topics which have resulted in the successful publishing of this book. They have passed on their knowledge of decades through this book. To expedite this challenging task, the publisher supported the team at every step. A small team of assistant editors was also appointed to further simplify the editing procedure and attain best results for the readers.

Our editorial team has been hand-picked from every corner of the world. Their multi-ethnicity adds dynamic inputs to the discussions which result in innovative outcomes. These outcomes are then further discussed with the researchers and contributors who give their valuable feedback and opinion regarding the same. The feedback is then collaborated with the researches and they are edited in a comprehensive manner to aid the understanding of the subject.

Apart from the editorial board, the designing team has also invested a significant amount of their time in understanding the subject and creating the most relevant covers. They scrutinized every image to scout for the most suitable representation of the subject and create an appropriate cover for the book.

The publishing team has been involved in this book since its early stages. They were actively engaged in every process, be it collecting the data, connecting with the contributors or procuring relevant information. The team has been an ardent support to the editorial, designing and production team. Their endless efforts to recruit the best for this project, has resulted in the accomplishment of this book. They are a veteran in the field of academics and their pool of knowledge is as vast as their experience in printing. Their expertise and guidance has proved useful at every step. Their uncompromising quality standards have made this book an exceptional effort. Their encouragement from time to time has been an inspiration for everyone.

The publisher and the editorial board hope that this book will prove to be a valuable piece of knowledge for researchers, students, practitioners and scholars across the globe.

List of Contributors

Kati Savolainen, Tapio Saukkonen and Hannu Hänninen
Laboratory of Engineering Materials, School of Engineering, Aalto University, Espoo, Finland

Reinhard Boons, Mark Van Iersel and Jean Louis Genicot
Radiation Dosimetry and Calibration-Environmental Health and Safety SCK-CEN, Belgian Nuclear Research Center, Mol, Belgium

Palliyakarany Thirumani Krishna Kumar and Toshikazu Takeda
Research Institute of Nuclear Engineering, University of Fukui, Fukuishi, Japan

Defu Liu, Fengqing Wang and Tao Zou
Disaster Prevention Research Institute, Ocean University of China, Qingdao, China

Guilin Liu and Huajun Li
College of Engineering, Ocean University of China, Qingdao, China

Adel Trabelsi
Research Unit of Nuclear and High Energy Physics, Faculty of Sciences of Tunis, Tunis El-Manar University, Tunis, Tunisia
Physics Department, College of Science, King Faisal University, Hofuf, KSA

Mansour Oueslati
National Center for Nuclear Sciences and Technologies, Sidi-Thabet, Tunisia

Kais Manai and Chiraz Ferchichi
Research Unit of Nuclear and High Energy Physics, Faculty of Sciences of Tunis, Tunis El-Manar University, Tunis, Tunisia

Vladimir D. Risovany, Konstantin V. Rotmanov, Genady I. Maslakov, Yury D. Goncharenko, Grigory A. Shimansky, Aleksandr I. Zvir, Irina M. Smirnova and Irina N. Kuchkina
Joint Stock Company, State Scientific Center — Research Institute of Atomic Reactors, Dimitrovgrad, Russia

S. A. Eman, S. H. Nageeb and A. R. El-Sersy
Department of Ionizing Radiation, National Institute for Standards, Giza, Egypt

Volnei Borges, Julio Cesar Lombaldo Fernandes, Bardo Ernest Bodmann, Marco Túllio Vilhena and Bárbara Denicol do Amaral Rodriguez
Department of Applied Math, Universidade Federal do Rio Grande do Sul, Porto Alegre, Brazil

Zengyu Xu, Chuanjie Pan, Xiujie Zhang, Xuru Duan and Yong Liu
Southwestern Institute of Physics, Chengdu, China

Mohamad Ali Shafii
Nuclear and Biophysics Research Group, Faculty of Mathematics and Natural Sciences, Bandung Institute of Technology, Bandung, Indonesia
Department of Physics, Faculty of Mathematics and Natural Sciences, Andalas University, Padang, Indonesia

Zaki Su'ud, Abdul Waris and Neny Kurniasih
Nuclear and Biophysics Research Group, Faculty of Mathematics and Natural Sciences, Bandung Institute of Technology, Bandung, Indonesia

Debabrata Datta
Health Physics Division, Bhabha Atomic Research Centre, Mumbai, India

Leila Abdolahi Shiramin and Seyed Farhad Masoudi
Physics Department, K.N. Toosi University of Technology, P.O. Box 15875-4416, Tehran, Iran

Mukhtar Ahmed Rana
Physics Division, Directorate of Science, PINSTECH, Islamabad, Pakistan

Sheng Liu
Qinghuangdao School, North East Petroleum University, Qinghuangdao, China

Xuanyu Sheng
Institute of Nuclear and New Energy Technology, Tsinghua University, Beijing, China

Nnamdi Norbert Jibiri and Tolulope Hadrat Abiodun
Radiation and Health Physics Research Laboratory, Department of Physics, University of Ibadan, Ibadan, Nigeria

Mohamed Abd-Elzaher and Abouzeid Ahmed Thabet
Department of Basic and Applied Science, Faculty of Engineering, Arab Academy for Science, Technology and Maritime Transport, Alexandria, Egypt

Mohamed Salem Badawi and Ahmed El-Khatib
Physics Department, Faculty of Science, Alexandria University, Alexandria, Egypt

Abdulaziz S. Alaamer
Department of Physics, Al-Imam Muhammad Ibn Saud Islamic University, KSA

Alessandro da C. Goncalves and Aquilino S. Martinez
Department Nuclear Engineering, Federal University of Rio de Janeiro, Rio de Janeiro, Brazil

Daniel A. P. Palma
Brazilian Nuclear Energy Commission, Rio de Janeiro, Brazil

Haikel Jelassi and Khaled Farah
Research Unit: Control and Development of Nuclear Technology for Peaceful Uses, National Center for Nuclear Sciences and Technologies, Tunis, Tunisia

Hichem Eleuch
Departments of Physics and Astronomy, Institute for Quantum Science and Engineering, Texas A & M University, College Station, Texas, USA

Ahmad Hichem Hamzaoui
National Institute of Scientific and Technical Research, Tunis, Tunisia

Arbi Mejri
Ionizing Radiation Dosimetry Laboratory, National Center for Nuclear Sciences and Technologies, Tunis, Tunisia
Research Unit: Control and Development of Nuclear Technology for Peaceful Uses, National Center for Nuclear Sciences and Technologies, Tunis, Tunisia

Xiu-Hua Liu, Yi Deng, Yin-Hang Zhou, Liang Xia, Lan-Lan Ding and Yu-Chuan Zhang
China Academy of Engineering Physics, Mianyang, China

Ayse Güneş Tanır, Mustafa Hicabi Bölükdemir and Rasoul Ghomi
Faculty of Sciences, Gazi University, Ankara, Turkey

Mukhtar Ahmed Rana
Physics Department, International Islamic University, Kashmir Highways, Islamabad, Pakistan
Physics Division, Directorate of Science, Institute of Nuclear Science and Technology, Islamabad, Pakistan

Farhad Zolfagharpour and Negin Sattary Nikkhoo
Department of Physics, University of Mohaghegh Ardabili, Ardabil, Iran

Nazli Hamdolahi
Department of Physics, University of Azad Tehran Markaz, Tehran, Iran

Abdulaziz S. Alaamer
Department of Physics, College of Science, Al-Imam Mohammad Ibn Saud Islamic University, Riyadh, KSA

Andrzej G. Chmielewski, Marcin Brykala and Tomasz Smolinski
Institute of Nuclear Chemistry and Technology, Warsaw, Poland

Shakeel Ahmad, M. Zafar and M. Irfan
Department of Physics, Aligarh Muslim University, Aligarh, India

A. Ahmad
Department of Applied Physics, Aligarh Muslim University, Aligarh, India

Adilson Costa da Silva, Aquilino Senra Martinez and Alessandro da Cruz Gonçalves
Nuclear Engineering Department, Federal University of Rio de Janeiro, Rio de Janeiro, Brazil

Kyeong Mo Hwang, Hun Yun and Chan Kyoo Lee
Power Engineering Research Institute, Yongin, Korea

Wolfram Westmeier
Gesellschaft für Kernspektrometrie mbH, Ebsdorfergrund, Germany
Fachbereich Chemie, Philipps-Universität Marburg, Marburg, Germany

Reinhard Brandt
Fachbereich Chemie, Philipps-Universität Marburg, Marburg, Germany

Reza Hashemi-Nezhad
Institute of Nuclear Science, School of Physics, University of Sydney, Sydney, Australia

Reinhard Odoj
Forschungszentrum Jülich GmbH, Jülich, Germany

Wolfgang Ensinger
Fachbereich Materialwissenschaften, Technische Universität, Darmstadt, Germany

Maria Zamani-Valasiadou
Physics Department, Aristotle University of Thessaloniki, Thessaloniki, Hellas

Andrei Sosnin
Joint Institute for Nuclear Research, Dubna, Russia

Yukino Imura, Hidenori Mimura and Toru Aoki
Research institute of Electronics, Shizuoka University, Hamamatsu, Japan
Graduate School of Science and Technology, Shizuoka University, Hamamatsu, Japan

Takuto Yanagida and Hisashi Morii
Research institute of Electronics, Shizuoka University, Hamamatsu, Japan